JEWS AND GENES

The Jewish Publication Society expresses its gratitude for the generosity of the following sponsors of this book:

Aaron J. Feingold, MD, and Brenda I. Liebowitz support this work in humble recognition of science and medicine's increasing revelation of G-d's universe and the nature of the miraculous.

Dr. Seymour and June Kessler, in memory of beloved grandmother Reva Goldstein.

UNIVERSITY OF NEBRASKA PRESS | *Lincoln*

JEWS AND GENES

*The Genetic Future in
Contemporary Jewish Thought*

Edited by Elliot N. Dorff
and Laurie Zoloth

FOREWORD BY MARK S. FRANKEL

THE JEWISH PUBLICATION SOCIETY | *Philadelphia*

Acknowledgments for the use of copyrighted
material appear on pages 157, 402, 417, which
constitute an extension of the copyright page.

Library of Congress Cataloging-in-Publication Data
Jews and genes: the genetic future in contemporary
Jewish thought / edited by Elliot N. Dorff and Laurie
Zoloth; foreword by Mark S. Frankel.
pages   cm.
Includes bibliographical references and index.
ISBN 978-0-8276-1224-2 (pbk.: alk. paper)
ISBN 978-0-8276-1192-4 (epub)
ISBN 978-0-8276-1193-1 (mobi)
ISBN 978-0-8276-1194-8 (pdf)
I. Dorff, Elliot N., editor. II. Zoloth, Laurie, editor.
[DNLM: 1. Genetic Phenomena. 2. Jews—genetics.
3. Biomedical Research—ethics. 4. Judaism.
5. Religion and Medicine. QU 500]
RB155.5
616'.042089924—dc23
2014034801

Set in Lyon by Lindsey Auten.
Designed by N. Putens.

CONTENTS

FOREWORD   MARK S. FRANKEL

As the world's largest general scientific society, the American Association for the Advancement of Science (AAAS) is committed not only to the advancement of scientific discovery but also to exploring the implications of scientific progress for the larger society. As part of this commitment, AAAS established the Dialogue on Science, Ethics, and Religion (DoSER) program in 1995 to promote collaboration and dialogue between the scientific and religious communities, to explore the ethical and religious implications of scientific research, and to deepen understanding of the cultural context in which science is embedded. This volume is a good example of that, with its focus on the intersection of molecular genetics and the Jewish tradition. The international collaboration and interdisciplinary discourse represented by the chapters in this book are a model for engaging religious scholars and scientific researchers on a topic of increasing importance for both communities as well as the larger society.

Although the project that led to this book began at the beginning of the decade, the scientific advances since then, many of which are described in this volume, have only made the need for such an effort more compelling. Advances in genetics present unique challenges for the Jewish community, which is not monolithic in its interpretation of how Jewish "law" can inform understanding of what these advances mean for Jews (and others) in the twenty-first century.

AAAS is pleased to have been involved in this project and delighted that its findings are now available to a larger audience. We are grateful to the editors and authors for producing this compilation of diverse views, insights, and wisdom, which will enrich discussions of these issues for years to come. We especially want to acknowledge the invaluable contributions of former DoSER directors Dr. Audrey R. Chapman and Dr. Connie Bertka for their dedication to and oversight of this multiyear project. We are also grateful to the Greenwall Foundation and the Walter and Elise Haas Fund for their grants in support of the project.

It is a rare day that goes by without some announcement of a new finding in genetics research. These breakthroughs hold great promise for improving human health and well-being, but many of them also raise questions about human dignity, the bonds among family members, and how the power of genetics can be harnessed for social good. The experience and wisdom embedded in religious traditions can help us sort through many of these questions, offering us informed ways of "knowing" as well as normative guides in our quest for answers. The discourse in this volume pushes our understanding to a new level. We invite readers to join the conversation.

INTRODUCTION ELLIOT N. DORFF AND LAURIE ZOLOTH

Why a book for nonscientists on genetics? Why a book on Jews and genes? Why *this* book on Jews and genes?

Answering the first question is easy. Genetics is playing an increasingly important part in our lives. Scientists have mapped the human genome, and now they are discovering linkages between genes and a number of specific diseases. That information is already helping to diagnose some diseases, and hopefully it will help us diagnose others and possibly prevent or cure genetic diseases in the not-too-distant future. The same is true for stem cell research, one branch of genetic research. Physicians are already using stem cell therapies to cure some cancers, and there are good grounds to believe that embryonic stem cell research, if not blocked by political forces and if given sufficient funding, will produce cures to some of humanity's worst diseases.

Genetics is also telling us a great deal about our identity. That is true for us as a species—the fact that 99 percent of our DNA is the same as that of monkeys should give us pause and maybe some humility—and also for us as individuals. Law enforcement officials and defense attorneys are using DNA to determine the identity of criminals as well as the innocence of those convicted long ago, and paternity lawsuits now have sound science to resolve them. Furthermore, as a section of this book demonstrates, genetic markers are helping scientists trace ancestry for generations—indeed, for thousands of years—adding to our knowledge of the movement of human populations long ago, the interbreeding of different cultural groups, and the ancestry of individuals living now.

Why a book on *Jews* and genes? Jews have a special interest in the new scientific developments in genetics for at least two reasons. First, for at least two thousand years the Jewish tradition has had a love affair with medicine. The Rabbis read the Torah to require us to set aside almost everything else in order to save lives (*pikkuah nefesh*), and they interpreted it as well to require us to avoid danger and to guard our health with proper diet, hygiene, exercise, and sleep. As a result, there is an illustrious line of rabbis over the last two thousand years who were also physicians, and Jews have honored doctors in large ways and small—including even a plethora of jokes depicting becoming a doctor as the pinnacle of success. ("Catholics believe that life begins with conception; the Supreme Court ruled in *Roe* that life begins at birth; and Jews believe that life begins when your child finishes medical school and moves out of the house!") Because genetic interventions hold out the promise of being the most effective line of curing diseases since the advent of antibiotics, it is no wonder that representatives of all Jewish denominations have enthusiastically endorsed embryonic stem cell research, genetic testing for diseases, and, if possible, the development of genetic cures. This is simply the new form of the Jewish mandate to heal.

Another factor makes a book specifically about Jews and genes appropriate. Because Jews were always a small population, and because historically they predominantly married only other Jews ("endogamy"), some genetic diseases are more prevalent among Ashkenazic Jews—that is, those whose ancestors lived in northern and eastern Europe—than among the general population. That is because if the breeding choices that a given group of people have are limited, any given person who has a random genetic change (mutation) is more likely to pass it on to others, and then they to their descendants; this is called "the founder effect." Such mutations are much less likely to be expressed or even to be passed on if the two members of the couple come from vastly different genetic backgrounds.

This is not an argument for intermarriage. Marriage, after all, involves serious and deep commitments to one's family, people, and faith, and those are not easily maintained in a marriage with someone from another faith. Furthermore, biologically, we can now detect the genetic changes that produce diseases and do things to prevent passing many of those diseases on.

It definitely is an argument, though, for testing couples for these genetic diseases and, if both partners carry the recessive genes that produce most of these diseases, to have their embryos or fetuses tested as well. Ashkenazic Jews are *not* a diseased population; in even the most prevalent of these diseases, only 3 percent of Ashkenazic Jews are carriers, which means that 97 percent are not. Still, the Jewish mandate to prevent diseases, if possible, and the personal interest of Jews to produce children free of these diseases make Jews especially interested in genetics.

Finally, why *this* book on Jews and genes? This book deliberately calls on the perspectives of Jews with all forms of Jewish commitments, from Orthodox to secular. Laurie Zoloth, one of its editors, identifies as Orthodox; its other editor, Elliot Dorff, is a Conservative rabbi; and the contributors to this volume include other Orthodox and Conservative Jews as well as Jews affiliated with the Reconstructionist and Reform movements and with no religious form of Jewish expression at all. Furthermore, this book summarizes the science in language understandable by nonscientists, and it presents a number of Jewish perspectives on how to respond to the new science in good, Jewish, argumentative fashion. It also includes Jews from North America and Israel, giving it an international scope. This plurality of approaches embraced by Jews from world Jewry's two largest communities enables this book to speak for all Jews, regardless of the nature of their Jewish identity.

Another factor recommends *this* book on Jews and genes—namely, its attention to methodology. Because nobody in biblical or talmudic times knew about genetics, the sources of the Jewish tradition can give us no direct guidance as to how to respond to our new knowledge. This means that one cannot simply decide what to do by quoting chapter and verse from one of the classical Jewish texts. Fortunately, though, Jews inherit a tradition that is amazingly sophisticated in its understanding of the multiple meanings that texts may have. The Rabbis, in fact, say that there are seventy faces to the Torah—that is, a large number of ways to read any verse.[1] The talmudic tradition that shaped what we know as Judaism also established powerful and wise ways to stretch the tradition to apply to new circumstances, including modes of interpretation (midrash), stories, principles, proverbs, and legal reasoning, including analogizing to previous precedents and distinguishing among them. Only with such open and honest attention to method can we create responses to the new genetics that are both practically wise and authentically Jewish.

This book focuses on four topics in the new genetics: stem cell research, genetic mapping and identity, genetic testing, and genetic intervention. Each part begins with a brief description in nonscientific language of the relevant scientific developments and then continues with several essays from differing viewpoints about how Judaism should be applied to this new research. Part 5 includes some reflections on the relationship generally between the approaches of America's many religions to national policy and between Judaism and genetic issues in particular.

This book derives from a project conceived originally by Laurie Zoloth and funded by the Haas and Greenwald Foundations. Their money enabled the authors of these essays to meet four times, to learn from the top scientists in the field on each of the four topics that this book includes, and to respond to first drafts of each other's essays so that they could be improved before publication. All of us who participated in this effort are immensely grateful to the Haas and Greenwald Foundations for making this project possible.

We were especially fortunate to have the American Association for the Advancement of Science sponsor the project and organize it for us. We want to thank Audrey Chapman, who first acted as our AAAS liaison, and then Connie Bertka, who sensitively and efficiently carried on with this project within the AAAS and brought it to fruition. We thank them for their direct personal interest and involvement in this project, their patience, and their skill.

We also want to thank the good people at the Jewish Publication Society who saw the importance of this book and helped us with publishing it. That especially includes Rabbi Barry L. Schwartz, director, and Carol Hupping, managing editor. We also want to thank our copyeditor, Mary M. Hill, for her skillful work in finding the issues that we still needed to address after Elliot Dorff and Carl Hupping together edited what they thought was going to be the final version of this book. We are deeply indebted to the sponsors of this book, Aaron J. Feingold, MD., Brenda I. Liebowitz, and Dr. Seymour and June Kessler, for financing the publication of this book.

Finally, we want to thank those who participated in this project. Their willingness to discuss these issues with Jews from many different religious and academic backgrounds speaks to their sheer humanity and their dedication to enabling Judaism to speak authentically and intelligently to these new genetic developments. They all brought their minds, hearts, and souls to this project, and we deeply appreciate each of their contributions to our discussions and to this volume.

NOTE

1. *Numbers Rabbah* 13:15–16.

ABBREVIATIONS

B.: Babylonian Talmud, edited by Ravina and Rav Ashi, ca. 500 CE

J.: Jerusalem Talmud, edited by Rabbi Yohanan ca. 400 CE

M.: Mishnah, edited by Rabbi Judah, president of the Sanhedrin, ca. 200 CE

M.T.: Maimonides's code, the *Mishneh Torah*, completed in 1177 CE

S.A.: Joseph Karo's Shulhan Arukh, completed in 1565 CE

T.: Tosefta, edited by Rabbis Hiyya and Oshaya, ca. 200 CE

JEWS AND GENES

PART 1 *Stem Cell Research*

CHAPTER 1   *Summary of the Science of Stem Cell Research*

ELLIOT N. DORFF AND LAURIE ZOLOTH

Both the goals and methods of scientific research are rooted in the historical and social terrain in which science seeks answers to its questions. This is true as well for the specific fields of science relevant to stem cell research, including molecular biology, genetics, and clinical medicine. Because scientific research is by definition an inquiry at the frontiers of the known world, its questions are often destabilizing ones. The researchers, then, are often brought into conflict with those aspects of human culture that express the traditional and familiar, including political structures, law, and religions. In such cases science represents the intrusion and insistence of the modern against the veracities of tradition. Sometimes traditions can adjust easily through new interpretations and applications of the tradition that can accommodate new scientific discoveries. Other times, though, the new science raises immense problems of self-understanding and of ethics, requiring everyone, scientists included, to assess the implications of the new science for how we understand our world, our place in it, and our duties to maintain and repair it. In every generation this is a persistent task for Jewish intellectual life.

Questions of ethics are especially challenging when the new science undermines our fundamental understandings of identity (who we are and who we choose to be) or changes our sense of the limits of our power (when it gives us tools to do things that our ancestors never were able to do). This leads us to ask whether we should assume the new identity and new agency that are available. This is the dilemma identified by Immanuel Kant in the late eighteenth century. As he pointed out, "Ought implies can." That is, if I cannot do something, then I never have to ask whether I should, because it is simply beyond my ability to do it anyway; but if I can do something, then I do need to ask whether I should do it, because there are all kinds of things that I can do that I should *not* do. This is obviously the case when new technologies are produced, and it is the case when new research presents us with dramatically new choices.

In the last two decades, one of the fiercest debates about our limits and our choices has been around the issue of human stem cell research—ever since James Thompson of the University of Wisconsin and John Gearhart of Johns Hopkins University first demonstrated that human embryonic stem cells, with their nearly infinite possibilities for differentiation and regeneration, could be grown in the laboratory from human embryos or gametes. Before we knew about stem cells and could manipulate them, we never had to ask whether or when we should do that, but now we must. In doing so we find ourselves not only addressing specific questions about the advisability and limits of using a particular technique but also reassessing the very nature of who we are, the connections we have to family, community, the environment, and God, and the limits, if any, we should impose on our capacity to do research and to heal.[1]

Although science has historically often been the cause of much of the unease about the new, the events surrounding the struggle over the research on human embryonic stem cells, beginning in the summer of 2001, have proven to be especially controversial.[2] In fact, this debate was one of the critical issues that generated the meetings that led to this book. This introduction to this section of the book about stem cell research begins with a definition of stem cells and a brief review of the emerging science and technology dealing with embryonic stem cells and the very newest research that has produced "induced pluripotent stem cells" from somatic or adult cells, something that was not possible at the time when many of our essays were written. The introduction then asks how this research challenges our understanding of the nature of knowledge and forces us to confront anew the moral limits, freedoms, and responsibilities of research. Following this introduction to the science of stem cell research are several chapters in which Jewish scholars consider how Jewish texts, laws, concepts, and values can be interpreted and applied to this new emerging science in order to gain wisdom about how we think about ourselves and our world and how we should act in it in light of this new science and technology.

The State of the Science

Stem cells are called "stem" because they are cells that can change into several different kinds of more specialized cells. They are undifferentiated; that is, they are not yet specifically one kind of cell. Stem cells can produce either any kind of cell in the human body (they are "totipotent") or at least several different kinds of cells (they are "pluripotent"). Of course, all living creatures begin as one cell, a zygote. In people, as in all mammals, this is the fertilized human egg, just after the sperm cell has entered it. The most flexible stem cells are those in the early embryo that is formed five to eight days after a sperm and an egg combine; these are "embryonic stem cells." The one hundred to two hundred cells produced in these first few days after fertilization ultimately mature into each and every kind of cell in the complex human organism.[3] That is, they "differentiate"—specialize—into specific kinds of cells so that some become the heart, others the brain, others the lungs, and so on, each group with its particular nature to enable the human body to live and function. At about five days of gestation, the form we call an early embryo looks like a small circle (the perimeter of which later in pregnancy becomes the placenta) with a clump of cells inside that circle called the "inner cell mass." The inner cell mass and the large circle that surrounds it are together called a "blastocyst." That is the name of the embryo at this stage. It is these inner cells that are extracted for purposes of embryonic stem cell research, and in the process, the blastocyst is destroyed. The cells are then placed in chemical solutions to enable them to develop; these are human embryonic stem cells in culture. As of this writing, sixteen years after the first stem cells were isolated, they are still replicating, which is why they are called "immortal" cell lines.

There are also stem cells in fully formed human beings; these are called "somatic stem cells" or sometimes "adult stem cells," whether they come from an infant or an adult. Somatic stem cells are not as flexible as embryonic stem cells, for they can change into only a few kinds of cells on their own, and they are not immortal. Still, the body uses them to renew blood, skin, and hair, for example, throughout an individual's life.

Biologists have long sought to understand how a single cell created when a sperm and egg combine ultimately creates a complex and highly differentiated system of intricate tissues and organs organized perfectly into a human being. How does the DNA program in the nucleus signal the cell to duplicate and differentiate? How does the small, microscopic mass of identical cells that have been formed in a woman's uterus or in a petri dish approximately five days after a sperm and an egg unite, the embryonic stem cells, ultimately form a human fetus?[4] Perhaps most intriguingly, if each of these cells of the early embryo has the capacity to develop into any and all cells of the human body, can that cell's mutability be used to create new cells in a person in order to heal him or her from a disease or to repair tissue that has been damaged?[5]

The process of embryology has long been studied through the use of animal models. Embryonic stem cells were first isolated in mice in 1981, and ever since then research has been conducted with embryonic mice, rats, and nonhuman primates. Much of the current success in understanding and using human stem cells, in fact, can be traced to the intensity of research in animal models, including the rapidly unfolding sciences of genomic mapping and molecular biology.[6]

Telomerase is an enzyme that enables genes to be flexible and to reproduce. In 1995 the genes for telomerase were cloned, enabling scientists to produce large numbers of them so that they could study them more easily. Biologists have used their expanding access to and knowledge about telomerase to search for ways to understand how the cells in the early human embryo maintain their plasticity and immortality.

Parallel work has studied stem cells that are found in some tissues of adults.[7] It had long been understood that some cells of the body, such as the lining of our intestines, blood cells, hair cells, and skin cells, are constantly renewed.[8] Researchers found that some of these tissues contain rare precursor or stem cells that are undifferentiated and that develop into mature and functional cells in the body.[9] These adult stem cells have been found, cultured, and used to treat some conditions. In bone marrow transplants, for example, harvested blood stem cells have been used to regenerate a new blood supply, and harvested stem cells in skin have been used to begin the process of creating new skin for skin grafts to repair the skin of burn victims, for example.[10] But these cells are limited in several ways.[11] They are rare and hard to find; they are not available for all tissue types; and, when cultured in the laboratory, they always cease dividing and lose their self-renewal properties because, with each division, the telomerase at the end of the nuclear chromosomes shortens.[12]

Because of these difficulties with adult, or somatic, stem cells, researchers have long been intrigued by the embryonic stem cells that are the precursors to these adult stem cells. In part, the interest is purely investigational. It has been the quest of some researchers to study precisely how the embryonic stem cell is programmed to do this, to understand what goes awry in genetic diseases, and to observe how the environment affects developing cells.[13] In part, the research has been driven by a therapeutic goal, not only to understand and observe the process but also to find ways to coax embryonic stem cells into specific uses. Researchers began to speak about creating banks of tissues to repair tissues damaged by illness or injury.[14]

The benefits of such an endeavor, if it succeeds, would clearly be enormous, for many of the diseases that beset us in modernity are precisely degenerative diseases. These include stroke (6 million people have one each year), congestive heart failure (6 million), neurological diseases (3 million with Parkinson's alone), diabetes (100 million), and liver failure

(5 million from hepatitis alone). It is important to note that, unlike much of the focus of high-technology-driven medical research, diseases of cell death and cell control are not limited to the elderly or to certain classes or groups, for degenerative diseases also plague children. Spinal cord injury repair is a target of this research as well.[15] Furthermore, unlike the search for medical treatment of the symptoms of disease, molecular-cellular medicine is in pursuit of permanent cures to disease by alteration or replacement of the genetic and cellular causes of the disease itself.[16]

Risks and Benefits in Transplantation Therapies

To make regenerative medicine work either financially or ethically, it must be scalable, biologically stable, safe, and universally usable. For human use, the problem of histocompatibility must be solved. That is, a way must be found to introduce stem cells into a person's body without its immune system reacting to them as foreign objects and attacking them, ultimately leading to the patient's death and thus frustrating the effort to cure the patient (the "graft-versus-host problem").

The science is thus interesting not only in its own terms but also because of the premise of *widespread* access to *significant* therapy, an essential health-care justice issue. The hope is that tissue transplantation, unlike organ transplants, would not be a boutique therapy for the lucky or wealthy few but could be widely available to large numbers of people worldwide and could be used without the terrible risk of graft-versus-host disease. The idea is to use human embryos to derive stem cells that can be used for tissue transplants—either by creating tissue-banking systems, or by finding a way to match donor and recipient, or by creating a universal donor cell—or to understand enough about the way that cells reprogram themselves to regulate this process within the human body itself. So far, reports of scientific progress are remarkable and swiftly appearing in peer-reviewed journals. It was only in 1999 that the first human embryonic cells were grown in the laboratories at the University of Wisconsin and Johns Hopkins University, and in 2001 they were still dividing, well beyond their six hundredth population doubling.[17] Researchers have already discovered that neural cells placed in animals with neurodegenerative disease migrate to the affected site, synthesize neurotransmitters, and extend neuronal processes. In laboratory studies at Johns Hopkins, John Gearhart has demonstrated that neuronal cells not only migrate to viral lesions in a living rat but also enable rats who have lost motor function to walk again. Liver cells make liver proteins; heart cells make contractile proteins, beat spontaneously, and respond normally to cardiac drugs; pancreatic cells make insulin. Blood cells have been made for all four blood groups. Still, with all this progress, the researchers were the first to admit that much of what they were seeking was a mystery—a terrain largely unknown. How do cells program and reprogram themselves? Can one create a universal donor cell?

The Ethical Problem of the Embryo's Moral Status

To procure embryonic stem cells, however, one extracts them from a five- or six-day-old embryo in a petri dish, thus killing it. Hence, the use and destruction of embryos for stem cell procedures immediately pose the significant moral question of the nature, meaning, and moral status of the human embryo. If it is already a human being, the killing of an embryo would amount to killing a person, so even though the goal of curing diseases is laudable, one may not kill one person in an effort to save another. On the other hand, if the early embryo is merely a clump of cells that would otherwise be discarded, then one may and arguably should instead use such cells to advance human knowledge and therapies.

Human embryonic stem cell research has been made possible by the technology first used in 1978 of in vitro fertilization (IVF), that is, bringing sperm and egg together in a petri dish. The resulting fertilized egg cell (the zygote) is cultured for a few days and then implanted in a woman's uterus. However, when a couple has produced several embryos in an effort to overcome infertility problems and either has had as many children as they want or has given up trying to have biological children of their own, the remaining frozen embryos that they produced in this effort are discarded. When the research first began, embryonic stem cells used by scientists came from embryos that such couples donated (and, in a Jewish side note, the initial embryos came largely from one

IVF clinic in Haifa, Israel). Typically, twelve embryos are created in the IVF process. If they are not transferred to a uterus, they will die. They can be frozen and stored (in the United Kingdom for five years, and in the United States indefinitely, as 100,000 of them are); they can be discarded (the most common course of action); or they can be donated for research. It is this creation and use of an embryo outside of a human body and in the hands of a largely unregulated marketplace driven by the deepest of yearnings for children that has reconfigured the moral landscape of reproduction in the developed world. For once one has created an embryo artificially, one is engaged in what has been a large but unstructured clinical trial without controls or even, in many cases, the careful consent of the people involved as required in other cases of medical practice and research.

How does one regard the central question of the moral status of the human embryo? As the National Bioethics Advisory Committee report on embryonic stem cell research clearly indicates, this is one of the key ethical disputes in society generally and among religions in particular.[18] It is also at the heart of American legal treatments of both abortion and embryonic stem cell research. Furthermore, does the status of the embryo in a woman's womb differ from that of an embryo in a petri dish?

In American law, the United States Supreme Court ruled in *Roe v. Wade* in 1973 that in a woman's womb a fetus is part of the body of the woman. Because American law grants adults the right to determine what may be done with their bodies, this means that a pregnant woman may choose to abort a fetus at will until such time that it can live outside her body—and even then, according to the Supreme Court's 1992 *Casey* decision, state statutes may impose some restrictions on the woman's right to abort but not to the extent that such statutes make it impossible for a woman to exercise this right. Recent state statutes that are being tested in the courts place significant restrictions on abortion, ultimately testing also the Supreme Court's determination that the fetus is a part of its mother's body and therefore subject to her will.

When couples create embryos outside a woman's womb for in vitro fertilization and have them frozen until they want to use them or discard them, however, what is the legal status of such embryos? Courts have faced cases in which after creating such embryos the couple divorces, and then later one or both of them want to use the embryos to implant either in the woman so that she can have a child on her own or in the man's new spouse when other attempts at pregnancy fail. To whom do the frozen embryos created by the couple belong? The courts have variously ruled that outside a woman's body embryos are either communal property, full human beings (only in state law, for the United States Supreme Court explicitly ruled in *Roe v. Wade* that fetuses are not to be treated as full human beings in federal law), or something in between but deserving special respect and protection.[19] Whatever the embryos' legal status, neither party may use them without the consent of the other for purposes of producing one or more children, because each partner has a right to refuse to become the biological parent of the children born through the use of frozen embryos produced with his sperm and her ovum.[20]

Although American courts and state laws vary in their determination of the legal status of the embryo, Catholics and some Protestants unequivocally see the early embryo as fully a person, and hence embryos, in their view, may not be destroyed in research, for that amounts to murder.[21] Roman Catholic authorities have maintained during the last several hundred years that a fertilized egg cell in a woman's womb is a full human being. They therefore prohibit abortion even to save the life of the mother, for one may not kill one person to save another. Catholic authorities object to the artificial creation of an embryo in an IVF procedure altogether, but if one is created, then that embryo is, according to decisions of Catholic leaders in the last several decades, also a full human being, even though the embryo has no chance of developing into a person unless it is implanted in a woman's womb. As a result, Catholic authorities have fervently argued against embryonic stem cell research, for removing the inner cell mass from an embryo to do research kills the embryo, and if one may not kill an embryo or fetus to save the mother's life, then certainly one may not kill the embryo to do research, which is a lesser good.

As the reader will see in the Jewish reflections in this section of the volume, the Jewish tradition takes a stance different from that of either American law or Catholicism. It sees the embryo during the first forty days of gestation (actually, the first fifty-four days, as Elliot Dorff's essay will explain) as "simply liquid" and thereafter until birth as being "like the thigh of its mother." The fetus does not become a full human being until birth, and only then does it attain all the attendant protections of full persons. This developmental view of the fetus in Judaism is shared by Islamic and some Protestant traditions.[22] Regarding the moral status of an artificially produced, microscopic blastocyst created entirely outside of a woman's body, halakhic (Jewish legal) experts have maintained that the small size of the blastocyst and its artificial location (outside of a womb) reduce even further the moral warrant for full status as a person.[23] The strong mandate within the Jewish tradition to seek to heal through both treatment and research and this developmental view of the status of the embryo have together led rabbis across the denominational spectrum strongly to endorse embryonic stem cell research, especially if the alternative is to discard the frozen embryos.[24]

Other Ethical Questions:
The Reconstruction of Creation's Tale

As ethicists struggled to understand and defend arguments about the moral status of the embryos first used in stem cell research, the research itself began to ask more questions about the mutable origins of the blastocyst. The next set of ethical problems included the essential issue of whether these "excess" embryos were the correct way to obtain the embryos needed to create stem cells or if other ways of stimulating gametes could also lead to a blastocyst and, if so, what the status of that newly made entity would be.

It was a deeply disturbing line of questioning. The narrative of human reproduction—one man, one woman, a meaningful cleaving of one to the other, as humans have done since Adam and Eve, leading to progeny that carry the story forward—is at the heart of all three Western faith traditions. Indeed, it is through this human story that the monastic Western traditions and several of the traditions of Eastern and indigenous religions as well tell of the creation narrative itself—it is a core narrative about the meaning, nature, and goal of being human.[25] Our understanding of ourselves as a part of this narrative, as children and then parents, also undergirds the religious imperative to procreate, the obligations inherent in the relationships within families and communities, and, in some religions, a commitment to natural law theory, according to which nature establishes not only physical laws but moral laws as well, so that nature tells us not only *how* things work but also how they *should* work.

However, since the early 1970s, the idea of the natural process of sexual reproduction has been disrupted by emerging scientific technology. Artificial reproductive technology has created many possible origins for any human embryo: it may be fabricated by mixing eggs and sperm or by injecting an egg with a selected sperm. The course of development may be altered as well. Sperm may be "spun" and separated by weight to select for gender; the egg may be altered to include extra mitochondrial DNA; embryos may be deselected by genetic trait so that only those embryos without a given trait (e.g., a disease like Tay-Sachs) are implanted in a woman's womb; the embryo may be implanted in a surrogate, the egg obtained from another woman or the sperm from another man, and the resulting child given to a family that may itself be constituted in a variety of genders and permutations. All of these disruptions in the core narrative have elicited considerable alarm initially and then extended social discourse about them and about yet other emerging possibilities. In many societies, the narrative has been reimagined, and retold, to account for these new possible origins of people.

But regenerative medicine offers not only another set of beginnings for the narrative of reproduction but also other possible ends for the embryo. Prior to artificial reproductive techniques (ARTs), a blastocyst conceived in a woman's body either implanted itself in the womb or, as in approximately 75 percent of cases, failed to implant and was naturally discarded during the woman's menstrual period. With the advent of IVF in 1978, a blastocyst fabricated in an IVF clinic might have any of at least four fates: it might be transferred to a human womb, where it might implant successfully; it might be transferred but not implant successfully; it might be frozen; or it might be discarded. Now there is a fifth possible fate for the embryo: it might be destroyed in a laboratory in the process of being used to make stem cells for research or for curing diseases. Now that our society has allowed for the first four outcomes, the last, use in a research laboratory, can be understood as an alternate ending or alternative goal.

For many, such a deconstructed narrative, with the possibility of origins other than monogamous union and ends other than reproduction, elicits a sense of moral repugnance, the ultimate horror of a scientific, desacralized world. But for others, the revised narrative elicits a curiosity and awe at the new possibilities for human understanding and of the possibility to alter other key aspects of what had been understood as moral fixities—the nature and scope of human suffering, the "natural" span of a human life, and the limits of our capacity to alter our existence. We have, in other words, gone well beyond our "ordinary reach," or, as Pascal and Roger Shattuck name it, our *portée*.[26] It should be noted that both responses—fear and awe—indicate that we are on the brink of reconfiguring the meaning of a core narrative of our existence, and this alone should create a moment for ethical pause. It was largely for this reason that one of this volume's editors (Zoloth) approached the American Association for the Advancement of Science (AAAS) for help in creating the place and format for scientists and Jewish theologians and ethicists to think together about this and the other pressing issues in the book you have before you.

Between the time that that first project ended and this book was ready to be published, much had happened in the debate. As ethicists, we fought about the issues of human embryonic stem cell research along the lines described above for nearly a decade, and much of the debate was about the question of the moral status of the embryo. In 2006 another debate emerged—about whether researchers should pay women to acquire their eggs for research. This began when scientists wanted to try to clone embryos to deal with the problem of tissue matching (histocompatibility). Moreover, by that time scientists understood that many donated embryos might be flawed because many came from infertile couples, thus diminishing the prospects of the success of the scientists' research. This led to a new round of debate both about cloning as a technique and about paying for human gametes. Meanwhile, an academic organization—the International Society of Stem Cell Research (ISSCR)—was founded, with ethicists (including Zoloth) on the board; the state of California allocated $3 billion to fund stem cell research and created ethical oversight committees (including Dorff) to create ethical guidelines for all stem cell research in the state; and President Obama changed previous federal policy so that the federal government would now support and fund stem cell research.

But in 2007, at the Toronto meeting of the ISSCR, the single most important change was announced, and it was a change in the science that promises to redefine the moral landscape. Shinya Yamanaka and his colleagues found a way to do what was thought to be impossible—to "reverse the arrow of time" and to change or deprogram an adult or somatic cell back to its original state as a pluripotent cell.[27] This meant that scientists had discovered a way to make adult stem cells behave substantially like embryonic stem cells by winding them back, as it were, to how they were before they differentiated themselves into specifically one type of cell (bones, fat, neurons, etc.).

Yamanaka's work was immediately understood as groundbreaking. He was awarded a series of scientific prizes and in 2012 the Nobel Prize for his research and its potential results. If human cells taken from a given patient could be reversed and reused to replace the patient's own damaged tissue, then there would be no problem matching the tissue, no graft-host incompatibility. The new cells would simply take over and begin to redifferentiate to a new cell type, thus hopefully curing the patient of his or her disease or disability. As a result, as of this writing, most of the laboratories that work on stem cells use iPS (induced pluripotent stem) cells from somatic (or adult) cells, discovering how they are alike and how they are different from hES (human embryonic stem) cells. However, iPS cells present significant safety issues. Yamanaka continues to work on solving the problem of how to prevent the implanted cells (he uses animal models) from forming tumors. But the advantages of tissue matching and of avoiding the need for a continual creation and destruction of embryos are powerful.

Serious questions remain about all aspects of the research. The time lag between the very large promises of research, made to, for example, the citizens of California in order to obtain their approval of the three-billion-dollar bond measure to fund this research, and the actual results, which are still pending, leads to yet other questions.

The essays in part 1 of this volume reflect on many of these issues from a Jewish perspective—one that was created by several years of discussion and one in which, of course, you will find not one answer but many. Some of the essays focus on the moral status of the embryo, and some wrestle with other moral issues.

Furthermore, some authors approach the moral issues raised by embryonic stem cell research and the other topics of this volume using the methodologies of Jewish law. That is the usual way in which moral problems, including completely new ones, have been addressed in the Jewish tradition, even though this requires stretching precedents to apply to topics their authors never contemplated, let alone sought to adjudicate.[28] However, precisely because of the novelty of many of the issues in stem cell research and other areas of modern genetics, other writers call upon other resources in the Jewish tradition to address these admittedly new issues—sources like Jewish theology, stories, text study (in the textual reasoning mode), maxims, and history.[29] These varying resources and methodologies provide a richly textured Jewish response to these important and groundbreaking moral issues that confront us now in the field of genetics with all their medical promise and all their moral difficulty.

Stem Cell Research Is a Work in Progress

In late January 2014, *Nature*, a highly regarded international science journal, published two articles written by researchers who claimed to have discovered how to convert existing, fully formed, mature skin cells into something "like" embryonic stem cells.[30] If ordinary skin cells were subjected to stress (e.g., an acid bath), the cells would be transformed into a totipotent cell, one capable of making any sort of tissue at all, including (unlike hES cells) placental tissue. This has been the long, elusive goal of stem cell biologists, of course: to create new tissue that would be a perfect genetic match with the person who needed the cells. While it had been discovered that mature cells could be reverse-programmed back to their earlier state, these induced pluripotent stem cells need tricky cocktails of viral vectors to rearrange the nuclear DNA, and so it had proven difficult to create such cells in the numbers needed for clinical use. This new research had been proven only in mice and using neonatal mouse cells, but the coauthor of one of the articles, Martin P. Vacanti, claimed to have replicated the study in human cells.

The story was made more appealing by its Cinderella-like aspects. The advance had been accomplished by complete outsiders at two of the major stem cell research facilities: Haruko Obokata, the lead author of one of the articles, a young female stem cell biologist at Japan's most prestigious lab, RIKEN; and Charles Vacanti, an anesthesiologist at Brigham and Young Hospital who was affiliated with Harvard but not working with the prominent Harvard stem cell biologists.

The reaction of the press was immediate; reporters were eager to discuss the fascinating news. But the reaction of stem cell biologists was more complex, a mixture of amazement and skepticism. One of the most interesting reactions came from the Knoepfler Stem Cell Laboratory, where researchers asked immediate questions: Can this work be reproduced by other laboratories? Will it work in human cells? Will it work in adult cells? (The trial only used neonatal mouse cells.) What are the molecular mechanisms? Do these cells possess significant rates of mutations or epimutations (the latter being abnormalities in the epigenome), and therefore will they prove to be unusable? Are these cells tumorigenic (i.e., do they produce tumors besides forming teratomas)?

Stem cell researchers began a poll on the project's believability and an open crowdsource blog where labs trying eagerly to reproduce the work posted their results. When, after eight weeks, none of the ten leading laboratories was able to replicate the work, some of the RIKEN authors (minus Vacanti this time) published a new methods paper that for some did not clarify the process at all. Knoepfler then released a poll of three hundred stem cell scientists who believed the results in week one but by week three were overwhelmingly skeptical. As of this date (April 2014), no major stem cell laboratory has been able to reproduce the reports, and it is unclear whether it is impossible to reproduce them or just very hard.

How did bioethicists react to the news? First, it has to be said that many, many outsiders have been the ones to transform medicine. One needs only to read the story of Ignaz Semmelweis, the Hungarian obstetrician who fought and lost the battle to prove that hand washing prevents sepsis against the entire weight of European medical science in the mid-nineteenth century, or remember that the cure for peptic ulcers was discovered by a plucky postdoc who later would win the Nobel Prize to raise questions about the wisdom of the crowd.

Bioethicists understood immediately that if this idea were proven true, it really could transform medical knowledge. And if it were true, it was a curious thing: Why, if cells could be so easily reprogrammed into stem cells, did the process not happen more often in actual mammalian life—for instance, when you crushed your finger or, for that matter, drank orange juice? What kept the cells committed to their little tenacious fates? What, in short, holds reality together? Would cells that turned back time in this way be the same thing as brand new tissue? Next, of course, there was the disturbing ontological problem that has haunted all stem cell research in the last decade: Could this process be used to make humans? If any of my cells can be reprogrammed into any cell, what would keep researchers from restarting life from a cluster of them surrounded by others coaxed into being a placenta? But really, despite all the initial promise, everyone who was familiar with stem cell research had another more disturbing question: *How can you know if any purported research result is true?*

There is something familiar about all of this. In 2004 one of the coeditors of this volume, Laurie Zoloth, stood in a Seattle Hotel theater next to a gentle and thoughtful young scientist from South Korea, an outsider who had just announced that he had cloned human embryonic stem cells.

Zoloth was the ethicist who had read his *Science* paper and, as it turned out, was just as duped by the faked signatures on the faked consent forms as the science reviewers had been duped by the faked data and photos. The researcher, Woo Suk Hwang, personally promised patients in wheelchairs, politicians with dying wives, and thousands of envious scientists that his research was valid and done ethically. It took two years, a court, a media storm, and an international scandal before the claim was disproven. The skittishness that affected everyone involved is an inexorable part of any new revelation. Ironically, it was the journal *Nature* that had uncovered the deception and relentlessly pursued the story.

Wary of the reactions, *Nature* announced that some irregularities had been found in the papers, a "mix-up" of images, the authors then claimed. PubPeer, a science blog, raised further issues, and finally RIKEN launched its own investigation. And so we are faced with a puzzle, what philosophers call a question of "testimony." What counts as true knowledge? Who can be trusted to tell it? How can I know if it is true? This last question is from Immanuel Kant, who believed that patience is needed for real knowledge: "Truth is the child of time; erelong she shall appear to vindicate thee." We will have to wait and see if, this time, she will arrive.

NOTES

1. Peter J. Donavan and John D. Gearhart, "The End of the Beginning for Pluripotent Stem Cells," *Nature* 414 (2001): 92–97.
2. United Nations Educational, Scientific and Cultural Organization, International Bioethics Committee, "The Use of Embryonic Stem Cells in Therapeutic Research," 6 April 2001. Also see George Bush, "Address to the People of the United States Regarding Stem Cell Research," 9 August 2001, http://www .washingtonpost.com (accessed 15 September 2001); "The Ethics of Human Cloning and Stem Cell Research," a report from "California Cloning: A Dialogue on State Regulation," Santa Clara, California, 12 October 2001, background paper for the Geron Ethics Advisory Board.
3. James Thomson et al., "Embryonic Stem Cells Derived from Human Blastocysts," *Science* 282 (1998): 1145–47.
4. Allan Spradling, Daniela Drumman-Barbosa, and Toshie Kai, "Stem Cells Find Their Niche," *Nature* 414 (2001): 98.

5. National Academy of Sciences, *Stem Cells: Scientific Progress and Future Research Directions* (Washington DC, June 2001), i, ES-1.

6. National Academy of Sciences, *Stem Cells*, ES-3.

7. National Academy of Sciences, *Stem Cells*, ES-2.

8. Ronald McKay, "Stem Cells in the Central Nervous System," *Science* 276 (1997): 66–71.

9. A. J. Friedenstein et al., "Heterotopic of Bone Marrow," *Transplantation* 6 (1968): 230–47.

10. National Academy of Sciences, *Stem Cells*, ES-32.

11. "Rarely have experiments that claim plasticity demonstrated that adult stem cells have generated mature, fully functional cells or that these cells have restored lost function *in vivo*" (National Academy of Sciences, *Stem Cells*, 38).

12. Irving Weisman, testimony to the National Academy of Sciences hearing, Washington DC (May 2001).

13. M. J. Shamblott et al., "Derivation of Pluripotent Stem Cells from Cultured Human Promordial Germ Cells," *Proceedings of the National Academy of Sciences* 95 (Washington DC, 1998), 13726–31.

14. Paolo Bianco and Pamela Gehron Robey, "Stem Cells in Tissue Engineering," *Nature* 414 (2001): 118–21.

15. Thomas Okarma, testimony to Congress, Committee on Science and Technology, Stem Cell Hearings (May 2001), 1. Also see Sally Temple, "The Development of Neural Stem Cells," *Nature* 414 (2001): 112–17; and "Stem Cells Help Paralyzed Mice to Walk," MSNBC report on John Gearhart (paper presented at the Bar Harbor Conference, Maine, 24 July). Also in videos seen by Zoloth, personal communication, 10 December 2001.

16. Tannistha Reya, Sean Morrison, Michael Clarke, and Irving Weissman, "Stem Cells, Cancer and Cancer Stem Cells," *Nature* 414 (2001): 105–11.

17. Reya et al., "Stem Cells," 106.

18. National Bioethics Advisory Committee report, "Religious Responses to Human Embryonic Stem Cell Research" (May 1999).

19. Roe v. Wade, 410 U.S. 113 (1973), at 162; Cynthia S. Marietta, "Frozen Embryo Litigation Spotlights Pressing Questions: What Is the Legal Status of an Embryo, and Can It Be Adopted?," http://www.law.uh.edu/healthlaw/perspectives /2010/marietta-embryolegal.pdf (accessed 20 October 2013).

20. National Legal Research Group, Inc., "Recent Case Law on Division of Frozen Embryos in Divorce Proceedings," http://www.divorcesource.com/research /dl/children/03mar54.shtml (accessed 20 October 2013), referring to Davis v. Davis, 842 S.W.2d 588 (Tenn.), on reh'g in part, 1992 WL 341632 (Tenn. 1992), cert. denied sub nom. Stowe v. Davis, 507 U.S. 911 (1993); Kass v. Kass, 91

N.Y.2d 554, 673 N.Y.S.2d 350 (1998); A.Z. v. B.Z., 431 Mass. 150, 725 N.E.2d 1051 (2000); Cahill v. Cahill, 757 So.2d 465 (Ala. Civ. App. 2000); J.B. v. M.B., 170 N.J. 9, 783 A.2d 707 (2001); and Litowitz v. Litowitz, 146 Wash.2d 514, 48 P.3d 261 (2002).

21. For the view of some Catholics, see United States Conference of Bishops Pro-Life Activities, "President Bush's Stem Cell Decision," 20 August 2001. For that of some Protestants, see General Conference of the United Methodist Church, Resolution 31530-CS-NonDIs-O., 2–12 May 2000.

22. For one Islamic view, see Aziz Sachindina, testimony to the National Bioethics Advisory Committee, Georgetown University Hearing, Washington DC, 2000; and CBS News presentation, 9 September 2001. For other Protestant views, see Ronald Cole-Turner, Karen LeBacqz, and Ted Peters, testimonies before the National Bioethics Advisory Committee, Washington DC, 2000.

23. Objects, even living ones, that cannot be seen by the naked eye have a different weight in Jewish law. Other factors, such as the motility of an object, also play a role in legal status. Avram Steinberg is developing an argument that follows from the *halakhah* that prohibits the murder of a "man inside of a man," one of the seven Noahide Laws. Because this prohibition has been understood to refer to abortion, if the "man" is not "inside of a man," then the act would be different. Using the legal authority to argue from one case to another, stem cell research would be allowable. This argument will be elaborated in further work by Dr. Steinberg (personal communication with Elliot Dorff, December 2001).

24. See Elliot Dorff's article in this volume for a full discussion of the classical sources of this view and the ways in which contemporary rabbis have used them to formulate this stance. In fact, the agreement on this one point across a wide spectrum of Jewish traditions is unusual.

25. Variants of the creation story include heroic or divine-human conceptions, but all are based on sexual union, gestation, birth, and rearing as a linked narrative.

26. Roger Shattuck, *Forbidden Knowledge* (San Diego: Harcourt Brace, 1996). This idea of "ordinary reach" is first discussed in the introduction (45) and is developed throughout the book.

27. K. Okita, T. Ichisaka, and S. Yamanaka, "Generation of Germline-Competent Induced Pluripotent Stem Cells," *Nature* 448 (2007): 313–17, doi:10.1038/nature05934/pmid 17554338.

28. For a discussion of the advantages and problems of stretching Jewish law to address issues that it never faced and for the methodology that Dorff suggests for doing that, see Elliot N. Dorff, "A Methodology for Jewish Medical Ethics,"

Jewish Law Association Studies 7 (1991): 35–57, reprinted in *Contemporary Jewish Ethics and Morality: A Reader*, ed. Elliot N. Dorff and Louis E. Newman (New York: Oxford University Press, 1995), 161–76; and see Elliot N. Dorff, "Applying Jewish Law to New Circumstances," in *Teferet Leyisrael: Jubilee Volume in Honor of Israel Francus*, ed. Joel Roth, Menahem Schmelzer, and Yaacov Francus (New York: Jewish Theological Seminary, 2010), 189–99.

29. For a discussion of how each of these resources within the Jewish tradition can be and has been used in providing moral guidance, even on issues that are completely new in kind or degree, see Elliot N. Dorff, *Love Your Neighbor and Yourself: A Jewish Approach to Modern Personal Ethics* (Philadelphia: Jewish Publication Society, 2003), appendix, 311–44.

30. David Cyranoski, "Acid Bath Offers Easy Path to Stem Cells," *Nature* 505 (2014): 555–608; Haruko Obokata et al., "Stimulus-Triggered Fate Conversion of Siomatic Cells into Pluripotency," *Nature* 505 (2014): 641–47.

CHAPTER 2   *Applying Jewish Law to Stem Cell Research*

ELLIOT N. DORFF

Until the late 1990s, when embryonic stem cells were first isolated, stem cell research was an arcane subject discussed by scientists alone. Since then, though, it has become the subject of intense political debates, with competing bills in Congress that would, on the one hand, criminalize it with ten-year prison terms and, on the other, encourage it with government funding. In the meantime, widespread media coverage has at least brought the subject to the public's awareness, even if most people do not understand it well.

In this chapter, which is based on my legal ruling adopted by the Conservative Movement's Committee on Jewish Law and Standards in March 2002,[1] I analyze embryonic and adult stem cell research from the point of view of Jewish legal sources and suggest a reading of those sources that would guide us in responding to this important new area of proposed research. I am a Conservative rabbi, but Orthodox rabbi Moshe Tendler and Reform rabbi Mark Washofsky, both of whom wrote the official rulings for their movements, agree with my conclusions and largely with my reasoning,[2] so on this issue—wonder of wonders!—most, if not all, Jews interpreting and applying Jewish law agree.

Readers of this volume, however, will find in the following chapters suggestions for using other aspects of the Jewish tradition to tone down the permissiveness and even the enthusiasm for embryonic stem cell research to which the legal sources of the Jewish tradition lead us, as I will explain below. The following chapters will make sense, however, only if readers understand what Jewish law would say about this issue, for they are all reactions to that position. Thus, it is important for readers to understand that legal position first.

Furthermore, although the Jewish tradition has used and continues to use multiple sources to guide Jews morally—including stories, proverbs, communal life, history, theology, prayer, and study—it has defined our moral duties most commonly and authoritatively through Jewish law.[3] In this way, in fact, it differs from other traditions that use conscience, authority figures (e.g., the pope), majority vote, or some other method to determine the right path on any given moral issue.[4] Thus, readers also need to understand the Jewish legal position on stem cell research—and particularly, embryonic stem cell research—because it is through interpreting and applying Jewish legal sources that the Jewish tradition most often makes its moral decisions.

A Jewish Perspective on Stem Cells: Fundamental Theological Principles

1. As stated above, the Jewish tradition uses not only law but also theology, stories, proverbs, communal life, study, prayer, and historical experiences to discern what God wants of us.[5] No legal theory that ignores the theological convictions of Judaism or these other resources of the Jewish tradition and people is adequate to the task, for such theories lead to blind legalism without a sense of the law's context or purpose. Conversely, no Jewish statement that ignores Jewish law can speak authoritatively for the Jewish tradition, for Judaism places great trust in law as a means to identify the moral issues involved in a given case or topic and to weigh differing moral claims in similar cases, thus giving us guidance. My understanding

of Jewish law's perspective on stem cell research therefore will, and must, draw on all these Jewish sources. Indeed, this is just one example of my general philosophy of Jewish law, which sees it as a living, organic system that integrates all of these various factors in formulating Jewish law and practice.[6]

2. God requires of us that we seek to preserve human life and health (*piqquah nefesh*). As a corollary to this, we have a duty to seek to develop new cures for human diseases. Saving a life directly is clearly demanded by the Jewish tradition; engaging in research to make that possible is not as clearly commanded in Jewish sources, but because the latter expands our ability to do the former, engaging in scientific research to develop cures, while not the duty of saving lives itself, is entailed in it.

3. The Jewish tradition accepts both natural and artificial means to overcome illness. Physicians are the agents and partners of God in the ongoing act of healing. Thus, the mere fact that human beings created a specific therapy rather than finding it in nature does not impugn its legitimacy. On the contrary, we have a duty to God to develop and use any therapies that can aid us in taking care of our bodies, which ultimately belong to God.[7] That said, Rabbi Avraham Steinberg, a prolific Orthodox writer on Jewish medical ethics and a participant in the project on which this volume is based, wrote this to me in private correspondence:

> Although, fundamentally, involvement with Creation is permissible, there are three preconditions:
> a. There is no inherent halakhic prohibition in the particular acts involved in the technological advancement.
> b. The efforts toward improvement of Creation do not result in an irremediable prohibition (as, e.g., murdering some people in an effort to save others).
> c. The benefit-to-harm ratio for humans is positive.

4. At the same time, all human beings, regardless of their levels of ability and disability, are created in the image of God and are to be valued as such.

5. Moreover, we are not God. We are not omniscient, as is God, and so we must take whatever precautions we can to ensure that our actions do not harm ourselves or our world in the very effort to improve them. A certain *epistemological humility*, in other words, must pervade whatever we do, especially when we are pushing the scientific envelope, as we are in stem cell research. We are, as Genesis says, supposed to work the world *and* preserve it;[8] it is that *balance* that is our divine duty.

6. Animals are part of God's world and deserve to be protected from pain as much as possible whenever we interact with them. (This is the principle known in Rabbinic literature as *tza'ar ba'alei hayyim*, [avoiding] pain to living animals.) Only human beings, however, are created in the image of God, and so we may and should use animals for medical research before we experiment on human beings. This is, of course, accepted medical practice in North America and elsewhere. This chapter will assume that scientists have done all their initial experiments of any proposed therapy on animal cells and then on full animals before they turn to using human beings to test the new therapy both because the Jewish tradition requires it and because the research methods scientists use demand it. Scientists test new therapies on humans only when experiments on animals suggest that there is a good chance that the therapy will work in humans.

   Whether human cell lines should be used to test therapies before full animals are used depends on the status of those human cells, as developed below; in fact, one of the promises of stem cell research is precisely that it can test therapies without harming animals or humans and, in addition, may produce more accurate predictions as to whether a given therapy will work in humans.

Using Aborted Fetuses for Research

Because human embryonic germ (EG) cells may be procured from aborted fetuses, the status of abortion within Judaism immediately comes into question. Its status, in turn, depends on the way that the Rabbis viewed gestation. Furthermore, because procuring embryonic stem (ES) cells requires taking out the inner cell mass of an early embryo and thus killing it, the status of the embryo and then the fetus is critical in determining the permissibility of using them for research.

According to the Talmud, during the first forty days of gestation, the fetus is "simply liquid." During most of gestation—specifically from the forty-first day until birth—the Rabbis classify the fetus as "like the thigh of its mother."[9] The Rabbis clearly determined this view of the early embryo and fetus based on what they saw when women miscarried at various points of pregnancy.

Neither men nor women may amputate their thigh at will because our bodies belong to God. We have them on trust during our lives, and hence we are forbidden to inflict injuries on ourselves.[10] On the other hand, if the thigh turns gangrenous and threatens a person's life, then both men and women have the duty to have their thigh amputated in order to save their lives. Similarly, if the woman's life or health is at stake, an abortion *must* be performed to save the life or the physical or mental health of the woman, for she is without question a full-fledged human being with all the protections of Jewish law, while the fetus is still only part of the woman's body.

When there is an elevated risk to the woman beyond that of normal pregnancy but not so much of a risk as to constitute a clear threat to her life or health, abortion is permitted but not required; under such circumstances (e.g., if the woman has diabetes), the woman should assess the risks of carrying the baby to term in consultation with the father, other members of her family, her physician, her rabbi, and anyone else who can help her grapple with the many issues involved in her particular case, and then she may decide to take the risks involved or to abort the pregnancy. This intermediate category, where abortion is permitted but not required, would include cases where the fetus poses serious threats to the mother's mental health, as, for example, if the fetus was conceived through incest or rape. In such circumstances, the woman may choose to abort, or, alternatively, because the child poses no physical risk to her beyond that of normal pregnancy, she may choose to carry the child to term and give it up for adoption or even raise it herself. Some recent authorities, including the Conservative Movement's Committee on Jewish Law and Standards, would also permit abortions in cases where testing indicates that the fetus is "severely defective," suffering from serious malformations or terminal diseases like Tay-Sachs.

Where no physical or mental risk exists beyond that of normal pregnancy, though, Jewish law would forbid abortion, not as an act of murder but as an act of self-injury.[11] Thus, Jewish law would forbid abortion on demand (i.e., because the couple simply does not want another child) or for economic reasons. Those are good reasons to use birth control but not to abort.

In sum, the official statement on abortion of the Conservative Movement's Committee on Jewish Law and Standards says this:

> Jewish tradition is sensitive to the sanctity of life, and does not permit abortion on demand. However, it sanctions abortion under some circumstances because it does not regard the fetus as an autonomous person. This is based partly on the Bible (Exodus 21:22–23), which prescribes monetary damages where a person injures a pregnant woman, causing a miscarriage. The Mishnah (*Ohalot* 7:6) explicitly indicates that one is to abort a fetus if the continuation of pregnancy might imperil the life of the mother. Later authorities have differed as to how far we might go in defining the peril to the mother in order to justify an abortion. The Rabbinical Assembly Committee on Jewish Law and Standards takes the view that an abortion is justifiable if a continuation of pregnancy might cause the mother severe physical or psychological harm, or when the fetus is judged by competent medical opinion as severely defective. The fetus is a life in the process of development, and the decision to abort it should never be taken lightly. Before reaching her final decision, the mother should consult with the father, other members of her family, her physician, her spiritual leader and any other person who can help her in assessing the many grave legal and moral issues involved.[12]

The upshot of the Jewish stance on abortion, then, is that *if* a fetus had been aborted for legitimate reasons under Jewish law, then the aborted fetus may be used to advance our efforts to preserve the life and health of others. In general, when a person dies, we must show honor to God's body by preparing it for burial and burying it as soon after death as possible. To benefit the lives of others, though, autopsies may be performed when legally required or when the cause of death is not fully understood, and Jews are urged to make their organs available for transplant to enable other people to live.[13] If we may and even should use the bodies of human beings to enable others to live, how much the more so may we use a part of a body—in this case, the fetus—for that purpose.

This all presumes, though, that the fetus was aborted for good and sufficient reason within the parameters of Jewish law. The law in the United States and Canada permits the mother to abort at will until the fetus is viable outside the womb, while Jewish law permits abortion only under the more restrictive conditions described above. Thus, undoubtedly some North American Jews abort their fetuses for reasons not justified by Jewish law. While in North America one can presume that the majority of aborted fetuses are not Jewish, the Rabbis understand the Noahide Covenant, given to all descendants of Noah, to forbid abortion altogether or, according to another opinion, to allow it only if the mother's life is at stake.[14] Non-Jews in North America also abort for many other reasons. Thus, one might think that doing research using embryonic stem cells from aborted fetuses would constitute a *mitzvah ha-ba'a b'aveirah*, a commanded act accomplished through a sin, and thus using the materials themselves would be forbidden.

The Talmud, though, restricts that consideration to prohibiting the people who committed the wrongful act from benefiting from it; after the fact, the Talmud specifically permits the community to benefit from such a sin in performing a commanded act of its own, a *mitzvah d'rabbim*.[15] Thus, even if Jewish law would not condone the particular abortion, once it has been done we may use the aborted fetus for a sacred purpose like curing diseases and saving lives. Using aborted fetuses to do *research* is not as directly and clearly permitted as using them for the cures themselves once they have been developed; but because aborted fetuses would otherwise just be discarded or buried, rabbis across the denominations extend the permission to use them for research that holds out the hope for curing diseases and saving lives.[16] What is critical here is what the Talmud states and what Orthodox rabbi J. David Bleich recognizes, namely, that the results of a prohibited act may be used for sacred purposes without in any way condoning the prohibited act. This is especially relevant to our case, for it is highly doubtful that a woman would agree to have an abortion solely for the purposes of providing a fetus for scientific research, and so using the fetus would not encourage abortion. Moreover, in our case, at least some, and perhaps many, of the aborted fetuses may have been aborted for reasons approved by Jewish law.

In sum, then, if a fetus was aborted in accordance with the dictates of Jewish law, we clearly have the right to use it for research purposes. Even if it was not aborted for reasons sanctioned by Jewish law, there are sufficient grounds in Jewish law to permit using it for research intended to produce cures for human ailments.

Using Donated Embryos Originally Created to Overcome Infertility for Research

Stem cells for research purposes, though, can also be procured from sperm and eggs mixed together in a petri dish and cultured there. In fact, in light of the controversial nature of abortion in the United States, scientists are much more likely to procure stem cells from embryos created by couples in the process of using in vitro fertilization and other treatments to overcome infertility than they are from aborted fetuses. These are typically embryos that have been frozen until such time that the couples decide to use them. Some couples ultimately decide never to use some of their frozen embryos, either because they have already had as many children as they want or because they have given up in their effort to bear a child. Some of those couples are, in turn, willing to donate their frozen embryos to research rather than simply discarding them. This, then, raises the question of the status of such early embryos in Jewish law.

According to the Talmud, during the first forty days of gestation, the embryo is "simply fluid (literally, water)."[17] That is because, as the Mishnah asserts, "a woman who miscarries [up to or] on the fortieth day need not worry that she has delivered a child [for which she has to observe the special period of impurity after the birth of a child, for] . . . the Sages say, the creation of both the male and the female takes place on the forty-first day."[18] Furthermore, according to the Mishnah, "Anything that does not have the form of a child is not a child,"[19] and thus an embryo before the forty-first day, which is without a form, is not a child. Maimonides calls upon his medical experience in saying that even on the forty-first day the figure of a human being is "very thin" and that within forty days "its shape is not yet finished."[20] Similarly, the Shulhan Arukh specifies that it is a vain prayer (tefillat shav) if a man prays after the fortieth day that his pregnant wife is carrying a boy, for by the forty-first day the gender of the child has already been determined.[21]

In our own day, when we understand that the fertilized egg cell has all the DNA that will ultimately produce a human being, we must clearly have respect for human embryos and even for human gametes alone (sperm and eggs), for they are the building blocks of human procreation. This is generally understood to entail a ban on abortion even during the first forty days of gestation except for therapeutic purposes. Indeed, the Jewish tradition demands that we have respect even for inanimate objects such that we refrain from destroying them unnecessarily (bal tashhit),[22] and if that is true, how much the more so must we respect animate and living substances such as human cells. Even the prohibition against destroying inanimate objects has its limits, though; we may, after all, use inanimate objects for our purposes, and we may even kill plants and animals for food. Thus, the question is this: Even if we may ultimately use embryonic stem cells for research, what level of respect should we ascribe to them, especially those outside the womb, where they have no potential for becoming a human being unless implanted in a woman's uterus, and how should that level of respect find expression in action?

We can, but we should not, say, in a positivistic mode, that because the sources of Jewish law never talk about embryos outside the womb, no law exists on the subject, and we may do with them whatever we wish.

This approach is at least honest—it states clearly where the tradition has nothing directly on point to say about a given issue—but ultimately it is an irresponsible way to approach Jewish sources, for it makes the Jewish tradition irrelevant to many modern issues not contemplated in the past. That does a disservice both to the Jewish tradition and to contemporary Jews trying to live by it.

On the other end of the methodological spectrum, we can, but we should not, try any device that will use the tradition to give us the moral advice we want. Some justify such an approach by citing the Mishnah, which says, "Turn it over, and turn it over again, for everything is in it."[23] That, though, misreads the Mishnah, which is describing the awe we should have for the Torah but not—certainly, not necessarily—prescribing the way that we should gain moral guidance for issues in our lives. Moreover, in the end, this approach involves doing eisegesis rather than exegesis—that is, reading whatever we want into the text rather than actually deriving guidance from it. This not only is a dishonest use of the text—interpreters are claiming that the text tells us something that it is not even addressing, let alone ruling on—but also is problematic from a practical point of view, for this methodology will result in good and wise advice on any contemporary issue only by coincidence. After all, deducing guidance from an ancient text that never even contemplated modern realities, let alone dealt with them, cannot possibly take into account the real issues that make the issue a question in the first place.

Where no relevant precedents exist, we must rather do "depth theology"—that is, we must seek to apply foundational Jewish concepts and values to the new case. People can, of course, disagree as to which concepts are relevant or how to apply them, but Jewish law is no stranger to disputes even when rabbis are reading and weighing precedents that are on point to the case at hand. Determining whether or not existing precedents are relevant is itself a matter of judgment. Still, when past rulings do not seem to give moral or legal direction, identifying Jewish concepts and values that can reasonably apply to the case at hand is the proper method to use, for it has the advantage of enabling the tradition to speak to new circumstances in a way that, while not a direct conclusion from the tradition, is strongly rooted in it.[24]

In this case, the Rabbis' classification of a fetus in the uterus up to forty days of gestation as "simply fluid" is a good precedent for us to consider in determining the status of such a fetus outside the womb—but only if modern science does not undermine the basis for seeing the embryo that way and, on the contrary, suggests some grounds for that talmudic perception. If, instead, that is only outdated science, we cannot reasonably rely on this Rabbinic precedent.

As it happens, modern science provides good evidence to support the Rabbis' understanding. As Rabbi Immanuel Jakobovits, former chief rabbi of the United Kingdom and author of the first full volume on Jewish medical ethics, noted long ago, the Rabbis' "forty days" are, by our genetic count, approximately fifty-four days, for the Rabbis counted from the woman's first missed menstrual flow, while geneticists today count from the point of conception, which is usually about two weeks earlier.[25] By fifty-four days of gestation the basic organs have already appeared in the fetus. Moreover, we now know that it is exactly around this time that the fetus begins to get bone structure and therefore looks like something other than liquid.[26] Indeed, the Rabbis probably came to their conclusion about the stages of development of the fetus because early miscarriages indeed looked like "merely fluid," while those from fifty-four days on looked like a thigh with flesh and bones. For that matter, even the Rabbis who proclaimed the embryo in the first forty days to be "simply water" clearly were announcing an analogy and not an equivalence, for they clearly knew that from that water a child might develop, unlike drinking water in a glass!

Thus, while we should have respect for gametes and embryos in a petri dish as potential building blocks of life, they may be discarded if they are not going to be used for some good purpose. If an embryo during the first forty days of gestation is "simply fluid," an embryo situated outside a woman's womb, where it cannot with current technology ever become a human being, surely has no greater standing; it is *at most* "simply fluid." Therefore, when a couple agrees to donate such embryos for purposes of medical research, our respect of such preembryos and embryos outside the womb should certainly be superseded by our duty to seek to cure diseases. Finally, because the embryo is "simply fluid" and "not a child" during the first forty days and all the more so within the first fourteen days, when stem cells would be removed for research, our duty to seek to cure diseases provides ample warrant, in my opinion, for removing the inner cellular mass in the first place so that stem cell research can go forward. In doing so, we are not killing a human being, as we would be if we were to remove a person's heart before death; we are rather taking a part of an object that has not yet even achieved the status of a formed fetus.

What, then, is the nature of the respect that we have for preembryos and even gametes? It is, in part, practical: we do not use them for creating cosmetics, cat food, or the like but only for human life and health. When they cannot be used for either, they may be discarded, but using them for stem cell research would be closer and more appropriate to their natural roles in life and would certainly not constitute a breach of the respect due them.[27] Moreover, our respect is attitudinal: we engage in this research with a certain fear and trembling, for we know that we are intervening in the processes of creating and sustaining life. As a result, we do not use frozen embryos with impunity: even if we decide to discard them, we do so after giving considerable thought to the matter, as I am doing in this essay, unlike most other things that we discard, and the same sustained moral reflection is required for using them in research.

What would happen, though, if we could gestate a human being entirely outside a woman's womb in some sort of machine? Infertile couples who can produce sperm and eggs but cannot carry a baby to term might indeed be highly interested in such a possibility. Would that change our perception of the fetus during its first days of gestation?

The problem is more theoretical than real, for it would take considerable time, effort, and money to develop such a machine. At this time, we do not know enough of what happens in the uterus even to know what we should try to reproduce artificially, let alone have the ability to do that. Moreover, given the options of surrogate mothers and adoption, and given the inevitably great cost of developing and then using such a gestation machine, it is not likely that such machines will be available for quite some time, if ever.

Second, it is important to note that the wisdom and authority of moral and legal decisions depend critically on their context. Sexual intercourse, for example, is both a good and indeed a commanded act in the context of marriage,[28] but it becomes one of the three things that we Jews are commanded never to do, even on pain of death, if it is in the context of adultery or incest.[29] Similarly, *at this time*, at least, we can and must say that an embryo outside a woman's womb is relevantly different from an embryo within a woman's womb. If and when we develop the ability to gestate a person outside a woman's womb, then the physical location of an embryo in a petri dish may cease to have as much moral import as it does now, but that would be a *different* context requiring a new weighing of the evidence.

The most important thing to note, though, is that I am not basing my argument for seeing the embryo as less than a person solely on the basis of where the embryo in a petri dish happens to be.[30] Rather, characteristics of the early embryo itself argue for assigning it the status of "mere fluid." Specifically, to procure stem cells, scientists can use only embryos during the first fourteen days of gestation, for then the neural streak, which later develops into the spine, appears. During that early period, the embryo in a petri dish can be distinguished from a human being not only according to its location outside of a womb and its resulting inability to develop into a human being (i.e., its lack of human potential) but also by *its low level of cell organization, the short period of time that it will remain in this state, and its incapacity to live on its own.* Thus, if very good scientific reasons support the talmudic precedent to classify an embryo of up to forty days to be "mere fluid," an embryo of fourteen days of gestation or less is even more justifiably classified as that, even if it were within a woman's uterus, and how much the more so outside one. At no point during those fourteen days, then, do stem cells become a human entity, and so embryonic stem cell research creates the possibility of enormous good at no human price. (In contrast, we regularly use full human beings in medical research because animal research alone cannot guarantee the safety and efficacy of medications for human beings. While we take safeguards to protect the human subjects in such research, people have died as part of such research, including the well-publicized cases of Jesse Gelsinger at the University of Pennsylvania and Ellen Roche at Johns Hopkins University. Stem cell research poses no such risks to human beings.) Thus, even if it were possible to gestate a human being mechanically, we would still have good reasons to classify an embryo during the first forty days as "simply fluid" and thus to use it for stem cell research.

In sum, then, frozen embryos originally created for purposes of overcoming infertility but that the couple no longer intends to use for that purpose may be discarded, but they may also be used for good purposes. One such purpose is to produce stem cells for medical research. Indeed, couples should be encouraged to donate their extra embryos—and any fetuses that they abort—to such efforts. Men or women are not duty-bound to do so, but such a donation is minimally an act of *hesed*, of loyalty and love, and possibly, given its goal of cure, even a mitzvah.[31]

Creating Embryos, Harvesting Eggs, or Doing Parthenogenesis Specifically for Research

Couples, though, are often reticent to donate their extra frozen embryos for research.[32] Moreover, scientists are increasingly wary of using embryos donated by couples having infertility problems for fear that the couple was having difficulty having children precisely because their embryos are defective in some way, and that might prevent those embryos from being good material for purposes of research and ultimately cures. This has led scientists to investigate other possibilities of obtaining embryonic stem cells. Creating embryos specifically for the purpose of doing medical research lacks the justification of using materials that would just be discarded anyway, but creating embryos specifically for research is nevertheless permissible under Jewish law with one condition.

Unlike the Catholic view, the problem for the Jewish tradition in doing this is *not* that it would amount to murder to destroy an embryo outside the uterus, for in that state an embryo has no greater claim to protection than an embryo in its first forty days in the uterus, much less than a person. Based, in part, on the story of Onan in Genesis 38, classical Jewish law forbids "wasting seed" (*hashatat zera*).[33] Even so, procuring the sperm through masturbation for "farmed" embryos would not constitute "wasting seed," for here the purpose of masturbating would be specifically to use the man's semen for the consecrated purpose of finding ways to heal illnesses.[34]

Procuring eggs from a woman for this purpose, however, does pose a problem. It is not so much that this requires subjecting her to an invasive medical procedure, for now eggs can be procured without surgery and with minimal risk or pain through laparoscopy. To produce the eggs, though, the woman must be exposed to the drugs that produce hyperovulation, and there is some evidence that repeated use of such drugs increases a woman's risk of ovarian cancer and other maladies.[35] Although such risks may be undertaken to overcome a woman's own infertility or even, I have held,[36] once or twice to donate eggs to infertile couples, assuming such risks for medical research is less warranted, especially because embryos can also be obtained from frozen stores that couples plan on discarding and possibly from some other methods. Still, the demonstrated risks for her to do this once or twice are minimal, especially if she is prescreened and deemed safe to undergo that procedure, and so a woman may donate eggs for this purpose with those limitations.

The same concerns about the risks in procuring human eggs would apply to using eggs to obtain stem cells from cloning procedures or from parthenogenesis, if that proves to be possible. Thus, while obtaining embryonic stem cells from frozen embryos that would otherwise be discarded is best, embryos may also be specifically created and eggs may be cloned or tricked into producing stem cells through parthenogenesis for purposes of medical research on the condition that the woman providing the eggs for such efforts is prescreened to insure her safety and even then does this only once or twice.

Some have raised two other objections to creating embryos intentionally for research. The first is that the embryo in a petri dish is, after all, potential life in that it could be implanted in a woman's uterus, and someday we may even be able to grow it in a machine. The second is that allowing the use of embryos specifically created for research creates a slippery slope in that human genetic materials will then be diminished in our estimation as just a means to a practical end, that human creation will lose its mystery and holiness. In Kantian terms, this smacks of violating the second version of the categorical imperative—that is, never treat a person merely as a means. Or, in the terms of more modern theorists (even though I am assuming that the donors are not paid), this seems awfully close to commodifying people in that we are looking at both men and women as (merely?) sources for genetic products. Without articulating it precisely this way, it is this concern that often underlies how the average person responds to the prospect of using embryos for research.

Embryonic stem cell research does, of course, entail the destruction of potential life, but one must remember that the embryo in a petri dish remains potential life only through considerable scientific interventions to provide an environment where the zygote will remain alive in that state. Its hold on life is, at best, tenuous; indeed, because as many as 75 percent of conceptions and approximately a third of implanted zygotes miscarry, usually in the first months of pregnancy,[37] the Rabbis were right in classifying such early embryos as "merely fluid"—and that is in a woman's uterus. In a petri dish, there is zero chance for the embryo to become a human being. Thus, it seems to me that we need to realize how weak the potentiality of that life is.

In contrast, the potentiality of stem cell research rests on a solid foundation of successful attempts to use adult stem cells in humans and embryonic stem cells in animals. Thus, when we speak of an embryo in a petri dish, we must remember that we are, at most, balancing *potential* life (and only if it is transplanted into a woman's uterus) against what we have good reason to hope will be actual treatments for serious diseases; that is much easier to justify than balancing *actual* lives against that hope, as we do whenever we use human subjects in medical research. If we do the latter—and we must, albeit under stringent controls, if we are ever going to have medications that are safe and effective—then we may do the former with yet greater warrant.

As for the second objection, I certainly agree that human creation must be honored and respected and that steps to protect that special status must be taken. The critical thing to note, though, is that we are not dealing with a *person* when we use embryos to advance stem cell research; we are dealing with genetic materials that, even in a uterus, have a long way to go before they become a person, with three-to-one odds that the zygote will miscarry and with an additional 33 percent chance that an embryo later in pregnancy will miscarry. *That is, especially in the first ten days or so of gestation (during which the inner cell mass would be removed in embryonic stem cell research), we are dealing with a thing, not a person.* That is what the classification of embryos as "merely fluid" entails. We surely are allowed—indeed, commanded—to use *things* to find ways to cure diseases. Moreover, in our case I do not see a serious danger of a slippery slope in the status of human genetic materials, for the use to which these embryos would be put is nothing less than another holy cause—namely, seeking to cure people of serious diseases. Thus, I do not consider the deliberate creation of embryos for purposes of stem cell research to demean the birth process in any way.

Removing a Cell from an Embryo for Research

Another suggested approach to obtain stem cells is to remove a cell from an early embryo (at, say, five to ten days of gestation) and then cultivate that cell to produce a line of stem cells, leaving the remaining embryo to develop. Because of significant experience in removing a cell from early embryos for preimplantation genetic diagnosis (PGD), we now have ample evidence that this does not negatively affect the resulting fetus or child.

This method poses no problems for the Jewish tradition whatsoever. The embryo itself outside the womb is at most "simply fluid," and, moreover, in this procedure the embryo from which the cell was taken can still develop normally if implanted in a woman's womb. The only disadvantage of this method from a moral point of view is that we are then using cellular material that would not otherwise be discarded. Still, given that we now know that no harm is done to the resulting embryo, it is perfectly permissible to do this.

Whether couples will be willing to have their embryos used in this way, though, is another matter. It is one thing to invade the embryo for PGD that might aid the couple in deciding whether to implant a particular embryo or not; it is another to take a cell from their embryo for purposes of research. Given this reality, frozen embryos that would otherwise be discarded are probably a more realistic source. Still, this method is as acceptable as using frozen embryos that would otherwise be discarded for the sacred purpose of trying to cure diseases.

Using Adult Stem Cells for Research and Therapy

When tissues or organs from another person are introduced into a patient, his or her immune system attacks the new tissues or organ as a foreign intruder. That is why organs and even blood must be typed in an attempt to prevent or at least minimize this problem. In the case of organ transplantation, even if the donor is a near match, usually the patient must take drugs for the rest of his or her life to prevent the immune system from attacking the new organ.

If needed tissues or organs could be produced from patients' own bodies, though, they would not need to take immunosuppressive drugs for the rest of their lives and endure the side effects of those drugs. As chapter 1 explains, that is precisely the hope of researchers and medical personnel. It would be wonderful if they could take cells from a patient's body, "reverse the arrow of time" by "winding them down" to an undifferentiated, pluripotent state similar to embryonic stem cells, and then reintroduce them into the patient's body, where they would differentiate into the cells needed to replace the unhealthy ones in the patient's body. The first attempts to do this proved fruitful in 2007, but the process, while curing the original disease, also produced cancers in patients. Furthermore, it is not yet clear that stem cells produced from adult cells are as flexible as embryonic stem cells so as to be able to produce any cell in the human body. Moreover, the pluripotent cells produced by winding down adult cells have not reproduced themselves over and over again, as embryonic stem cells have, and therefore this method may not produce a sufficient supply of healthy cells to cure the disease. So this method of using stem cells to investigate and ultimately cure diseases has yet to be perfected.

If scientists do find ways to avoid these problems, though, using cells from a patient's own body to cure his or her disease is clearly the best possible alternative, both pragmatically and morally. Pragmatically, as explained above, this would avoid the need for the patient to use immunosuppressive drugs for the rest of her or his life. Moreover, patients with diseases would likely be very willing to use some of their own cells to cure their disease, so procuring cells for the cure would not be a problem. Morally, the use of adult cells means that we do not even have to ask about the moral status of the embryo or aborted fetus because we are not using either one. So the only question is whether doctors may "assault" the patient's body by taking some of his or her cells in order to save his or her life or vastly improve his or her health. The answer to that question is a resounding "Yes!" by all Jewish writers and by all writers in other traditions that approve of medical interventions in the first place.

Other Factors in the Use of Embryonic Stem Cells

Given that the materials for stem cell research can be procured in permissible ways, the technology itself is morally neutral. It gains its moral valence on the basis of what we do with it.

The question, then, reduces to a risk-benefit analysis of stem cell research. The articles in a *Hastings Center Report* raise some questions to be considered in such an analysis.[38] I will not repeat them here but only note two things about them from a Jewish perspective.

First, the Jewish tradition sees the provision of health care as a communal responsibility,[39] and so the justice arguments in the *Hastings Center Report* have a special resonance for Jews. That is, when and if this technology becomes available, poor people as well as the middle class and the rich should be able to benefit from it. That is especially true because much of the basic science in this area was provided by public funds. At the same time, the Jewish tradition does not demand socialism, and for many good reasons, we in the United States have adopted a modified, capitalistic system of economics. The trick, then, will be to balance access to applications of the new technology with the legitimate right of a private company to make a profit from its efforts to develop and market applications of stem cell research.[40]

Second, the potential of stem cell research for creating organs for transplant and cures for diseases is, at least in theory, both awesome and hopeful. As a result, in light of our divine mandate to seek to maintain life and health and its corollary to develop therapies so that we can do that, I would argue that, from a Jewish perspective, proceeding with such research is not only permissible but *mandatory*. Moreover, given the immense potential for cures in stem cell research, we should pursue it aggressively. As difficult as it may be, though, we must draw a clear line between uses of this or any other technology for cure, which are to be applauded, as against uses of technology for enhancement, which must be approached with caution.

As I shall explain below, enhancement and therapy do not present a neat and clear dichotomy; rather, they lie on a spectrum, where the ends are easy to define but the middle is murky. Research to cure cancer, neurological diseases, and the like is clearly therapeutic. But because Jews have been the brunt of campaigns of eugenics in both the United States and Nazi Germany,[41] we are especially sensitive to creating a model human being that could be replicated through technologies now available and others to come. Moreover, when Jews see a disabled human being, we are not to recoil from the disability or count our blessings for not being disabled in that way; rather, we are commanded to recite a blessing thanking God for making people different.[42] Contrary to Nazi policy, then, we clearly should not kill the disabled but value them as much as we do the (temporarily) able-bodied while at the same time striving to prevent and cure disabilities.

Defining exactly where the category of disability (and therefore therapy) ends and where that of enhancement begins, though, is very challenging, especially because people's expectations change continually as medicine develops. Thus, what looks like enhancement today may look like expected therapy tomorrow. Eyeglasses, for example, might have been considered enhancement at some point in the past, while now they are clearly therapy, covered by most medical insurance plans. Similarly, abortions to prevent the birth of malformed fetuses are now justified as "therapeutic abortions," even though a generation ago we had no idea of the fetus's status in the uterus and would have considered an abortion based on the possibility of malformation unwarranted.

Contemporary philosophers have addressed the therapy/enhancement divide. Some suggest a definition of disease that would sharply distinguish the two, while others, like Erik Parens, maintain that the distinction simply cannot be neatly defined and that we should therefore consider each situation on its own merits.[43]

Although genetic engineering poses the problems of enhancement much more starkly than does stem cell research, it is important to underscore that this chapter addresses only stem cell research for purposes of what are clearly medical cures. Discussion of the use of this or any other technology for purposes of enhancement requires further consideration.

What We Should Do Now

In sum, then, after scientists have accomplished all that they can toward a given therapeutic goal through animal experiments, we may and should take the steps necessary to advance both adult and embryonic human stem cell research and its applications in an effort to take advantage of its great potential for human healing for two reasons. First, we have a duty to heal and, as a corollary to that, to develop our abilities and methods to heal. Second, genetic materials, including early stage embryos, lack the status of a person or even part of a person (e.g., a thigh). So while embryos and even gametes deserve our respect, for they are the materials that have the potential of creating human beings, that status is outweighed by the duty to seek to cure.

In accordance with Jewish law, stem cells may be procured from a variety of sources, and the following list ranks these sources from the most desirable to the least desirable:

a. Adult cells that are "wound back" to a pluripotent state, especially those from the patient who needs the therapy. As indicated above, this method avoids the need for using immunosuppressive drugs and the moral problem of the status of the embryo, but there are some serious drawbacks to overcome in using adult cells that are reversed into pluripotent cells.

b. Aborted fetuses.

c. Frozen embryos originally created for overcoming infertility that the couple has now decided to discard but has agreed to donate for stem cell research instead. In both (b) and (c), researchers are not responsible for the abortion itself or for creating the frozen embryos, and they are using materials that would otherwise just be discarded, but option (b) avoids legal fights over frozen embryos as well as the frequent unwillingness of couples to donate their frozen embryos just in case they decide to use them later. At this stage of research, we do not know, but it may be the case that more can be done with embryonic stem cells than with embryonic germ cells because the latter have already differentiated into reproductive cells, while the former are totipotent. If stem cells indeed turn out to be more malleable, then (b) and (c) may be reversed in order, or it may be that both sources are equally acceptable, each with its advantages and disadvantages.

d. A cell taken from an embryo and grown independently. This technique would avoid the extra dangers to the woman involved in the methods listed in option (e) below, but it does not have the advantage of using materials that would otherwise be discarded, as in (b) and (c). So couples may be more reticent to cooperate with this method of obtaining embryonic stem cells than they would be using method (b) or (c) above.

e. Embryos created specifically for medical research by combining sperm and eggs donated for that purpose, by cloning (somatic cell nuclear transfer, or SCNT), or by parthenogenesis. These are the least desirable because of the increased danger to the woman donating her eggs, but they are permissible sources of stem cells if the woman donates eggs for this purpose only once or twice after being prescreened to insure that it is safe for her to do this. A man does not violate any Jewish laws by masturbating to contribute to stem cell research.

Although the use of adult cells, especially a patient's own, is clearly the easiest to justify morally, fixing the practical problems with this method should not replace nor even slow down our attempts to develop healing methods from embryonic stem cells, for at this stage of the science, we do not yet know whether the pluripotent stem cells produced from adult cells hold out the same promise as do embryonic stem cells.

All stem cell research, adult and embryonic, should carry with it restrictions to enable access to its applications to all who need it. This is part of the broader Jewish mandate for the community to provide needed health care for those who cannot otherwise afford it, but it applies to stem cell therapies as well. In fact, the more the government funds such research, the more the public will have a right to demand that the cures that we hope will result from it are available to all who need them, with appropriate provisions for those who cannot afford them.

As difficult as the distinction between therapy and enhancement is to define, and as much as the line may change over time, here I am endorsing embryonic stem cell research only for purposes of therapy. Much more must be considered in evaluating the possible use of this and other techniques for purposes of enhancement. When our goal is curing disabilities or diseases, though, the Jewish tradition would impel us to pursue both adult and embryonic stem cell research aggressively. May God grant us success in pursuing this promising new methodology to bring healing to many who suffer.

1. Elliot N. Dorff, "Stem Cell Research," http://www.rabbinicalassembly.org
   /sites/default/files/public/halakhah/teshuvot/19912000/dorff_stemcell
   .pdf (accessed 8 August 2012), and printed in *Conservative Judaism* 55, no. 3
   (2003): 3–29.
2. For Rabbi Moshe Tendler's position, see his testimony to President Clinton's
   National Bioethics Advisory Commission at http://bioethics.georgetown.edu
   /nbac/stemcell3.pdf (accessed 8 August 2012), H1–5, and see official Rabbinical Council of America statements based on Rabbi Tendler's views at http://
   www.rabbis.org/news/article.cfm?id=100553 and http://www.rabbis.org/news
   /article.cfm?id=105421. For Rabbi Washofsky's position, see his book *Jewish Living: A Guide to Contemporary Reform Practice* (New York: UAHC Press,
   2001), 237, 245.
3. For a discussion of how Judaism uses all of these sources to make its moral
   decisions, see Elliot N. Dorff, *Love Your Neighbor and Yourself: A Jewish Approach
   to Modern Personal Ethics* (Philadelphia: Jewish Publication Society, 2003),
   appendix, 311–44.
4. For more on the use of law versus other methods to address moral issues, see
   Elliot N. Dorff, *Matters of Life and Death: A Jewish Approach to Modern Medical
   Ethics* (Philadelphia: Jewish Publication Society, 1998), appendix, 395–423.
5. For more on these theological principles and other fundamental assumptions
   of Jewish medical ethics, and for the Jewish sources that express these convictions, see Dorff, *Matters of Life and Death,* chap. 2.
6. Elliot N. Dorff, *For the Love of God and People: A Philosophy of Jewish Law* (Philadelphia: Jewish Publication Society, 2007). Also see Dorff, *Matters of Life and
   Death.*
7. When I shared an earlier version of the responsum on which this article is
   based with Rabbi Avraham Steinberg, MD, a prolific Israeli Orthodox writer
   in matters of medical ethics, he wrote me this:

   > It seems to me right that you did not use the popular term "playing G-d,"
   > because from our theological point of view any and all the innovations
   > of science deal with the unearthing of preexisting factors in the nature
   > of Creation; the utilization of knowledge of nature for varying uses in no
   > way constitutes a new creation. We are speaking of creation of humans
   > "extant from extant," which diverges from nature in technique alone, not
   > in substance. Only the Creator of the universe is "able" to create a world
   > *ex nihilo.* Similarly, these technologies do not solve the mystery of life or
   > disclose life's basic, fundamental essence. On the contrary, these wondrous

genetic discoveries can strengthen one's faith in the Creator of the world because where there are laws of nature, there is a Creator. It is a confirmation of the biblical verse (Psalms 104:24), *"How abundant are your works, O Lord, with wisdom You made them all."*

Dr. Steinberg cites the following sources articulating this conviction that we are to see ourselves as partners with God in improving the world: B. *Bava Kamma* 85a and Rashi there, based on Exodus 21:19; *Genesis Rabbah* 11:7; *Pesikta Rabbati* #23; *Midrash Terumah* #2; *Tanhuma Yashan*, cited in *Torah Shelemah* on Genesis 2:3 (comment #58); *Ramban* on Genesis 1:28.

8. Genesis 2:15.
9. B. *Hullin* 58a, where the status of the fetus is a dispute between Rabbi Eliezer and Rabbi Joshua; B. *Sanhedrin* 80b, where the position that the fetus is the thigh of its mother is just assumed; and elsewhere (e.g., B. *Gittin* 23b; B. *Bava Kamma* 78b).
10. See M. *Bava Kamma* 8:6 for the prohibition on self-injury. For more discussion, together with sources, on God's ownership of our bodies and the implications of holding them in trust, see Dorff, *Matters of Life and Death,* chap. 2.
11. Dr. Avraham Steinberg (see note 7 above) pointed out to me that some rabbis suggest reasons other than self-injury as the grounds for prohibiting abortion—specifically, (1) the destruction of divine creation (*Zohar* on Exodus 3:2; *Responsa Yaskeel Avdi*, part 6, on S.A. *Even Ha-Ezer* 85:1); (2) unwarranted destruction of seed (*Responsa Chovot Yair* #31; *Responsa Bet Yehudah* on S.A. *Even Ha-Ezer* #14); and, most importantly, (3) those who see abortion as murder (*Responsa Tzofnat Paneah*, part 1, #59; *Responsa Maharam Schick* on S.A. *Yoreh De'ah* #155; Rabbi I. Y. Unterman, *Noam*, vol. 6, pp. 1ff.; *Responsa Iggrot Moshe, Hoshen Mishpat*, part 2, #69–71; Rabbi Y. B. Zolti in his approbation to the book *Harefuah Le'or Ha-Halakhah*, vol. 1, 5740 [1980]; *Responsa Yabi'a Omer*, part 4, *Even Ha-Ezer* #1; and others).

    Even on the basis of injury, he asks how the Conservative Movement's Committee on Jewish Law and Standards (CJLS) can justify abortion when the fetus is judged by competent medical authority as severely defective: "If the only reason to prohibit abortion is the prohibition of *Habalah* (injury) to the mother, why may you injure her in order to abort the defective fetus?" As he points out, "Rabbi Waldenberg in *Responsa Tzitz Eliezer* is one of the few Orthodox *Poskim* [authorities] I know who permits abortion of a defective fetus, but he bases his view on his assumption that abortion of a Jewish woman altogether is only rabbinically prohibited and 'where the woman suffers pain, the Rabbis did not decree their prohibition.'"

The answer to his question, I presume, is that the CJLS permits abortion in such cases for the same reason that one may undergo any surgery—namely, that the injury to the person is undertaken to accomplish a greater good, in this case, to preserve the mother's mental health. The CJLS position can also be justified on alternative grounds (1) and (2) above in a similar way—namely, that the mother's mental health takes precedence over these concerns in such a case—but it could not hold if the ground for prohibiting abortion is that it is tantamount to murder. That is, the CJLS position is clearly based on the view that abortion is prohibited as an act of self-injury, an injury that is justified when the fetus is severely defective in order to preserve the mother's mental health.

12. *Proceedings of the Committee on Jewish Law and Standards of the Conservative Movement, 1980-1985* (New York: Rabbinical Assembly, 1988), 37, with supporting papers for that stance on pp. 3-35; also available at http://www.rabbinical assembly.org/sites/default/files/public/halakhah/teshuvot/20012004/07 .pdf (accessed 9 August 2012). For more on the Jewish stance on abortion, together with the biblical and rabbinical sources that state that stance, see Dorff, *Matters of Life and Death*, 128-33; and David M. Feldman, *Birth Control in Jewish Law* (New York: New York University Press, 1968), reprinted under the title *Marital Relations, Abortion, and Birth Control in Jewish Law* (New York: Schocken, 1973), chaps. 14 and 15.

13. For classical sources on this, see Dorff, *Matters of Life and Death*, chap. 9. According to a rabbinical ruling by Rabbi Joseph Prouser, approved by the Conservative Movement's Committee on Jewish Law and Standards in December 1995, postmortem donation of vital organs and tissue constitutes *pikku'ah nefesh* and is actually obligatory, not optional. See http://www.rabbinicalassembly .org/sites/default/files/public/halakhah/teshuvot/19912000/prouser_chesed .pdf (accessed 9 August 2012).

14. B. *Sanhedrin* 57b; M.T. *Laws of Kings* 9:4. "Another view is that this extension of the Noachide laws was intended, on the contrary, as a protest against the widespread Roman practice of abortion and infanticide" (Immanuel Jakobovits, *Jewish Medical Ethics* [New York: Bloch, 1959], 181, based on Weiss, *Dor, Dor Ve-Dorshav* 2:22). This creates a problem, though, for elsewhere in the Talmud the presumption is stated that Noahide Laws may not be more stringent than Jewish law; see B. *Sanhedrin* 59a. Tosafot therefore seek to show that Noahides (i.e., non-Jews) may also avail themselves of the permission in Jewish law to abort to save the mother's life or health; Tosafot, B. *Sanhedrin* 59a, s.v. "*leyka*"; cf. Tosafot, B. *Hullin* 33a, s.v. "*ehad*"; J. *Shabbat* 14:4 (14d); and J. *Avodah Zarah* 2:2 (40d). On all of this, see David Novak, *The Image of the Non-Jew in Judaism* (New York: Edwin Mellon Press, 1983), 185-87, with the endnotes at 197.

15. B. *Berakhot* 47b. This exception is also asserted in B. *Gittin* 38b; cf. M.T. *Laws of Slaves* 9:6; S.A. *Yoreh De'ah* 267:79. See also B. *Sukkah* 30a and B. *Bava Kamma* 94a, although in those places the exception for a *mitzvah d'rabbim* is not developed. The owner of the aborted fetus that was to be discarded has also despaired of getting it back (*yai'ush*), and if researchers are now in possession of the aborted fetus, a change of location and ownership (*shinui reshut*) has also taken place, and so one also could argue that the fetus no longer bears the taint of its origins.

    The Talmud's permission for the community to use ill-gotten gain for sacred purposes, however, seems to me to be the more appropriate category here—assuming that the abortion was not done for reasons condoned in Jewish law in the first place. For a discussion of Judaism's approach to ill-gotten gain, see my rabbinical ruling, approved by the Conservative Movement's Committee on Jewish Law and Standards, "Donations of Ill-Gotten Gain," http://www.rabbinicalassembly.org/sites/default/files/public/halakhah/teshuvot/20052010/Dorff_Donations%20of%20ill-Gotten%20gain.final.062909.pdf (accessed 9 August 2012).

    Dr. Steinberg (see notes 7 and 11 above) argues that we do not need to invoke *mitzvah d'rabbim*, because in the case of abortion, the transgression (assuming it was one) has already been completed, and therefore those who use tissues from the aborted fetus are not thereby participating in a sin. That is true, of course, but I worry about gaining from a sin (*hana'ah*), just as many oppose the use of any results from Nazi medical experiments on Jews and others. So the Talmud's exception for a *mitzvah d'rabbim* is, I think, still important in providing warrant for using tissues from aborted fetuses.

16. Even Rabbi J. David Bleich, who objects to federal funding of research on fetal tissue lest that encourage abortion, permits using organ tissue obtained from a homicide victim because "utilization of the body of the victim for scientific purposes could not conceivably be construed as an endorsement of the antecedent homicide" (*Contemporary Halakhic Problems* [New York: Ktav and Yeshiva University Press, 1995], 4:201). He also permits the use of scientific data obtained through immoral experimentation, as in the case of the Nazi experiments (4:171–202 and 218–36, esp. 234–35). I would like to thank Rabbi Aaron Mackler for drawing Rabbi Bleich's positions to my attention. I frankly think that the possibility that a woman would be persuaded to abort her fetus so that it could be used for stem cell research is very remote.

    Rabbi Moshe Tendler, in his testimony to the National Bioethics Advisory Commission, invoked a different source to justify the use of aborted fetuses—namely, M.T. *Kilayim* 9:3. Even though biblical law prohibits cross-breeding of any two species of animals, such as a horse and a donkey, the product of such an illicit mating, the mule, may be used for the benefit of the owner, even though a biblical prohibition was transgressed (Rabbi Moshe David Tendler, "Stem Cell Research and Therapy: A Judeo-Biblical Perspective," in *Ethical Issues in Human Stem Cell Research*, vol. 3, *Religious Perspectives* [Rockville MD: National Bioethics Advisory Commission, June 2000], H-4 [see note 2 above for the website link to his testimony]). Rabbi Avram Reisner, however, points out that the laws of cross-breeding have been treated by Rashi and other authorities in Jewish law as a special case—a *hukkah* and a *hiddush*—and cannot legitimately be extended to other areas of the law ("Curiouser and Curiouser: Genetic Engineering in Nonhuman Life," in *Life and Death Responsibilities in Jewish Biomedical Ethics*, ed. Aaron L. Mackler [New York: Jewish Theological Seminary, 2000], 506–22, esp. 511–14; also available at http://www.rabbinicalassembly.org/sites/default/files/public/halakhah/teshuvot/19912000/reisner_curiouser.pdf [accessed 9 August 2012]).

17. B. *Yevamot* 69b. See note 25 below.

18. M. *Niddah* 3:7 (30a).

19. M. *Niddah* 3:2 (21a).

20. M.T. *Laws of Forbidden Intercourse* 10:2, 17.

21. S.A. *Orah Hayyim* 230:1. The Mishnah on which this is based, however, does not mention the limitation on this to after forty days; see M. *Berakhot* 9:2 (54a).

22. B. *Shabbat* 67b, 129a, 140b; B. *Kiddushin* 32a; B. *Bava Kamma* 91b; B. *Hullin* 7b.

23. M. *Avot (Ethics of the Fathers)* 5:22.

24. For more on these methodological issues, see the appendix of my book *Matters of Life and Death*; and see my article, "Applying Jewish Law to New Circumstances," in *Teferet Leyisrael: Jubilee Volume in Honor of Israel Francus*, ed. Joel Roth, Menahem Schmelzer, and Yaacov Francus (New York: Jewish Theological Seminary, 2010), 189–99.

25. See Immanuel Jakobovits, *Jewish Medical Ethics* (New York: Bloch, 1959, 1975), 275, for his estimation that "forty days" in Rabbinic counting amounts to just under two months in modern obstetrical count.

26. The forty-day marker comes originally from Aristotle (*History of Animals*, Book VII, Part III), who believed that embryos were originally vegetable, then animal, and human ensoulment of males occurred at forty days and of females at eighty days. The forty-day marker was adopted by none other than Augustine (*On Exodus* 21, 80)and Aquinas (*De Potentia*, q. 3, a. 9, ad 9 [reply to the ninth objection]. See also *Summa Contra Gentiles*, lib. 2, cap. 88, n. 3, and *Summa Theologiae* 1a q. 118, a. 2, ad 2). Thus abortions before forty days were prohibited

as contraception but did not qualify as homicide. In fact, the Catholic Church itself did not hold that a fertilized egg immediately became a person until 1869, when the members of the First Vatican Council wanted strongly to affirm the virgin birth of Mary, and so they needed to see her as a person immediately upon conception by the Holy Spirit. That change did not occur in canon law until 1917. See David M. Feldman, *Birth Control in Jewish Law* (New York: New York University Press, 1968), 268–71; Aaron L. Mackler, *Introduction to Jewish and Catholic Bioethics* (Washington DC: Georgetown University Press, 2003), 127–31; and "Roman Catholicism and Abortion Access: Pagan and Christian Beliefs, 400 BCE–1983 CE," http://www.religioustolerance.org/abo_hist.htm (accessed 3 June 2014).

27. I would like to thank my friends and colleagues on this project, Dr. Jeffrey Burock and Professor Dena Davis, for raising this issue and for Professor Davis's suggested response to it, as recorded here. Professor Alan Weisbard suggested that in some ways, our use of stem cells is similar to our use of body parts for transplant: we respect the body and therefore normally bury it with appropriate rituals; but we also use bodies for organ transplants, for our respect for bodies is superseded by our duty to save life and health.

28. Genesis 1:28; Exodus 21:10; M. *Yevamot* 6:6; M. *Ketubbot* 5:6. For a discussion of these commandments regarding sex for both procreation and companionship, see Dorff, *Love Your Neighbor*, chap. 3.

29. B. *Sanhedrin* 74a.

30. Interestingly, Rabbi Dr. Steinberg (see note 7 above) sees the issue of forty days as "not so simple in Jewish law" (which he does not explain in his letter to me) and prefers to base the permission to use preimplanted fertilized eggs for stem cell research on their existence outside the womb of the woman. He points out that the Talmud (B. *Sanhedrin* 57b) understands Genesis 9:6 to say that one has a human being (*adam*) only within a person (specifically, a woman's womb), and so the whole notion of murder does not apply to fertilized egg cells outside a woman's womb.

I agree with that analysis entirely, and therefore part of my insistence that we should aggressively do stem cell research on fertilized egg cells in petri dishes is precisely because they currently have no prospects whatsoever to become a human being while they remain outside a woman's womb. At the same time, a day may come when we can indeed gestate a human being outside a woman's womb, and then I think that using the Talmud's interpretation of Genesis 9:6 would be an anachronistic justification of stem cell research on the basis of where the gametes happen to be. It is for that reason that I have proceeded further to analyze, in light of contemporary science, the strength of the Talmud's claims about the status of the fetus during its first fifty-six days even in utero in order to justify stem cell research on gametes outside a woman's womb.

31. I am here assuming that the donors are not compensated financially. As Rabbi Joel Roth has discussed with regard to kidney donation, monetary compensation for organs raises a host of halakhic problems ("Organ Donation," 60–61, 116–25, at http://www.rabbinicalassembly.org/sites/default/files/public/halakhah /teshuvot/19912000/roth_organ.pdf [accessed 9 August 2012]). Even though stem cells have developed even less toward human status than a full organ has, the same difficulties and arguments that Rabbi Roth raises would, in my judgment, apply to stem cells as well. I would like to thank Rabbi Susan Grossman for pointing out this wrinkle to me.

32. Gina Kolata, "Researchers Say Embryos in Labs Aren't Available: Few Couples Have Agreed to Allow Frozen Specimens to Be Used in Experiments," *New York Times*, 26 August 2001, 1, 21.

33. The Mishnah and Talmud forbid a man from touching his penis lest he induce it to become hard and ejaculate (B. *Niddah* 13a–13b), and later Jewish sources use the phrase *hashatat zera*, "destruction of the seed" (Tosafot on B. *Yevamot* 12b, 32b, and B. *Ketubbot* 39a; S.A. *Even Ha-Ezer* 23).

34. Rabbi Steinberg (see note 7 above) does not agree with me here. He thinks that the prohibition of emitting semen for other than procreative purposes would not be superseded by the promise of saving lives through stem cell research. In this we simply disagree about the relative weight to give to the prohibition of emitting semen and the duty to do research to save lives.

35. Robert Spirtas, Steven C. Kaufman, and Nancy J. Alexander, *Fertility and Sterility* 59, no. 2 (1993): 291–93. Still, after the 1992 Stanford study, on which this article is based, suggesting that fertility drugs might raise the risk of ovarian cancer, "later research cast doubt on that finding—but only after thousands of women were terrified" (Michael D. Lemonick, "Risking Business? Do Infertility Treatments Damage Babies' Genes? Doctors Used to Think Not. Now They Are Not So Sure," *Time*, 18 March 2002, 68–69; the quotation is on 69). Still, the 1988 congressional report listed a number of other possible complications caused by commonly used drugs to stimulate the ovaries, including early pregnancy loss, multiple gestations, ectopic pregnancies, headache, hair loss, pleuropulmonary fibrosis, increased blood viscosity and hypertension, stroke, and myocardial infarction; see U.S. Congress, Office of Technology Assessment, *Infertility: Medical and Social Choices*, OTA-BA-358 (Washington DC: U.S. Government Printing Office, 1988), 128–29. The demonstrated risks are thus not so great as to make such stimulation unwise for a woman who

needs to do this to overcome her own infertility or even to donate eggs once or twice to infertile couples, but they are sufficient to demand that caution be taken and that the number of eggs donated be limited. Here, where the eggs will be used not for producing a child but for medical research, undertaking such risks seems even less warranted.

36. Elliot N. Dorff, "Artificial Insemination, Egg Donation, and Adoption," *Conservative Judaism* 49, no. 1 (1996): 48-50, reprinted in Mackler, *Life and Death Responsibilities*, 81-84; and in Dorff, *Matters of Life and Death*, 106-7.

37. As many as 75 percent of fertilized eggs miscarry: http://pregnancyloss.info /statistics/ (accessed 8 August 2012). Approximately a third of all pregnancies miscarry: http://www.americanpregnancy.org/main/statistics.html (accessed 8 August 2012). See also Allen J. Wilcox et al., "Incidence of Early Loss of Pregnancy," *New England Journal of Medicine* 319 (1988): 189-94.

38. *Hastings Center Report 29*, March-April 1999, 30-48. The Hastings Center is the first, and still one of the most prominent, institutes for the study and dissemination of scholarship on issues in bioethics.

39. See Dorff, *Matters of Life and Death*, chap. 12 for the sources on this duty and for the variety of approaches traditional sources take as to who should get medical care and who should pay for it. See also the responsum on the distribution of health care written by Rabbi Aaron Mackler and me and approved by the Conservative Movement's Committee on Jewish Law and Standards, printed in Mackler, *Life and Death Responsibilities*, 479-505; and in *Responsa 1991-2000 of the Committee of Jewish Law and Standards of the Conservative Movement*, ed. Kassel Abelson and David J. Fine (New York: Rabbinical Assembly, 2002), 319-36, and at http://www.rabbinicalassembly.org/sites/default /files/public/halakhah/teshuvot/19912000/dorffmackler_care.pdf (accessed 9 August 2012).

40. For more on Judaism's view on communal responsibilities in the distribution of health care, see Dorff, *Matters of Life and Death*, chap. 12.

41. See Stephen J. Gould, *The Mismeasure of Man* (New York: W. W. Norton, 1996); and George J. Annas and Michael A. Grodin, *The Nazi Doctors and the Nuremberg Code: Human Rights in Human Experimentation* (New York: Oxford, 1992).

42. For a thorough discussion of this blessing and concept in Jewish tradition, see Carl Astor, "*. . . Who Makes People Different*": *Jewish Perspectives on the Disabled* (New York: United Synagogue of America, 1985).

43. For one attempt at a definition, see Bernard Gert, Charles M. Culver, and K. Donner Clouser, "Malady: A New Treatment of Disease," *Hastings Center Report* 11 (1981): 29-37; reprinted and expanded in their book, *Bioethics: A Return to Fundamentals* (New York: Oxford University Press, 1997), chap. 5,

93–130, esp. 104; and in Charles Culver's article on the definition of malady in *Morality and the New Genetics: A Guide for Students and Health Care Providers,* by Bernard Gert et al. (Sudburg MA: Jones and Bartlett Publishers, 1996), chap. 7, 147–66. On page 147 of this last book, Culver states the definition as follows: "A person has a malady if and only if he has a condition, other than his rational beliefs and desires, such that he is suffering, or is at increased risk of suffering, a harm or an evil—namely, death, pain, disability, loss of freedom, or loss of pleasure—and there is no sustaining cause of that condition that is distinct from the person." Culver then proceeds to explain four elements of that definition: (1) harms or evils; (2) distinct sustaining cause; (3) rational beliefs and desires; and (4) an increased risk. The essay by Ronald M. Green in this volume explores the implications of this definition in depth. In contrast, for a position maintaining that the distinction between therapy and enhancement cannot be clearly drawn, see Erik Parens, "Is Better Always Good? The Enhancement Project," *Hastings Center Report* 28, Special Supplement, January–February 1998, s1–s17, and see the extensive bibliography on this subject on s16 and s17.

CHAPTER 3 *Divine Representations and the Value of Embryos*

God's Image, God's Name, and the
Status of Human Nonpersons

NOAM J. ZOHAR

An important issue in the debate over stem cell research is the moral status of the very early, preimplantation embryo, which is necessarily destroyed in the process of deriving embryonic stem cells. For some ethicists, even these very early embryos are "one of us," as one of the members of President George W. Bush's Bioethics Commission described it, and have equivalent moral status to born human beings. For others, these embryos are definitely not people—yet they have some kind of moral status or value above that of (say) cells we routinely wash off in the shower.

It is clear that the Jewish tradition does not give full moral status even to a nine-month-old fetus, which can and must be sacrificed if the pregnant woman's life is at risk.[1] The tradition has less to say about very early embryos and nothing at all to say, for obvious reasons, about embryos existing outside of the womb. Several authors—representing a wide range of Jewish perspectives—have concluded that the very early embryo has little if any moral weight and that the hoped-for medical and scientific benefits to be obtained from embryonic stem cell research easily outweigh any concerns about the destruction of such embryos.

Insofar as those benefits amount to "life saving" (*pikku'ah nefesh*), this conclusion seems indisputable. As I have argued elsewhere, however, it is highly problematic to conflate the long-term therapeutic benefits of medical research in general with the more specific category of concrete "life saving" with its unique, compelling urgency.[2] Life saving indeed overrides nearly all precepts and prohibitions. But projects of basic medical research as such do not necessarily count as "life saving"; rather, they must each be examined individually, weighing the expected benefits against the values infringed.

Against this background, it is hard to rest satisfied with the aforementioned consensus. We still must ask: Do embryos have any moral weight at all? Are there *any* limits to what may be done with them? To put the question most starkly: Is there any ethical reason why very early human embryos should not be collected and used as cat food?

This (perhaps outlandish) suggestion is made in order to underscore the seriousness of the question about the value we attach to such tiny embryos. It can connect to some practical concerns that may well come up in the future as stem cell research moves from its present state of basic research to more narrowly targeted practical uses. One might, for example, want to make the argument (parallel to one often used in the animal welfare debate) that early embryos could morally be destroyed in research aimed at medical progress but not in research aimed at shinier lipstick. Also, if early embryos carry any weight at all, then perhaps we should maintain a distinction between creating embryos for the purpose of research, on the one hand, and using "leftover" embryos that are available from fertility clinics, on the other. In this chapter I shall offer a basic framework (along with some tentative applications) for a halakhic-theological discussion of these questions.[3]

Divine Representations

Judaism contains a strong tradition of religious humanism—that is, a supreme valuation of human beings anchored in religious, theistic conceptions. The central idea of this tradition is that human beings are created in the divine image. Some medieval Jewish philosophers endeavored to reduce the divine image to our noncorporeal aspects—our rational intellect alone. But within the Rabbinic tradition as a whole—and *halakhah* in particular,the divine image clearly pertains to the human body as well. This is evident, for example, in the requirement to respect a human cadaver: it is not a person, but its human form still represents, to some degree, the image of God.

Unlike a human cadaver, however, an early stage embryo—certainly one at the extremely early stage that would be used in stem cell development—lacks not only personality but also any semblance of human form. This simple fact was known long before the age of molecular biology, as attested in the talmudic statement that, prior to forty days of gestation, a fetus is "mere fluid" (B. *Yevamot* 69b). This statement has been plausibly quoted in contemporary discourse to allow abortion easily at such an early stage. More recently, it has been adduced in the context of stem cell research in denying any protected status to such an early stage embryo. Proponents of this view acknowledge that a fetus at much later stages of gestation, though not a person, may already reflect the divine image. They assert, however, that at this early stage it is "mere fluid" and has not yet attained human form; hence, its use or destruction should not be subject even to the restrictions relating to the use or destruction of a cadaver.

But as suggested by my question about cat food above, things are not so simple. The discomfort many of us feel when such a use of embryos is suggested reflects an intuitive sense of their value. What we need is a conceptual language to express this intuition. Can our tradition provide terms for defining the value of such an entity? True, such an embryo lacks human form and is therefore not in the *image* of God. Even with the aid of great magnification, it is not a visual representation of the divine. But the embryo has something that a cadaver lacks—namely, the full genetic code with the living capacity to develop into a unique human being.[4]

Hence, I want to suggest that it is emblematic of the divine in another way: not as an image but as a name. In the halakhic tradition, inscriptions of the divine Name are endowed with holiness and require great respect. Perhaps, by analogy, similar respect might be due toward an early stage embryo.

Obviously, the formal halakhic conditions that define "defacing a [divine] Name" do not exist in the case of an embryo. For the formal prohibition to apply, the Name must be written in Hebrew in one of the seven forms recognized as direct representations of the divinity (see the citation from Maimonides's code below). Yet in light of the unique status of human beings as the divine image, it seems plausible to regard the genetic inscription of this image as requiring respect akin to that commanded by divine Names inscribed in letters. Within a traditional Jewish universe of discourse, this seems a plausible way to express our intuition that this embryonic entity is not devoid of value.

One helpful feature of halakhic discourse is its specificity: religious and theological concerns are worked out in refined detail. Pursuing our basic analogy, we should seek to learn from some of the detailed rulings that have been elaborated by rabbis interpreting *halakhah* regarding the prohibition of defacing divine Names. Following are two illustrations of such proposed analogies in detail, both relating to the fundamental puzzle of symbolism, that of form versus meaning.

Form and Meaning

1. Objective Criteria: Meaning in Context

As noted above, sanctity only attaches to specific language forms, those that are regarded as God's proper names or direct appellations. Where other linguistic forms are employed—even if the reference is unambiguously to the divinity—their meaning (referent) alone is not enough to endow them with sanctity: "Anyone who destroys one of the holy, immaculate names by which the Holy One, blessed be He, is called—is subject to flogging under Torah [law]. . . . These are seven Names: the name written *yud, he, vav, he*—which is the explicit name—or that written *adonay, el, elo'ah, elohim, elohei, shadday* and *tseva'ot* . . . . The other

appellations by which God is praised, such as *hanun, rahum* [merciful] ...
are like any other [words in] Scripture and may be erased."[5]

This does not imply, however, that meaning does not matter. True,
the fact that a phrase refers to God is not enough to endow it with
sanctity: it must also have the precise form of one of the seven divine
Names. But although meaning is not sufficient for a symbol's sanctity,
it is still necessary. Even a perfectly inscribed "name of God" may have
no sanctity at all if it does not signify God's name. This is due first of
all to the inherent ambiguity of certain linguistic forms, most notably
the word *elohim* (but also, as we shall see below, to the importance of
subjective intention). Grammatically, this is the plural form of "god"
and in some cases, in fact, denotes the many gods of the pagans (as in
the Ten Commandments, Exodus 20:3: "You shall have no other gods
[*elohim aherim*] besides Me"). More often in the Hebrew Bible it constitutes a respectful-plural reference to the one God but sometimes also
to other suprahuman entities.[6] In places it may even refer to human
judges (e.g., Exodus 22:7). Similarly, *adonay* means also simply "sirs"
or "masters."

Normally the meaning can be determined (the ambiguity resolved)
objectively by looking at the context. Hence, the Talmud (B. *Shevu'ot*
35a–35b) provides guidance with respect to particular appearances of
the forms *adonay* and *elohim*, pronouncing some "sanctified" (*kodesh*)
and others "mundane" (*hol*). The same talmudic section also records
several Rabbinic disputes regarding certain instances in which the context allows more than one interpretation.[7] The fact that there are such
disagreements does not show, of course, that the words' meanings—and
status—are determined subjectively by individual readers. Rather, the
objective meaning, as determined by the scriptural context, can be read
in more than one way.

What can this suggest, by analogy, with regard to human embryos?
First of all, I think, it indicates that the same physical entity need not
always have the same status. When an item's hallowed status derives from
its quality as a symbol, the possible ambiguity calls for an examination

of context. Two human embryos that are quite alike in their stage of development may have very different status on account of a crucial difference in setting. Specifically, I have in mind the difference between an embryo implanted within a woman's womb and an embryo in a laboratory container or in a deep-cold freezer.

For these three contexts in which early embryos might be found, I propose the following tentative distinctions. When a woman is pregnant, her embryo is a divine representation, a living code of a future human being. In the context of the pregnancy—her project, within her body—it is analogous to a sanctified divine Name within Scripture.[8] By contrast, an embryo in a laboratory container as part of a scientific project—whose explicit goal is to produce not a human being but, say, stem cells for therapeutic purposes—is analogous to the word *elohim* in a context that establishes its lack of sanctity. Finally, an embryo in deep freeze is of uncertain status. Normally, it will have been produced as part of a fertility treatment, but it might now be redundant for that project, its future undetermined. Like the instances of *adonay* or *elohim* disputed among the talmudic sages, such an embryo's context allows for either a sacred or a mundane significance.

Thus, context can sometimes fail to resolve the ambiguity of a symbol's significance. In such instances, we might look beyond objective context to subjective intention. Moreover, even where the context seems univocal, its determination of meaning can be taken as no more than prima facie. Individual intention can sometimes pull conclusively away from the direction suggested by objective context.

2. Subjective Criteria: Intention and Meaning

Maimonides states that "if one writes a Torah scroll, or *tefillin*, or a *mezuzah*, and while writing lacked intentionality, and wrote even a single instance of God's name without proper intent [*shelo lishmah*]—they are void" (M.T. *Laws of Tefillin* 1:15). Things are even worse if an illicit intention is suspected or presumed; hence, a Torah scroll written by "a heretic" must be burned. Whatever the historical identity of this "heretic," the main point is explained by Rashi: "For undoubtedly, his intention in writing the name was aimed at an alien deity" (Rashi on B. *Gittin* 45b, s.v. "*yisaref*").[9]

The objective setting is not sufficient to determine a symbol's sanctity as a divine Name. Prima facie, God's name within a scriptural scroll refers to the true God and is sanctified. But the word's sanctified status can be contravened by the subjective intent of the person who produced it. The analogy suggests itself: an embryo within a woman's womb might not be deemed a divine representation if her intention at the time of intercourse was opposed to its coming into existence and she plans to negate its developing into a human being through an abortion.

This seems right with regard to a pregnancy that was definitely unwanted. Examining intention is less helpful in reducing the ambiguity in the status of frozen embryos. Insofar as they were created in the course of fertility treatment, they were meant—at least tentatively—to become human beings. Yet since more were knowingly produced than were expected to be implanted, their status as divine representations remains uncertain. Similar uncertainty may, indeed, occur also within the womb, where the embryo's creation was accompanied by no clear intention either for or against procreation. In order to shed a bit more light on the question of uncertain divine representation, it is worth looking further at halakhic discourse regarding "uncertain" instances of the divine Name.

The basic requirement is that the writing of every instance of God's Name be explicitly accompanied by proper intention. This is, of course, an extremely demanding rule, and a less demanding alternative was endorsed by Rabbi David Halevi (Poland, seventeenth century) (S.A. *Yoreh De'ah* 274:1), author of a leading commentary ("*Taz*") on the Shulhan Arukh, Rabbi Joseph Caro's code of Jewish law. Halevi maintains that at the beginning of the entire project of writing a scroll of the Torah, the scribe should pronounce a general intention to sanctify each and every instance of the divine Name. This general pronouncement will then extend to cases in which, in the course of writing, he forgets to articulate the proper intention prior to writing the Name.[10] Now, this naturally covers all instances where the Name is indeed meant to be sanctified, and it excludes all instances where the same word-form has a definitely mundane referent.

But what about the instances of disputed, ambiguous meaning? First of all, the question arose regarding the intention a scribe ought to have in writing them. The earliest advice on this, from a student of Rashi, is to "play it safe"—that is, "when in doubt, adopt a sanctifying intention."[11] Another way of "playing safe" was to rely on the initial, general pronouncement: if the true referent is God, this instance (like the less ambiguous ones) is covered by the general intention articulated at the outset.

A second question, of more immediate interest for our present discussion, pertains to the status of these ambiguous instances: Are they "sanctified" (*kodesh*) or "mundane" (*hol*)? Normally, this would not matter very much. A word-form such as *elohim*, if its referent is uncertain, should be regarded as *safek* (in doubt), and one ought not deface or destroy it, for it might be holy.[12] But a unique case in the nineteenth century gave rise to an intriguing discussion of the relation between objective uncertainty and subjective intention.

Near the very end of the Torah scroll, at Deuteronomy 32:37, the scribe forgot that the word *elohemo* (an archaic form of "their deity") was to be regarded as ambiguous. Instead, he thought of it as referring definitely to an alien deity and wrote it with the intention that it not be sanctified.[13] Rabbi Abraham Bornstein was consulted about the status of this word and indeed of the entire parchment. He first mentions the Tosafists' minority opinion that would give priority to the aforementioned initial, blanket pronouncement of intent over the actual intent in writing the particular word. This opinion is insufficient, however, to render the scroll kosher. After all, a Name that might require sanctification was written with explicitly contrary intent! What, then, can be done to correct the matter? The author moves on to consider whether the name may be erased in order to rewrite it with appropriate intent. Normally, this would be barred by its ambiguous status under the objective criteria: on one reading, the context confirms this as a divine Name. Is the scribe's contrary, subjective intention enough to exclude the possibility of sanctification in the face of the Tosafists' emphasis on objective meaning? Rabbi Bornstein concludes:

> It seems to me that it is permitted to erase this *safek* [doubtful] name. . . . First, because the Tosafists' opinion is a lone opinion; and arguably even they attribute [sanctity based on objective meaning alone] only to the Tetragrammaton, which refers exclusively to God, blessed be He, but not to other names such as *elohim*. Since angels and judges too are called *elohim* . . . it stands to reason that sanctifying meaning is not be attributed without explicit intent [*lo amrinan staman lishman*]—especially here, where it is in fact ambiguous. Second, even if we take account of the Tosafists' view, in this case [the scribe] explicitly intended the mundane [meaning].[14]

Note the crucial combination: the scribe's subjective intention might not be enough definitely to exclude the sanctity of the word *elohim* where it clearly refers to God. Because, however, the objective context leaves the word-form in ambiguity, the scribe's intention can define it as lacking sanctity, and it may be erased.

By analogy, then, the ambiguous status of a frozen embryo might be resolved by an explicit pronouncement of intention by those who produce it. If, at the time the sperm and ova are collected and then fertilized, the individuals involved declare that any surplus embryos are designated for use in medical research and/or therapy, this would remove the ambiguity. The surplus embryos, not intended as beginnings of human beings, are defined as not being divine representations. The possibility of these embryos being divine representations is allowed by the context, but, as in the case discussed by Rabbi Bornstein, it is ruled out by subjective intention. Hence, they may be "erased"—manipulated and destroyed—for the sake of medical progress.

This analysis suggests that—other things being equal—it would be even better to use embryos for which the status of divine representation is excluded by the objective context as well. Creating embryos explicitly for research purposes would rule out any vestige of sanctifying intent that might attach to embryos created in the context of fertility treatments.

In fact, of course, things are not equal, since the surplus embryos often exist anyway and stand to be discarded and destroyed even if not put to scientific use. Avoiding such waste might be a reason to prefer the use of such embryos above production of additional ones solely for research purposes. Still, from the perspective of my discussion here, there is nothing inherently wrong with such intentional production. On the contrary: conceiving the value of embryos in terms of divine representation makes that value depend on each embryo's significance as defined by both context and subjective intention. From this perspective (and contrary to an often-encountered position), absence of any plan for use in human reproduction is a marked advantage.[15]

Divine Representations and Other Values

Finally, we must consider what sort of countervailing values might override the duty to respect and preserve divine representations. Most instructive in this matter is the Rabbinic observation about the ceremony of the woman whose husband suspects her of adultery, the *sotah*, in which, according to the Torah, God orders that His Name, written in holiness, be erased to restore peace between husband and wife.[16]

In its original context, this refers to actually erasing an entire biblical passage (more precisely, scraping the text off the parchment into the water the woman is made to drink)—including several instances of God's name—in the course of the *sotah* ceremony. The suggested logic is one of necessity: without the potent scrapings, the "bitter waters" will not be trusted by the husband, and the woman's ordeal will not suffice to quiet his suspicions and hence restore marital peace. These instances of the divine Name are unquestionably sanctified, and their destruction is formally a grave offense. Yet it must be committed for the sake of something even more important.

The lesson seems clear. Even though an embryo might have value akin to that of a properly inscribed divine Name, it may and should be destroyed if that is necessary in order to achieve a goal comparable to the restoration of marital peace. This would apply even to an embryo within a woman's womb, and even if it was produced intentionally for procreation.

Certainly the mandate would extend to embryos in vitro—even those whose status is defined by context and intention as divine representations.

The crucial question is, then, how to characterize "restoring peace between husband and wife" so as to facilitate meaningful comparisons. One way to go about that is to analyze the *sotah* context: the manner in which one recounts that narrative will affect one's understanding of its telos. There are, of course, other ways too, and they can be complementary. I will not attempt this assessment here, although—if one accepts the basic framework of my argument—that would be the logical next step. By way of illustration, though, I can say with some confidence that alleviating Alzheimer's disease has enough (or, perhaps better, *the right kind of*) value, whereas cheaply feeding cats does not. As for developing more effective cosmetics—that would depend more closely on the way we understand "restoring peace between husband and wife" and on the value (whether positive or negative) we ascribe to contemporary practices of enhancing body image.

These thoughts about the relative weight of particular values are presented as mere suggestions, meant to indicate the kind of discourse that should arise once it is recognized that embryonic stem cells may require some measure of respect. My main purpose in this chapter has been, however, to explore the grounds and the parameters for such respect from a halakhic-theological perspective. I believe that this perspective facilitates a nuanced view of the symbolic value inherent to representations of the Divine—a value attached to the symbolic entity in its objective form yet also highly dependent upon the intentions of the human agents who brought it into being.

My thanks to Dena Davis for extensive help in thinking about these issues and sources.

1. See M. *Ohalot* 7:6; and D. Schiff, *Abortion in Judaism* (Cambridge: Cambridge University Press, 2002), 27–57.

2. Noam Zohar, *Alternatives in Jewish Bioethics* (Albany: SUNY Press, 1997), 130–35.

3. By this hyphenated appellation (halakhic-theological), I mean to emphasize the method adopted here. The technological novelty makes halakhic discourse of the more standard kind less feasible. In such discourse too (where it can work), I would hope to find theological and philosophical dimensions (see Noam Zohar, "Developing Halakhic Theory as an Essential Foundation for the Philosophy of Halakhah: A Theoretical Discussion and Three Examples," in *New Streams in Philosophy of Halakhah*, ed. A. Ravitzky and A. Rosenak [Jerusalem: Magnes Press, 2008], 43–63 [in Hebrew]); here, however, I am knowingly venturing beyond formal halakhic argumentation, providing instead theological discourse and analogies grounded in halakhic teachings.

4. Note that this is not an argument from potentiality alone. I am not suggesting that the value, or the moral status, that will attach to a person who will come to exist in the future somehow attaches to the entity that now has (merely) the potential to evolve into that person. Rather, value attaches to this entity in its present state, being emblematic of the divine.

5. Maimonides, M.T. *Laws of the Foundations of the Torah* 6:1, 2, 5.

6. According to some interpretations, certain occurrences of *elohim* refer to human beings such as judges or rulers, but none of these is unequivocal. See Francis Brown, S. R. Driver, and C. A. Briggs, *Hebrew and English Lexicon of the Old Testament* (Oxford: Clarendon Press, 1907 [1966 reprinting]), 43.

7. Most famous, perhaps, is Genesis 18:3. The first (1917) Jewish Publication Society translation is "My Lord," while the new (1985) Jewish Publication Society translation offers "My lords," adding in a footnote "or 'My Lord'"; Maimonides's ruling fits the earlier version.

8. Note that the woman provides the context in two complementary senses. Biologically, the embryo is carried and nourished within her body, like a word within a book. But she is also the author or scribe, insofar as this act of procreation is her project. In this, she may not be alone: often the project will be a shared one with her spouse, and the relevant intentions will be his as well. Still, the woman's involvement has primacy, and I will continue to refer simply to her; in actual situations, this should be amended as appropriate.

9. Rashi describes the individual as "a devout idolatrous priest," which renders somewhat puzzling his project of writing a Torah scroll. In the *novellae* to *Gittin*

attributed to Ritva (in a section more likely written by Ramah in thirteenth-century Spain), an alternative is put forth: "'A heretic'—this refers to evil Israelites, disciples of Jesus the Nazarene, who combine the divine Name with another entity and reinterpret the entire Torah according to their view as referring to Jesus. Such an individual certainly wrote [the name] intending an alien deity" (*Novellae Attributed to Ritva, Gittin,* vol. 1, ed. Eliyahu Lichtenstein [Jerusalem: Mosad Ha-Rav Kook, 1980]).

10. This suggests that a woman's general intent to reproduce may be enough to endow her embryo with the status of a divine representation: there is no need for specific intention during intercourse.

11. Rabbi Simhah of Vitri, *Mahzor Vitri* section 517.

12. By analogy, there would be a prima facie case based on doubt (*safek*) against destroying a frozen embryo that is preserved toward possibly producing a human being.

13. All translations I was able to check concur, rendering this as "gods"—the polytheistic plural.

14. *Responsa Avne Nezer*, Yoreh De'ah 359.

15. Incidentally, this line of reasoning would support a distinction between so-called therapeutic (i.e., research) cloning and reproductive cloning not because the latter is inherently evil but because of the special respect due to any embryo produced for procreation. Whatever one thinks of reproductive cloning, a cloned embryo produced explicitly for research would—on account of that purpose—not be deemed a divine representation.

16. *Sifre Numbers*, Piska 16; also T. *Shabbat* 13:5. For an interesting application, see Rabbi Meir's comment cited in J. *Sotah* 16d.

CHAPTER 4 *"Like Water"*

Using Genesis to Formulate an Alternative
Jewish Position on the Beginning of Life

YOSEF LEIBOWITZ

When does real life begin? That is one of the key questions in stem cell
research and medicine today. The dominant Jewish view is that life begins
at birth, that from forty days to birth the fetus is "like the thigh of its
mother," and that before forty days the fetus is "like plain water." Some
have taken that statement to mean that early embryos may be destroyed
to advance research aimed at finding cures for diseases. I think that
this last talmudic statement has been taken out of context, that it is a
purposeful exaggeration in order to make a point about fetal development. It strikes me as a callous disregard for the mystery of life to use it
to form a policy that allows us to destroy embryos for stem cell research.

One of the most difficult aspects of the study of Rabbinic thought
is that we do not know the origins of particular ideas that came from
that remarkable period of Jewish discourse and found their way into
the Talmud. Were they biblical, did they come from the nations among
whom Jews lived, or are they particularly Jewish, articulating the sages'
own worldview? And what do they mean for us today? Should we replace

definitions and decisions recorded in the Talmud with new knowledge gained from science? And if so, what is the process for such replacements?

Let me try to put the question of when life begins into another context, namely, a biblical text that, I will argue, presents an alternative Jewish view.

Creation in the Divine Image

The first chapter of the book of Genesis concerns the nature of the human soul and the image of God within it as articulated in the verse "Let us make man in our image after our likeness, and let him have dominion over the fish in the sea, the birds in the air, the animals and over all of the earth and over the things that crawl upon the earth."[1] A short study of this text will, I hope, help us understand a Jewish view of stem cell research and genetic engineering based on the Bible rather than on later Rabbinic texts. We will look at both the nature of the soul and the beginning of life, as well as the unique responsibilities of the human being toward the sanctity of the world.

The Bible is the primary document that reflects the very first stages of Jewish thought. It is accepted by all as such, but I believe that it is often misread. In some learned circles, the words of the Bible are filtered through the lens of Rabbinic and later interpretations.[2] Some readers see Jewish theology as starting with the medieval Jewish philosophers after they adopted and adapted Greek linear thinking and writing.[3] For others, the Bible is interpreted in a fundamentalist way, without any attempt to penetrate its nuances. In academic circles until recently, it was studied in terms of either the Documentary Hypothesis or parallel Near Eastern sources. In more recent years, there has been an attempt at what is called "a close reading of the text." The methodology I take here follows this last approach.

Although the creation story in the first chapter of Genesis is the best-known one, it is not the only description of God's creation in the Bible. There are others—in the second chapter of Genesis, in Psalm 104, and in chapter 38 of the book of Job. While each one presents a different view, they are complementary, not contradictory.

Let me digress for a moment to explain why contradictions in the Bible can be complementary. The Bible is not a book of science. Science is interested in the *mechanics* of creation, but the Bible is interested in the *meaning* of creation. In describing the mechanics, we use only the most accurate description that we know of at the time.[4] But when trying to explain the unexplainable—here, the meaning of creation—language is not enough. The writers of the Bible knew this, and in their attempt to communicate meaning, they told the story of creation in several "languages," or variations, to convey its many meanings.

This difference between the description of the mechanics and the meaning can be illustrated in a very simple way. If I woke up in the morning at sunrise and I said to my family, "Come look at the sun rising," one of them might very well say that the sun is not rising but rather that the earth is revolving around its axis. So I should really say, "Come look at the sun as the earth reaches the specific degree of its revolution." The statement that the sun is rising does not describe the mechanics of what is happening but the impact that it has on the soul of a person standing on the surface of the earth. Even an astrophysicist can still marvel at a sunrise.

In Genesis 1 man is the last to be created, and his creation is preceded by the question "Shall we make man in our Image?"[5] It implies, perhaps, that you could have a universe without human beings and that their absence would not make a difference. But in Genesis 2 man is in the center of the story. There is no need for a world with plants and animals if there is no human. Each one of these two chapters depicts a different view of man's position in the world, but they are complementary because we, so to speak, feel both ways at different times. Sometimes we wake up in the morning and feel that the whole world revolves around our being, and at other times we feel insignificant in the vastness of the universe.

What characterizes the description of Creation in Genesis 1? It is a description of a world that is "mechanical" in its nature; Creation is presented as orderly, repetitious, and evolutionary. The repetitious element of "And God said: 'Let there be . . .' And there was. And God saw that it was good. And there was evening, and there was morning" forms the backbone of the story as it moves in an evolutionary direction from the most basic elements to the most complex. It is this very nature of the description, without poetry or emotion, that suggests that this miraculous event was an orderly scientific process.

For our purposes the crucial verse is the description of the creation of the human being. "Let us make Man in our Image after our likeness, and he will have dominion over the fish of the sea, the birds in the air, the animals, the entire earth, and all those things that crawl upon earth."[6] Two different terms are used in the verse: "our image" (*b'tzalmenu*) and "our likeness" (*ki'dimutenu*). When it comes to the language of the Bible one must always be sensitive to the use of two similar but slightly different words in the same passage. Are they synonymous or complementary? Does the second term add to the meaning of the first and help define it?

In his great philosophic work *The Guide for the Perplexed*, Maimonides explains that there are two words in the Torah that mean very similar things: *tzelem* and *to'ar*. The word *to'ar* refers to physical image. The term *tzelem*, in contrast, refers to the metaphysical. How would Maimonides justify this in terms of idols being called images? In order to understand this, we have to ask about the functions of idols. Although the prophets ridiculed the idolaters for believing that idols were actually gods, these idol worshipers were really much more sophisticated than that. They understood that there was a realm beyond the physical, and it was there that the gods resided. Their man-made idols were a means of bringing the gods down to earth, so to speak, to reach them or perhaps to try to manipulate them. In this same sense, Maimonides suggests that the word *tzelem* denotes a physical object that carries with it a metaphysical presence.[7] The passage "Let us make Man in our Image after our likeness" asserts that the metaphysical element found in this new creation, man, is also connected to something that belongs uniquely to God.[8]

Understanding "the Image of God" in the Torah's Account of Creation

The nature of this entity, called "the image of God," can be understood only by parsing the entire chapter and the message that it conveys to us regarding the nature of the world from the perspective of the human being. One of the qualities of the world that is depicted by Genesis 1 is orderliness. The cadence of the words and of the repetition conveys this feeling. The phrases recur: "And God said: 'Let there be . . .' And there was. And God saw that it was good. And there was evening, and there was morning." This repetition mimics nature's own orderliness and rules. The physical world around us is repetitious and follows patterns.

There is a second element that I believe is also present. It is that there are different levels of being in this world. Genesis 1 posits four of them, the four that science sees as well: the inanimate, plant, animal, and human. The Torah sees these as distinct levels. They are also in evolutionary order, from the simplest to the most complex.

Let us begin by looking at the first two levels, the inanimate and the plant. The verse that described the new entity called "plants" says, "Let the earth bring forth green things, grass that seeds seeds, fruit trees that make fruit after its kind whose seeds are in it."9 Here we see the introduction of the concept of reproduction. The inanimate soil and water of level one do not reproduce, but plants do. The Bible sees this as extraordinary. The text not only states this element of reproduction as the major theme of the verse but also reflects this new phenomenon using a literary tool. In Hebrew, verbs and nouns are often derived from one another. The concept of reproduction is mimicked in the verse by using verbs and nouns in a way that is characteristic of Hebrew. The verbs give rise to nouns; words are born or reproduced. This phenomenon of verbs and nouns giving rise to one another is a central characteristic of the Hebrew language. So the English translation, "bring forth green things," is captured in the Hebrew by the doubling of the verb and the noun.

Here the Torah is emphasizing the fact that plants reproduce exact duplicates of themselves. This is actually the core of what we are talking about in cellular biology and the concept of genes and stem cells. The phenomenon of what we call life is built on the ability of cells to reproduce themselves and to gather together into an organism. The cell itself is a "living" entity. It contains the elements of birth, death, and nutrition. Within the cell, however, is also a mechanism that allows one cell to divide and eventually differentiate. What that mechanism is is still unknown. In some mysterious way what was once one cell divides into many like cells with the same genetic makeup. Suddenly these cells are organized to form an entity. Each cell becomes a different part, and each cell "knows" where its place is, and from this one cell emerges a complete plant. This entity has a life of its own beyond the life of each cell.

We could in fact describe this as saying that the plant has within it a mechanism that allows it to become what it is. This mechanism is not found in any one cell or group of cells but in the constellation as a whole. It pervades the entity, gives it its identity, but is not found in any given physical element in any of the cells. This sets the stage for the next level.

(We should note again that the Torah does not attempt to explain what is happening; it is not interested in the mechanics of this process. That is for science to do. The Torah does, however, want us to stand in wonder at the occurrence itself. And this modern science has helped us to do.)

Now we come to the animal level. This is the description in the Bible: "Let the waters swarm with swarming things, a living *nefesh*, and let the birds fly across the firmament of the sky."[10] The verse here is adding two things in describing the newest level of creation. It first begins with the same doubling of the verb and noun: "Let the waters swarm with swarming things." Now it calls the new phenomenon "a living *nefesh*." The word *nefesh* is usually translated as soul, but that is misleading. The word "soul" in later Hebrew and in English has a different meaning. It usually refers to that which is unique to the human being. Here the word *nefesh* is being used to describe that element in the animal that is not found in the plant world.

What does the *nefesh* bring to animals to make them different from plants? How does this difference manifest itself? The Bible gives us hints. There is the verse that says the birds fly over the face of the heavens. Animals have willful motion; plants do not. There is another sophisticated use of language: "swarm with swarming things [*yishretzu sheretz*]." As in the plants, the expression here has the verb come before the noun, for reptiles are somewhere in between plants and birds in their ability to move intentionally. In the second part of the verse, however, when it describes the birds flying (*'of ye'ofef*), it uses the same technique but has the noun before the verb, "birds fly." It is as if to say that in the plant world it is the action that controls the noun, while in the animal world it is the noun that controls the action.

Once again the Bible, unlike science, is not interested in the mechanics that make this work, only in the manifestation that we all see. Both the plant and the animal are alive. They contain those elements that define life—birth, nutrition, reproduction, and death. In the plant world, these work automatically. In the animal world, the animal has control over them and exhibits motion. I take this to mean that the Bible says that animals have souls.

Now we come to the human, which is unique because it is created in the image of God. What is this image? Once again the Bible does not describe it. It states that there is something in us that accounts for the differences between human beings and the other animals, and it is called the image of God—which is to the human what *nefesh* is to the animal. This point must be emphasized. What confuses the issue sometimes is that the word *nefesh*, translated as "soul," changed its meaning over the ages. In later Hebrew, it came to mean "soul" in the sense of that unique element that characterizes the human being. In the Bible, animals have souls, while in later Hebrew literature, only humans do. This was not a change in theology but a change in language.

When the Bible talks about what we call "soul," it sometimes uses the phrase "the image of God." It uses this term to describe the unique aspects of the human that pervade its being. Notice that Genesis 1 does not speak at all as to when this element called "the image of God" enters the body,[11] nor where it is located, nor what it consists of. It also does not do this with the concept of life and reproduction in the plant or the concept of *nefesh* in the animal. It does, however, indicate things that are brought about by this image of God. These manifestations of behavior that make the human unique are hinted at in the verse. These elements are going to be important for us as well in understanding the conflicts we face in dealing with the manipulation of life.

In the description of the human in the image of God, the Torah points to unique elements, the first of which is speech. We already are aware that God creates the world through speech: "And God said: 'Let there be light.'" Why does the Torah use this to describe creation? Do we have to imagine God speaking as we do? The verse depicts the power of speech in the human being as a power derived from the image of God within us. Speech in humans is significantly different from that of animals. Animals can communicate, but they lack the sophisticated apparatus of abstract thinking that characterizes humans.

In fact, we human beings use speech as one of the most creative tools that we have. Real human speech begins when the infant learns that some words are not designative but abstract. The word "mama" might at first denote a particular object, but at some point after hearing other women being called "mama" as well, the child suddenly realizes that the word does not designate an object but a concept. In Hebrew the root d-b-r means both "speech" and "object."

The power of speech gives us the power of thought. It also gives rise to our realities. We function in a world that consists not only of physical elements but of concepts as well, such as family, nation, history, progress, and morality. This last element is most crucial for us. The human power of speech gives us the power to provide moral guidelines for our actions and to set and debate the parameters. This is not found in the animal world.

On the second level, the language "used by God" to create the human is different from all the other levels of creation. Until this last creation, speech was the tool of creation: "And God said: 'Let there be . . .'" Now it becomes the tool of deliberation: "And God said: 'Shall we make . . . ?'" This once again mimics the human being. Our power to deliberate is unknown in the animal world. Should we build or destroy, consume or protect? These questions and deliberations are very special in the human. The Torah ascribes our abilities to question and think to the image of God within us.

What the Torah is telling us is that humans are a leap above the animals. One of the things that makes this clear is the Torah's use of language. Look at the second part of the verse: "And let him have dominion over the fish in the sea, the birds in the air, the animals and over all of the earth and over the things that crawl upon the earth." It depicts the uniqueness of the human being through the process of control. "Dominion" means total control. There is no animal outside of the human that exercises control over the world the way we do. All other animals live in nature, and we impose our will upon nature.

There is, however, an ambiguity in the verse that is missed in the translation. It can be understood three ways: as prediction, permission, or command. Prediction would see it as telling us something about this new, unique creature in the image of God. He will end up having dominion over the world because he will control it. In fact, the verse hints in its own way, of something else. Look at the order of the nouns: fish, birds, animals. It is the order in which God created the world. The verse is in fact hinting at the fact that by our power of creation we become godlike, or we think of ourselves as God. We shape the world into what we want it to be.[12]

Let us for a moment, however, note another aspect of the verse that ties into this. The order of the nouns is in fact the order of creation, except for the last one. The verse says: fish, birds, animals, earth. The earth should be first. This, however, is based on another theological message of the Bible. The word 'aretz here, which is translated as "earth," has in fact two complementary meanings in biblical Hebrew. It can mean the earth of nature,[13] or it can mean human civilization.[14] This reflects the same concept of man taking control of God's world and shaping it to be his own.

The second way of reading the verse is permission. What characterizes the human with all our power to create is also our power to refuse to destroy. Man can become a vegetarian, refusing to destroy animal lives. Man can also refuse to pollute the waters and risk destroying life in them. We make such decisions and act on them through thought and deliberation.

This brings us to the last possibility, command. Because of our reluctance at times to control what we do with the world, religious ideology can tell us to remove ourselves from temptations. Therefore, we might read the verse to say that we are created by God with our ability to control the world, and we are meant to do it, but within limitations. What emerges from this is a different understanding of what the image of God is meant to tell us. This is the beginning of the description of the human in the Bible. We are a being that has abstract knowledge, inquisitiveness, the power and need to shape the world around us, the capacity to think, and the fear of being powerless and alone. We alone of all the animals stand before the concepts of life and death with awe, trembling, and fear. We can set up moral frameworks, and these frameworks will be debated and honed.

This part of us is described as "the image of God." As mentioned at the outset, the biblical description is not interested in the mechanics of how this comes about. It is also not interested in when this image is formed or where it is found. It is interested only in the significance of "the image of God" being in us and how it influences how we act. The easiest reading of the biblical text, then, is that "the image of God" in the human cannot be separated out from the material part of the human at all. We are a duality. In contrast, a plant is a plant from its inception, as is an animal.

The complex makeup of the human being is part of the dilemma that we face in stem cell engineering. It raises for us all those crucial conflicts that lie at the core of our humanity. On the one hand, we have the power to control our lives, at least to some degree. Doing so draws upon the very best of human abilities—of understanding, abstract knowledge, inventiveness, persistence, and caring. Stem cell engineering also, however, raises specters of the other side of human qualities: the need to control, the fear of being helpless, the fear of the unknown and of death.

Understanding "the Image of God" in the Story of the Flood

The phrase "the image of God" appears again in the Flood narrative. In this story, after the world has been destroyed, God begins to build it again with Noah. He informs Noah of the following rules, which will govern human existence. They are very much connected to the blessings given to Adam in Genesis 1, but here they are in an expanded form.

> And God blessed Noah and his children and He said to them: Be fruitful and multiply and fill the earth.[15]

And your fear and awe will be upon all of the animals of the fields and upon the birds of the sky, and all that crawls upon the earth, and all the fish of the sea shall be placed in your hands.[16]

Every crawling thing that is alive will be for you as food; I have given you all just as the vegetables of the fields.[17]

However, meat, whose *nefesh* is in its blood, you shall not eat.[18]

And also I will requite your blood of your *nefesh*, from every animal I will requite it, and from the hand of man, from the hand of man and his brother I will requite the *nefesh* of the man.

He who spills blood of a man, through man his blood shall be spilled, because man has been made from the image of God.

The verses expand upon the themes that we saw in Genesis 1: the power and domination of man over the world while at the same time the need to be aware of the miracle and sanctity of life. In this last verse, there is again the phrase "the image of God." The verse talks about the responsibility of humans toward life around us. We control the world, but at the same time we have to relate to it through moral responsibility.

The Implications for Stem Cell Research

So the question remains: Does the taking of the stem cells from a fetus at any stage violate this image of God? As I have attempted to show in this chapter, the answer is closely connected to the biblical concept of the image of God depicted in Genesis. The image of God is part of the life force within us, one that makes us different from plants and animals and it is actually present in the genetic makeup of the first cells. It does not "enter" the body at any single moment; it is there from the very beginning to the very end.

If God's "image" is within us so long as our physical bodies are here to embrace it, are we to nurture and protect it from the moment of the first living cell until the last? Are we to be sensitive to this power within us, responsible for how we use it to control and distort, to sanctify and defile? Should such an awareness give us pause in our decisions about killing the inner cell mass to do stem cell research?

NOTES

1. Genesis 1:26.
2. By Rabbinic we mean the period between 100 BCE and 500 CE. Throughout the ages, there have been great differences in how the Bible has been approached.
3. The first of these was Saadia Ga'on. The most famous and widely read is Maimonides.
4. Even this is not entirely true. Einstein's theory of relativity is not used by humans in "telling time" in the relatively confined world of human existence. We relate to time as if it is uniform and not relative.
5. Genesis 1:26. There are a number of ways to read this verse. Either he was consulting with others, and that would account for the plural, or he was declaring an order, or he was debating with himself. I opt for either the first or the third.
6. Genesis 1:26.
7. See Maimonides's *Shemoneh Perakim* for a more detailed view of his understanding of the nature of the soul.
8. Or according to one interpretation, to God and the angels.
9. Genesis 1:11.
10. Genesis 1:20. I am purposefully not translating the word *nefesh*. It is usually translated as "soul," but, as we shall soon see, this is misleading.
11. Compare with chapter 2, which describes God breathing life into the human. This act was taken in later Jewish thought to be describing the entrance of the *neshamah*.
12. In fact, this is the basis of the Sabbath. Just as God ceased to create on the seventh day, so do we.
13. As in the opening verse of Genesis: "God created the heaven and the earth."
14. As in the story of the Tower of Babel: "And all the earth was one language and one deed" (Genesis 11:1).
15. This is a repetition of the blessing given to Adam.
16. Here it is fear and awe of God that imbues them with this great responsibility.
17. Adam and Eve were vegetarians until now.
18. Here *nefesh* should be translated as "being." Blood is the symbol of the life force, which pervades the whole body.

CHAPTER 5 *Reasonable Magic*

Stem Cell Research and Forbidden Knowledge

LAURIE ZOLOTH

I am a Jewish ethicist, working within a traditional textual perspective and practice; hence, I accept the halakhic (Jewish legal) parameters regarding embryonic stem cell research as determinative. I am not unconcerned by the position of my Catholic colleagues, and, while it is not appropriate for me to comment on the internal history of the Christian theology about embryos, I have learned much from it despite my differences, which are essentially theological. It is the Catholic concern for abortion that has allowed us, as a society, to think carefully and gravely about the issue of stem cell research, to reckon the losses carefully, and to reflect prudently about how we deal with public science policy in the face of deep moral dissension. It is a debate about issues of the self, power, and providence not unlike the debates over the use of slave labor in the 1860s. However, even if one thinks that the early human embryo may be used in the service of healing, as do Jewish thinkers, and even if one has permissible and contextually mutable norms about reproduction and abortion, as do Jewish thinkers, there are still issues, at least for this Jewish thinker, that ought to give us pause as we contemplate the promise of this therapy and what it demands of us morally. Thus, I shall now turn to the topic of how we understand the burdens and the limits of forbidden, and dangerous, knowledge and how we allow for freedom and for duty in the conduct of such powerful and intricate science.

The Nature of the Limit Itself

Much of our knowledge about the molecular biology of human embryonic stem cells derives from the new field of genetics and its uses in genetic medicine. It is not a surprise, therefore, that we might now conceptualize the nature of the self and the worth of the self in terms of medical deficit or medical prowess. In fact, for a moral philosopher, the most interesting issues in genetics are precisely these; they are not only ones about regulation, disclosure, insurance, or privacy but also ones about the power and the *portée*, or "reach," of the human self. The subjects of the self and the power that the self wields relative to others have been theological, philosophical, or political in earlier eras, and the nature of the self is described as a philosophic ontology of discipleship or citizenship. But we as moderns are now intrigued by the nature of the self as the wielder of health. The power to control history and destiny is now often in the hands of research science more than in the hands of priests or politicians. We are led to ask: How does this kind of biological and genetic knowledge shape our ideas of moral agency? How does it address the capacity for good? And how does it address the temptation and enactment of true evil?

One method employed in science is to understand power, motive, and causality by understanding ever smaller component parts of the process itself. It is the nature of science, as Gilbert Meilaender has noted, to observe and seek to understand the workings of the world by examining ever smaller pieces of it—kingdom, species, phyla, family, self, cell, DNA, proteins, and genes. We see this point made even more vivid by the way that computers and other digital recorders break knowledge into pixels of data, which are easier to store and to manipulate. But is a self merely a set of basic units making other basic units, stem cells coaxed into specific cell fates? If so, and if we understand these units to be relatively interchangeable, what can we hope for except functionality?[1] What is the nature of the human self as a self? What do we now mean by identity? Is it merely cumulative steps in a causal process, one that theoretically could be as clearly done as undone, if the cells were able to be rewound?[2]

At stake in how we respond to these questions are two issues. First, what is the weight of the duties and obligations we have toward each other, particularly in caring for the suffering of illness, in the face of this transformed knowledge of the cell and hence of the self? Second, what is the risk to the structure of our ontological and normative constructs if we act fully on this knowledge? I am not thinking here of the usual sense of the danger to our rights as persons or workers but rather of the nature of our increasing obligations, including, perhaps, some obligations regarding the study of dangerous knowledge.

We often debate issues about the use and regulation of research pertaining to genetic information and medicine in a binary way: Should we ban it? Should we fund it publicly? I am asking whether we as a society might find some middle ground, a moral location that is fully aware and wary of the dangers of genetic reductionism, cognizant of the temptation toward mere curiosity and the hubris of science and yet also aware that cowardice in the face of danger and fear of the dark present can themselves cause problems. Could memory exist with forgiveness of the past? Could radical awe and radical irony engage in a reasonable discourse over what we collectively do when we are faced with the terrible choices of science? Has such a moment come before, and what resources—textual, social, and ethical—were mobilized to respond to it?

A Narrative and Historical Argument

We need to become more nuanced in our thinking about the new molecular genetic medical research and knowledge, and to do this, I turn to societies and texts in which the regulation of knowledge and of dangerous knowledge is at stake. Further, since we often talk, on the negative side, about the terrible seduction of genetic science, the charlatanry and the false illusions of medicine, and the desacralization of the world or, on the positive side, about the magical healing promises of genetic science, I turn also to social discussions and texts in which magic is both believed in and problematic. Hence, I turn to the world of the Talmud—definitely a place of law, story, irony, and faith. For readers not familiar with this text, I offer a brief description.

Jewish ethics as a discourse exists in reference and relationship to and within the halakhic legal system articulated in the multivolume work of the Talmud, a compilation of biblical exegesis, precedents, extrapolated rulings, and the controversies of interpretation and narratives attached to discussion of these laws. It is to the consideration of the ethics of daily life, ritual activity, and responsibilities that the language games of the Talmud are applied. It is not only in the realms of prayer, citizenship, and discipleship that the sacred is revealed but also in the mundane and the daily exchange of worth, materials, and meanings that the marketplace represents. Further, this attention to every detail of human encounter not only creates a broad discourse of action but also allows for a broad inclusion of moral agents from a diverse set of social locations and geographies, including the moral agency of women, whose business acumen, obligations, and leadership have traditionally been part of the economic life of the Jewish community. Hence, in reflecting on Jewish ethics, one considers both the whole of human activity and the whole of the community as well. This discourse is primarily contained in the extensive literature of debate and exegesis of the Rabbinic literature, a discourse of contention and casuistic narrative ethics that both determines and discusses the Torah's 613 commanded acts named as the "mitzvot" by the Rabbis of the talmudic period (and compiled over a seven-hundred-year period, from approximately 200 BCE to 500 CE) as they commented on and argued about the interpretive meaning of the words and the narratives of the biblical text.[3]

Magical Knowledge and the Order of the World/Word Itself

In the first century of the Common Era, as recorded in the Mishnah, the earliest level of these debates, Rabbi Akiba and Rabbi Jose are discussing magic and its implications. For these men, lovers of knowledge, is there any inquiry that is forbidden? Can it be said that there is a category of knowledge that is dangerous, even prohibited? Is there a kind of perception and use of perception that by its linguistic enactment can threaten the social order? In the talmudic paragraph below, they are in the middle of a prolonged debate about magical power and sorcerers, what can be created and what cannot. The Rabbis of the Talmud depend on the interpretive and interrupted task: textual reasoning, imagination, discernment, definition, and norm setting.

> The seducer is he who says: Let us go and worship the stars. The sorcerer, if he performs an act, is subject to penalties, but not if he merely creates illusions. Rabbi Akiba, in the name of Rabbi Joshua, has said: Two people gather zucchinis, one of them is subject to penalties, the other is exempt; the one that performs the act is subject to penalties, the one that gives the illusion of it is exempt.[4]

At stake are the apprehension and the control of nature, the power of a text and of human language, and, finally, the nature and the place of nature in contrast to the place of human creativity. The penalties mentioned alert us to the depths of the concern for the issue of sorcery: violators will be stoned to death. In fact, the conversation breaks off for several paragraphs to discuss the precise, and quite grisly, modes of execution (a lot worse than a ban on cloning!). The Rabbis drastically limited the offenses that would lead to capital punishment and introduced evidentiary rules that made it almost impossible to be applied. Even so, why is sorcery so dangerous?[5] Why is it subject, even if only theoretically, to the death penalty?[6]

The Gemara, or Talmud, which is the later exposition and record of an ongoing debate about this passage, gives us details: it is worse if the sorcerers live in a gullible idolatrous city (presumably one with no worried ethicists), and it is particularly noted that it is more likely that women need to be warned, "because most women know sorcery." It is undoubtedly true that most women know something about the creation of living beings and its mysteries. It is this discussion, which alerts us to that over which women have a secret knowledge and power —namely, creation—that is one of the key tropes in the text. At stake is the regulation and comprehension of creation itself, perhaps even the key narrative of Creation, and hence the questions of the limits of Knowledge and the relationship to the forbiddenness of Knowledge, especially at the boundaries of good and evil (the Tree encountered) and of life and death (the Tree unencountered).

Let me be clear: there is no doubt in this text that, first, magic is possible; that, second, while generally forbidden, it is also at times permitted; and that, third, it can have actual effects on the natural world. In fact, this is the troubling problem. The text is not concerned with games or illusions or with those who merely pretend to gather zucchinis but with the category of human knowledge that, by its "magical" nature, challenges the "authority of God." Seduction is thus clearly the deepest problem, and the seduction about which they are worried is the misunderstanding of place and law, not the magic outcome. The real evil is not in knowing nature, or even in the manipulation of nature by this magic, but in worship of the thing.

In the discursive text that follows, we are taken first through a discussion of the kinds of magic that are not permitted: essentially, these are things that render the user of magic foolish or that, like the zucchini trick, alter the process unfairly or alter the marketplace unfairly. Hence, zucchini, in its huge abundance, flooding the marketplace, overturning the rules of the agrarian system, and a silly trick besides, as all gardeners know,[7] cannot be made by magic. Stories are then told, by first one and then another authority, in which Rabbis are deceived, tricked into buying a magical disappearing donkey or riding donkeys that turn into women,[8] or barely avoiding drinking water that turns into scorpions, with many of the stories taking place during visits to Egypt, a (textually) tricky sort of place. All are stories in which the borders and limits blur, limits between animals and humans, women and men, or animate and inanimate objects. Some stories hint at bestiality. In fact, the place of sorcery is precisely Alexandria, the greatest of all cities of Egypt and, as Emmanuel Levinas notes in his discussion of the talmudic passage, the place of "high civilization, a metropolis, one of our great capitals,"[9] thus the place of Arabic science. All of this sort of magic is clearly of the prohibited genre.

There follows a long discussion about the moral status of different living creatures, references to the different categories of being. These rules refer to some complexities of moral status, in which the size of the entities determines largely how we come to understand our duties toward them, and whether they are to be feared as demonic life forms, or not feared, and hence not subject to prohibitions. Things "smaller than a barleycorn" cannot be made by demons; large animals can be so made. Forbidden knowledge is like the magic of Pharaoh's sorcerers in Egypt; to remind us of the link, a reference is made to the Exodus from Pharaoh's magic Egypt and the plagues. This sort of magic is made as a direct challenge to the God of Sinai, as the text adds: "Rabbi Yohanan said: Why is sorcery called *keshafim? Because it challenges the Assembly on High, [and] The Holy One is God; there is no other.*"

Permitted Magic

But there also are cases of permitted magic, reasonable magic. We are first told about the limits and the rules of use of this reasonable magic: it is only linguistic; it cannot use any specific "paraphernalia." So the reasonable sort of magic is the manipulation of physical reality by linguistic power, a power of reasoning and scholarship at the core of the Rabbinic worldview for which in this first century CE, as in so many centuries, the state—here, the Romans and all their technology, weapons, and things—waits just outside the study house door to pounce in case there is an infraction. A clear case is offered as an example: one may use words to make large animals, indeed even a calf (with all that this suggests to the careful reader).[10]

Magic, we are told, is like any other kind of "work" in that it can be described in the way that the all work is categorized, by whether it is permitted or not on the Sabbath. There are three such categories: some work is forbidden in the biblical text and punishable by death, some is merely discouraged by the Rabbinic community, and some is allowed. Hence, the argument itself that the work of magic is "like the Sabbath" itself allows for moral judgments: by linking reasonable magic to the centrality of the Sabbath, it is linked to the system of Rabbinic order. This is a complex ethical move, for the ordering of the normative rules for behavior on the Sabbath is both antithetical to the order of technology—it is precisely this cessation, this limit on technology, that marks the Jew on the Sabbath—and an imposition of the activities of the Law on the world of nature.[11] May one work on the Sabbath? Yes, for the work of the moral takes precedence: "Some actions are entirely permissible, like the one of Rabbi Hanina and Rabbi Oshaia, who, every Sabbath eve, studied the doctrine of Creation, by means of which they created a half-grown calf and ate it."[12]

Now why is this act of creation of this "half-grown calf" permissible? It is especially curious because an earlier narrative mentions that the Rabbis not only can sometimes hear what the dead are saying but that if they wanted to, truly righteous ones could create the world. Indeed, in an earlier part of this long and complex passage describing magic, it is told that Torah scholars have indeed created a person by the use of the manipulation of the letters of the divine and secret name of God. This creature looks like a perfect man but cannot answer or argue, so the Rabbis know he cannot be a real Jew, and they send him "back to dust." He cannot speak because of the flawed moral agency of his creators. It is our distance from God that makes our power to create in this way fail. But against this classic story of the first golem, the story of the Friday-night meal of roast veal from the magically created calf is recounted. So how is it different? The Talmud does not stop to explain.

Perhaps this story is different because magic that is performed under certain circumstances is reasonable. It is not reasonable if you are making a whole person, but it might be if you are making something living but solely for the use of persons. Magic is reasonable if a living thing is not made privately or secretly, as the golem is made, but openly and in the familiar companionship of the talmudic scholar, *b'hevruta*, in the fellowship of two (one to do it, one to worry about it). Magic is done on the Sabbath to honor that day, and it is for the immediate benefit of particular hungry people, not made in huge, socially disruptive categories, like zucchini in large piles in a public field. Hence, it is reasonable only within the confines of discursive community, made by rationality and textual study, using the letters of creation that scientists now call ACTG, and it is not for the procurement of wealth and power. It is a limited and permitted reasonable magic.

The Nature of the Naming of Things

As an academically trained ethicist, I would raise the question here of whether our use of magic can ever be reasonable. Modernity is in some way based on the acquisition of things, on the leisure and the surplus to have extra things and then to yearn for them and to feel the lack of things that are not dinner tonight. Think of Vermeer and the Delft school of seventeenth-century Dutch artists and the way that acquisition is linked to the beginning of the kind of science that categorizes things. Here we might see a link between the rationality of science, as in Francis Bacon and then Darwin, and the development of the lists of our stuff. Science is the ordering of the language that defines the natural world, and hence not only do we see it as existing in genus and species, we see it as a series of things to possess. The possession and the control of things, the control of naming and ordering, setting the order precisely, is what we do in science. Our selves are then shaped and defined by the stuff we possess and hold, and this includes medicine in the modern, postantibiotic sense.

Bioethicists Adrienne Asch, Daniel Callahan, and Stanley Hauerwas, like Gilbert Meilaender, have raised some of these questions.[13] If medicine is reduced to a series of entitlements, it can look rather like getting more of something—getting quantities of life units without questioning the premise or value of the units themselves. Several of these commentators have struggled with the fact that living entails agony. If we are to understand and make meaning from a human life, ought we not to expect some degree of suffering? In Asch's words, "Isn't the point of life that it is hard sometimes?" An argument that arises here is that the magical promises of medicine would take from us some of what it is to be human, a sort of epidural for every pang of birth, an antidepressant for every loss. The promises would take from us the very thing that makes the human person human and not a perfect golem—the ability to resist, to answer back, to face adversity with courage and creativity, to learn about failure, and to rise after defeat. It is the end of the book of Job, in which the prophet struggles to answer the problem of a terrifying universal. Here the most human response to evil is not to eradicate it but to be in conversation, prayer, with God. If we understand that accumulation is not the goal, such a conversation can exist. However, as these ethicists note, the language of science can easily blur toward a language of redemption.

This slippage occurs, I would argue, in part because Americans confuse "bigness" and "length" with "goodness." It is reflected in the steady medical answer to the problem of death. We understand health care as working correctly when we get "more,"—for example, when we compare the life expectancies of persons in the first century to our own. Mary Winkler observes that we think about genetics and our bodies in the same way that we have been taught by Martha Stewart to think about our houses, namely, that perfect acquisition is "*a good thing*."[14] It is a life, as Winkler notes, of perfection, a fictionalized naturalness, cleanliness, and purity, that is the only moral good we can agree on.

This has become the answer to illness and mortality—since one wants more of life, it makes sense to worry primarily about the safety of interventions rather than their moral point. Hence, the "precautionary principle" has become the moral touchstone for new technology rather than a thoroughgoing debate on meaning and telos. This occurs with regularity: if we cannot agree on moral status, we can agree that neatness counts. Medicine, to the extent that it promoted this idea, goes from being a model that cures specific illnesses to the intellectual project of acquisition of more things—having a catalog of beautiful parts and beautiful capacities. The self is in part the accumulation of experiences (of the right sort) and attitudes (coaching) and quotas of happiness. "New" assumes "new and improved"—improvement is built into the newness, and it is the seduction of this sort of magic that points us toward mistaken worship.

Returning to the conversational narrative: "The seducer is one who says to the gullible city: Let us go and worship the stars!" But any human knowledge or certainty raises these concerns: How do we know how best to live in the face of real, not theoretical, illness; the grief and the loss that underlie the diseases with which I began this article; the very real suffering that motivated the quest for stem cell research in the first place, including diabetes, cancer, Parkinson's disease; and the suffering of children with genetic diseases like familial dysautonomia? Speaking now not as an academically trained ethicist but as a clinically engaged one, how do I walk around the real pediatric unit with my philosophical, ironic critique? How do we address the real child with the real degenerative illness? What can the wariness of the bioethicist offer to guide the scientific inquiry in a way that is more complex than a simple prohibition? We might better answer these questions after examining a subtlety and a story that is found in the Talmud directly after the business with the magic dinner of the calf created on Friday night, linking the problem of death with the problem of magic.

From the Certainty of Magic to the Uncertainty of Healing

The argument about sorcery gives way in the talmudic text to a scene at the bedside of a dying man. It is not just any man—it is the great Rabbi Eliezer, teacher to the hero of the Talmud, Akiba, who was the Rabbi who first began the project that ultimately produced the Mishnah. Rabbi Eliezer is famous for being the Rabbinic authority with the greatest sort of expertise about Rabbinic details, the one who taught the nature of specific limits about what is pure and what is impure yet whose authority on the purity or impurity of an oven was challenged by his colleagues of the Assembly in a famous scene that appears in another part of the Talmud. He is outraged (he is the expert, after all), and in the narrative of the Talmud he calls on various forces in nature to support him. But although the rivers run backward and the walls lean in, and a voice from heaven is heard in support of his position, the Assembly of the house of study rejects his opinions in favor of the ones of the discursive community. They declare: truth is interpreted here, in the public discourse of study, not called out from heaven and not derived from nature.[15]

Rabbi Eliezer is forced to leave and forced out of teaching, his knowledge—or perhaps his certainty about his method—so threatening that even as he lies dying his students must sit four cubits away. Even Rabbi Akiba shuns his teacher, although we are told that he could have continued to learn even from this distance. Somewhere else, on this Sabbath evening, two sages are "making a half-grown calf," making a cow to eat, but here the teacher faces the last night of his life, and he is angry, defiant, and yelling at his son about the rules of this Sabbath, his last. His own family members are doing things out of order, preparing for the Sabbath incorrectly, confusing biblical with Rabbinic norms! Then he yells at his lost disciples, Why have you only now come?! Rabbi Eliezer is so outraged by the abandonment of his students that he rebukes the way they have turned away from knowledge, and he mourns that loss of all the things he has discovered and cannot live to pass on; he compares his very crippled body to a closed and rolled scroll of the Torah. He prophesies a violent death for his students to punish them for their abandonment. We, the readers, know that this prophecy will indeed come to pass, for they are famous as martyrs in the Jewish tradition for teaching Torah despite the Roman emperor Hadrian's prohibition of that after the Jews' revolt of 132–35 CE, and Eliezer is describing their deaths as Jews will come to read about every single year in the liturgy of Yom Kippur. On that day, in which Jews pray against illness and the capacity for evil and death, we recount the death of Akiba and his colleagues, the worst conceivable torture at the hands of the Romans, the masters of technology against the magic of the word. Paradoxically, the Romans mockingly will wrap Rabbis in the now unrolled scrolls of the Torah and set them aflame, thinking that they are winning the argument about power at last.[16]

Eliezer struggles to teach his disciples one last thing. It is about the measures of purity and impurity, the same sort of argument that ostracized him. He is the one who taught Akiba about the limits of magic and the bit about the zucchini. He is correct, but he is not able to teach, not because he is wrong about the content, but because of something in his method of transmitting knowledge. His exclusion from the textually interpretive community meant that he was not available to students to ask questions of his teachings, and not to be queried and interrupted by dissent meant that the knowledge could not be properly transmitted. He dies and is forgiven, and his student Joshua can now come close to him. It is Joshua who stands over him at the end and says, at last, "The prohibition is annulled." And this may be why it is compassionate Joshua, the one who speaks of forgiveness, who teaches Akiba the limits of forbidden knowledge at last.

Eliezer's student Akiba mourns him: "I have a lot of money but no money changer to change it for me." Without an interpretive community the entire process of teaching and learning, question and answer, is broken down, and without this process, knowledge is again forbidden, unreasonable, and, ultimately, lost. The Gemara ends in a paradox by asking whether Eliezer was a sorcerer and answering "No." Yes, he could make a field of zucchini and then make them disappear, but doing so was a warranted use of magic, akin to basic research, and so he was not to be stoned to death. Every kind of knowledge, even the forbidden sort, is finally permitted in order to study it. Thus, the narrative ends with this last claim: "It is different when it is in order to learn. You must not learn to do. You should not learn them [the magical acts] in order to practice; but you must learn to do everything in order to understand and to teach."[17]

Hence any prohibition on magic is lifted for teaching. To teach and to research and, most importantly, to teach in a way that allows for questions to be asked and even for utter certainty and absolute authority to be queried, prohibited magic is permitted. Even if sorcery does challenge essential orders of creation, its study, as suggested by the conclusion of the text, is permitted. A distinction is made between learning and practice, understanding and application.

Where does this leave ethics in the struggle to understand whether the "magic" of the new cells, as it is described in the literature of the most enthusiastic supporters of stem cell research, is warranted or forbidden magic? And if forbidden, forbidden for what purpose in a world that "*must learn everything*"? How might such an imaginative narrative be useful? I have suggested that moral philosophy is changed by new scientific knowledge and, as in this story, that the idea of how philosophy is shaped and what knowledge is permitted is linked to the content and the context of its use, but how might genetics be altered by moral philosophy? Can we do more than watch scientists make the "half-grown calf"?

Such a story is about many things: magic, epistemic certainty, the nature of nature, the distinction between the moral status of different kinds of being that seem alive to us, the power of the state, the tension between technological power and the word, the seduction of idolatry and sacredness, and the mortality that we know frames the discourse. Such a story, I think, can do some work for us, suggesting that knowledge can be reasonably used and that scientific freedom is not only permitted but in many cases necessary, just not in every conceivable case. Since some ends justify some means, we will need to assess carefully our limits on the seeking of risky and dangerous knowledge. In contrast to proposing a ban, as a solution I believe that we are capable of creating a discursive community with overlapping ethical norms that will be sensitive to need as well as to social limits. One of the first tasks of such a moral community is the framing of categories and the discernment of our duties relative to the entities in those categories, including the way we will need to think about embryos in the first two weeks of life, whether donated or fabricated. The norms suggested by my texts are a way to begin: that dangerous knowledge is secondary to study, not only to the marketplace; that core elements of a society, such as the labor of agriculture, are to be respected; that creative activities, even necessary ones, should be within the parameters of a shared moral universe, like the rules of Shabbat that govern our narrative; and that the point of the activity should be directly useful, not frivolous, not merely acquisitive. There will be other texts from other sources, as well. On the Sabbath, learning itself never ceases to be an activity—it is a kind of activity that is always permitted and done ideally in the company of another, hence witnessed.

The Use of Metaphor: Alchemy and the Rules of Creation

The core narrative of science is informed by many such tales. Roger Shattuck has suggested that the tension between human reach, human curiosity, and the danger of forbidden knowledge is a narrative that can be said to drive many of our most essential Western canonical texts.[18] In Chaucer's story "The Wife of Bath," in Milton's *Paradise Lost*, in *Faust*, in *Frankenstein*, in Emily Dickinson, in *Billy Budd*, in *The Stranger*, we can see the struggle to embrace or to withdraw from the forbidden knowledge employed as literary and philosophic strategies. Shattuck's insight is that this tension informs how we think of both the struggle to contain nuclear energy and the struggle to direct genetic medicine.

The ideas presented in this chapter may lead us in several directions. One fruitful discussion would be in how the ancient, medieval, and early modern world understood the nature and pursuit of alchemy. Alchemy is an interesting metaphoric parallel. The alchemist stood at times within and at times apart from the science that developed into chemistry. The pursuit of the alchemist was in part to turn "baser" or "more primitive" metals into gold. Metallurgy was understood as having a processional, hierarchical pattern of increasing refinement, and the goal was to push it along by means of mystical and practical arts. Although wealth was one obvious goal, another was finding the "philosopher's stone," which was the secret elixir of life and of healing.[19] It is a task regarded as possible and interesting in both the Talmud and the Midrash.[20] As an example of this in the Midrash, the following is from an aggadic narrative in which a Rabbi believes in and yearns for the magic:

> There was a spice seller who used to make the rounds of the villages near Sepphoris . . . and say "Who wants to buy the elixir of life?" The people crowded around him. R. Yanni [a Rabbinic sage] called out to him, "Come over here and sell it to me!" The seller said to him, "Neither you nor those like you [scholars] need it." But R. Yanni persisted, whereupon the spice seller went over to him, took out a copy of the Book of Psalms, and pointed out to him the verse, "Who is the man who desires life [*mi ha'ish*]?" and asked "What is written here? Keep your tongue from evil . . . and pursue good (Psalms 34:13–14)."[21]

Here, the sorcerer sets up the tension between the truth of morality, revealed in the psalms the Rabbi is supposed to know and trust, and the healing elixir the sorcerer is hawking. Ironically, it is a premodern pharmaceutical retailer who reminds the ethicist that his task is to query the marketplace pursuit, not to buy the goods. Later alchemists are reported as understanding that the art of manipulation of substances was an important part of healing, of commerce, and of moral order. Early Hellenistic accounts of alchemy describe "Maria the Jewess," among other alchemists, and early Islamic philosophers practiced alchemy as part of a general search for the origins of metals.[22]

Raphael Patai, like Shattuck, notes that the divisions between art, spiritual seeking, philosophy, and science were differently drawn in the premodern and early modern eras from how we define those fields today. But then as now, the distinction between purpose and curiosity, experience and prohibition of experience, can be difficult to hold stable. Alchemy itself became discredited relatively late, but the idea of an inquiry that does what alchemy attempted to do for fifteen centuries—healing, life extension, transmutation of the material world from base to refinement—remains a powerful motive.

Stories and metaphors and aggadic midrashim extend our understanding of both the halakhic system and the ways in which particular religiolegal systems can operate as heuristic frames for a larger social response to issues such as the compelling research on human embryonic stem cells. If bioethics is the scholar reading in the village square when the alchemist comes with his wares, the elixir of life, for which we too still yearn, will we remember the text that reminds us of the moral imperatives as well? How will we be reminded to ask the sharp questions of students to their teacher that the expert needs to ask in order to keep his or her work a part of the world we share? In this, we need to ask questions about our responsibilities. They are first of all the duties of ethical inquiry toward all science and, in particular, toward stem cell research, research that needs our civic witness to proceed.

First, are we willing to proceed with this work in spite of our understanding that all of our medical research occurs against a background of the deepest social inequities? Stem cell research is not unique in this, and, in fact, it is especially tempting because it holds the promise, unlike cloning or esoteric interventions, of actually being scaleable, cheaper and more accessible to large numbers of patients, and distributable. It is a commitment of Jewish text and practice to stress the justice issue and to be as insistent on the necessity for a just distribution of medical interventions as my Catholic colleagues are on the matter of the sanctity of life. The world of oppression surrounds the work of creation, but, as in our text, it cannot go unchallenged. It must be said in every effort to distribute the medicines we are so driven to create.

Second, what "moral images," in the words of Hillary Putnam, will guide our work?[23] Is the self particular? How particular? As Cynthia Cohen has warned, will the very understanding of our interchangeability at the cellular level lead to commodification of these parts of the self?[24]

Third, what of our duties to the ill? Can we ever turn away from knowledge, or must we learn everything, and are we obligated to lose nothing of what can be learned?

Fourth, what are our duties to the future? Do we struggle with frail and imperfect technology today in the hope that, as in the case of stem cell research, we might attain the very thing that would allow us to move beyond the necessity to destroy embryos?

In thinking about these questions, I suggest a preliminary and partial answer, queried, not directed, but interrogated, by the texts I have offered. We can develop a policy of what I call civic witness, allowing for the fullest freedom for basic research and careful apprehension about any applied clinical experiments. In this, I argue for accountable, witnessed, and discussed research, for public discourse, and for scholarly argument, beginning at the bench of basic science. Maximal freedom can be fully supported—*all knowledge in order to learn*—but fully learned in order to be fully taught and explained and held to the duty of justice. This means support for basic research, including the free pursuit of basic research with stem cells, but a research reconsidered prior to human clinical trials and with careful public oversight for both public and private ventures: to learn, but not yet to do. I am arguing for a witnessed and burdened freedom, not unlike, as Shattuck reminds us, Aristotle's notion of the distinction between curiosity and studiousness: "The former, driven by pride, vanity, the impulse to sin, or superstition, leads us astray. Studiousness, on the other hand, falls under the virtue of temperance and leads us to the knowledge of sovereign truth."[25]

As we struggle to understand the emerging medical promise of genetics, and it is certain that there are many promises we would hope to have realized, we stand in a world clamorous with need, not only for genetics, but also for justice, for access to health care, basic public health, and nursing care. It is a world fraught with a troubling history of eugenics and with the eager pharmaceutical marketplace, and some are happy to privatize the research. On some level, faced with the reality of human suffering, pediatric disease, and devastating degenerative illness, gene therapies cannot come too quickly. On another level, faced with our propensity for mistakes, our history of eugenics, and our already unfair social practices, these therapies need a far slower pace and the steady interlocutors of the respondent community. If stem cell researchers cannot answer us on these latter issues, we are alerted to trouble. We will be the civic witnesses to what we hope will be a transparent process of work. Precisely now, at this moment, is when the questions are still ours to consider: What is sorcery? What is research? What magic is permitted to us?

NOTES

This chapter was originally presented as part of my Isaac Franck Lecture at Georgetown University. The Isaac Franck Lecture presents a privileged opportunity to raise reflective questions about the nature, goal, and meanings of Jewish tradition, Jewish practice, and bioethics and specifically to reflect on Jewish thought as it confronts the challenge of medicine. I am grateful for the opportunity to discuss the emerging controversy about stem cells in light of the Jewish tradition. My thanks to the Georgetown faculty for the invitation and for their comments on an earlier draft of these remarks. Further thanks are due to a thoughtful group of colleagues who offered suggestions on this chapter in its embryonic form, namely, Daniel Callahan, Gilbert Meilaender, Roger Shattuck, Adrienne Asch, Mary Winkler, Carl Elliott, Leroy Walters, Robert Gibbs, Aaron Mackler, Lewis Newman, Al Jonson, Ted Peters, Karen Lebacqz, Ernle Young, Lenn Goodman, and Alan Brandt. For full disclosure,

I am a member of the Ethics Advisory Board of the Geron Corporation and of the Howard Hughes Medical Institute, both of which fund stem cell research, and a member of the Executive Board of the International Society for Stem Cell Research, the academic society of stem cell researchers.

1. This line of reasoning is the focus of arguments best made by Gilbert Meilaender in his work on the conceptual problems of stem cells and the use of human embryos. See, for example, "The Point of a Ban. Or, How to Think about Stem Cell Research," *Hastings Center Report* 31, no. 1 (January–February 2001): 9–16.

2. Brigid Hogan has raised this question repeatedly in her work, querying whether, if we could learn to understand the molecular switches that regulate the programming of cytoplasm, we could theoretically deprogram adult cells and dedifferentiate them (speech at Vanderbilt University, October 2001).

3. This paragraph is taken from Laurie Zoloth, "Her Work Sings Her Praise: A Framework for a Feminist Jewish Ethic of Economic Life," in *Spiritual Goods: Faith Traditions and the Practice of Business*, ed. Stewart Herman with Arthur Gross Schaefer (Charlottesville VA: Philosophy Documentation Center, 2001), 381–401. Within the period defined as "Rabbinic Judaism" are three separate divisions: (1) the *tannaitic* stratum (ca. 30–200 CE), including materials recorded in the Mishnah and Tosefta and alluded to as a "berita," or extra source outside these collections in the Gemara; (2) the *amoraic* stratum (ca. 200–600 CE), including materials recorded in the Palestinian Talmud (edited ca. 450 CE); and (3) the Babylonian Talmud (edited ca. 600 CE). "Talmud" is the Hebrew name for the material developed after the Mishnah: "Gemara" is the Aramaic name.

   The Bible, the Mishnah, and the Gemara create a structure for Jewish response, but it is a fluid and mutable one. Jewish law has evolved, and continues to develop, in the fourteen hundred years since the redaction of the Talmud by an ongoing series of responsa, or written legal rulings by respected scholars and learned community leadership on the laws codified in these texts. Throughout the medieval and early modern periods, difficult cases of social crisis were brought before authorities and scholars who then ruled on the facts of the cases, on the methodological principles of logical discourse, and on certain key principles of relationships in the civic and commercial spheres. Relationships between and among the producers, sellers, craftspersons, consumers, traders, and managers of trade were expected to be regulated by certain essential principles intended to stimulate and regulate commerce and to protect the reasonable interests of all parties to the encounter. Key precedents and case exemplars were given in order to find parallel moral and legal appeals, and from this latticework of cases, some general principles emerged.

Major works of the medieval period developed and codified Jewish ethics. Moses Maimonides (known by the acronym of his Hebrew name, the RAMBAM), the authoritative philosopher of the twelfth century, writing in Muslim Spain and North Africa, authored critical halakhic interpretations and a full code of Jewish law, the *Mishneh Torah*. His rulings on matters of civic responsibility continue to be referenced in this ongoing debate. The Shulhan Arukh (Set Table), written by Joseph Caro in 1564, is a later Jewish code that is regarded as centrally authoritative from that date to the contemporary period because it includes the glosses of Rabbi Moses Isserles, who indicated where the practices of northern European (Ashkenazic) Jews differed from those of Mediterranean (Sephardic) Jews that Caro had recorded. Caro built his codal structure on authoritative scholarship from the tradition heavily influenced by interactions with Islam, including such Jewish figures as Maimonides, Asher ben Yehiel, and Alfasi, all of whom show ample evidence of this influence.

For centuries, most Jews lived in enclosed communities, subject not only to civic procedures but also to a system of rabbinical scholarly authority and rabbinical courts that adjudicated disputes. Hence, the meaning and intent of Jewish law had both an aspirational and a practical power. Since 1945, when it first became true that most of the world's Jews live in democratic countries as full citizens with the rule of law, the rabbinical courts have been limited in their authority but not in their intellectual processes of analyzing and resolving contended issues for their community. This process continues in the contemporary period, with responsa (interpretations meant to be directive) about ethical issues published and debated—for example, in the work of philosopher and Conservative rabbi Elliot Dorff or Orthodox rabbi Moshe Tendler. Complex cases involving communities and individuals routinely are debated in all branches of the Jewish community. In many Orthodox circles, rabbinical decisions in the business world carry authoritative power.

4. M. *Sanhedrin* 7:10–11 (67a).
5. Emmanuel Levinas also has written extensively on this passage, exploring the importance for the Rabbinic authorities of the act of teaching. But here, Levinas notes that to have Rabbi Akiba, noted in the Talmud as the most gentle of authorities, lavishly recounting the gory modes of death is particularly odd.
6. Note here that the death penalty was discussed, but there is dispute about whether it was ever enacted.
7. Zucchini is extremely easy to grow in abundance. In the American context, it has a special place in the garden joke lexicon based on the reality that even one zucchini plant will produce more than a family can eat, and hence the problem is finding enough places to give it away. I, being a gardener, am familiar with

both the joke and the phenomenon. The talmudic commentators were also rooted in an agrarian economy and hence created this reference deliberately. In some translations, the word is translated as "cucumbers," also easily grown, but for American readers, the amusing aspects of the issues are more fully revealed by the use of "zucchini."

8. Hence, the Rabbi is suddenly "riding," or embracing, a woman, presumably a sorceress.

9. Emmanuel Levinas, "Desacralization and Disenchantment," in *Nine Talmudic Readings* (Bloomington: Indiana University Press, 1990), 151.

10. To one not familiar with the Hebrew Bible, the reference alludes to the making of the Golden Calf, the prohibited form of idol worship that led to the destruction of the first tablets of the Law in Exodus 32–34.

11. At precisely this moment in the textual argument, another, seemingly unrelated, narrative breaks in. The reader is referred back to an earlier section of the debate in which the same point is at stake in a rather violently argued debate between Rabbi Akiba and the Roman tyrant Tineus Rufus, called Tyrannos in the Talmud.

12. B. *Sanhedrin* 67b.

13. Adrienne Asch, personal communication, June 2001.

14. Mary Winkler, unpublished seminar paper, June 2001, Cambridge.

15. B. *Bava Metzia* 59b.

16. B. *Avodah Zarah* 18b.

17. B. *Sanhedrin* 68a.

18. Roger Shattuck, *Forbidden Knowledge: From Prometheus to Pornography* (New York: St. Martin's Press, 1996).

19. Also called the "sorcerer's stone," as millions of *Harry Potter* readers can explain.

20. B. *Yoma* 44b–45a; J. *Yoma* 4; *Exodus Rabbah* 35:1; *Numbers Rabbah* 12:4; B. *Shabbat* 88b; B. *Yoma* 72b.

21. *Leviticus Rabbah* 16:2.

22. Raphael Patai, *The Jewish Alchemists: A History and Source Book* (Princeton NJ: Princeton University Press, 1994). Raphael Patai extensively details the entire history of alchemical practice and provides in this work (61–99) a detailed account of Maria the Jewess and her contemporaries.

23. Hillary Putnam, speech to the American Society for Bioethics and Humanities, Philadelphia PA, 1999.

24. Cynthia Cohen, personal communication, 2001.

25. Shattuck, *Forbidden Knowledge*, 308. He is referring to Aquinas, *Summa Theologiae* 2 qq. 166–67.

PART 2 *Genetic Mapping and Identity*

CHAPTER 6 *Summary of the Science of*
Genetic Mapping and Identity

ELLIOT N. DORFF AND LAURIE ZOLOTH

The study of the unit of heredity, genes, is called genetics, and the study of the larger units of all the genes of an entire organism, the genome, is called genomics. Genomics is both a young and a growing field that may significantly change the practice of medicine. In a sense, of course, all disease is "genetic," because all diseases reflect some basic change in the molecular biochemistry of the cell, or groups of cells, and all such chemistry is driven and organized by the combining, folding, and interactions of the proteins that are created by gene patterns. The genes work differently in different environments. Learning how this genetic chemistry works is the puzzle of genomic science.

The Human Genome Project

The Human Genome Project (HGP) was an international collaboration to map, chromosome by chromosome, the genetic code of the human species. The HGP promised much. It intended to expose the origins of human difference and of human individuality; to lead to understandings about how evolution transformed our species; to change the concept of disease, medicine, and medical intervention; and to understand the risk and range of genetic mutations across populations.

In the last decade, the entire DNA "map" of humans and then of mice, mosquitoes, rice, chimpanzees, and other species has been decoded. This work leads in many directions with many applications.

One direction is a new form of anthropological science. If we can trace how populations are similar and how they are different on the genetic level, we can understand the relationships among human populations in both the past and the present in new ways.

Another direction is medical science. Understanding how the genome is created and the places in which it is defective is a subtle art. Diseases form a spectrum of conditions. Some are what are called "Mendelian" after Gregor Mendel, an Austrian monk and abbey director in the early 1700s who studied the effects of crossbreeding on peas. It was his insight that traits are carried over from one generation to another and tend to be inherited intact. In other words, a short plant and a tall plant did not produce medium-height offspring but instead produced an arithmetic mix of short and tall plants. This pattern was then repeated in subsequent generations. This was the foundation for Darwin's theory.

Genomics, although a very new science, is revealing how random mutations play roles in human history. Thus, one mutation that has been fortuitous for European populations, the *CCR5* mutation, enables individuals to survive epidemics, for example, the Black Death. This mutation, the single most common mutation in the white population, affecting one in one hundred people, is largely missing in African populations. It has also been protective during the AIDS pandemic, for people with this mutation resist infection with HIV.

Although some diseases can be attributed to a mutation in a single gene, such as Tay-Sachs and Huntington's chorea, most are a result of complex, multifactorial gene and protein interactions. These are far more common and thus, in terms of public health, much more important to understand.

Another possible implication of this new research is personalized genetic medicine (PGM), an idea based on an older conception of medicine: that the nature of a disease varies from patient to patient. More recently, for prevention, diagnosis, and therapy, health care has relied on models that view patients as representatives of humanity in general or at best some subcategory of humanity. However, each patient is an individual with a unique biology, and so treating a patient as some biological Everyman limits the efficacy of such models. Genomic-based knowledge may give us the tools to learn the individual genome of each patient and discover how pathological or advantageous mutations call for more or less aggressive care, for specific sorts of care, or for drugs to manage the person's disease. These treatments may not make sense for the entire population but would be efficacious for the individual patient.

The Use of Genomics to Understand and Cure Diseases

For genomics to benefit us, we must know several things, whether we are consumers of health care or providers of that care. First, it is important to understand modes of inheritance and the role of family history. Knowing a patient's health-care history across three generations is important for physicians who wish to understand and treat individuals. (This is often difficult in Jewish families, given the annihilation of six million Jews in the Holocaust.)

Second, both the provider and the patient must know what genetic tests mean and how to calculate risks. This process requires a new language and the skill to communicate risk, a task made difficult by the levels of both fear and expectation in the lay community and the lack of training in health-care programs.

Genomics seeks to know about the identification and characterization of genes and their arrangement in chromosomes, in contrast to the traditional method of study of genetics, which is devoted to the study of the origin of human uniqueness, usually using diseases carried in human families. The HGP was devoted to the juxtaposition of genomics and human genetics. It was largely devoted to understanding, first, the coding regions of the human genome. In fact, when the project that produced this book began, scientists thought that much of the genome is "mere junk," residue codes that carried little meaning, a view that had changed by the time the essays of this volume were assembled. It is hoped that understanding the body not as a nineteenth-century machine with "breaks" but as an organism ordered by an evolving linguistic code will allow research into the processes of development, maturation, and aging, as well as the interactions of organisms with complex environments, themselves shaped by an organic matrix composed of variable genomes.

Molecular genetics has changed medicine because it has introduced the idea that if we are to understand any disease, it will be through recognition of variations in the elements of the biological and physiological apparatus of the cell, and these elements are protein products of the genes that produce and regulate them. The HGP is the next step in the history of the examination of the human body for medical purposes. One of the first tools of science, the use of autopsy, allowed the study of organ systems, and the use of the microscope enabled researchers to shift their attention to the cell. As genes began to be understood as the basic unit of disease and of the specificity of the molecules that are the engines of the cell, the logical next step was to create a complete list of the genes. It is the concept at the heart of biology: if you understand the parts and the structure of each part, then you can understand the whole and the function of the whole.

The HGP was fully federally funded, and it set aside 3 percent of its budget to study the ethical, legal, and social implications (ELSI) of the HGP. The HGP had several benefits. The first was that diagnosis and prognosis would be helped if a test was developed for conditions known to be genetically caused, such as cystic fibrosis, Tay-Sachs, and Huntington's. Next, the concept of disease, in terms of how it can be generalized and theorized, was altered by insights into the details of structure and function pathways. Disease began to be understood in its adaptive and dynamic forms, related to context, yet it can be reduced to models that can be manipulated. Genomics allowed biologists more fully to capture the role of proteins and to develop a new field—proteomics—to describe the "business end" of the cell. Protein gene products are both the cause and the response to disease, translating structure into functions that then create phenotypical traits. Functional and evolutionary biology meet at this level.

There are important implications of this knowledge. If the HGP promises to expose the origins of individuality, it shows how different narratives of place and time shape the traits of human populations. Thus to understand disease in individuals, we must characterize it in terms of three time scales at once: that of phylogeny, that of the lifetime, and that of the moment.

Scientific Questions That Remain

As one might expect of a new field, questions remain. First, little is known about the fact that there is such a variance, called a polymorphism, between the alleles, or functional sequences, of a gene. The polymorphic alleles, in most cases, create difference but not disease. But which ones, and how many, are associated with disease? Is this heterozygotic advantage important to maintain? What role do such different alleles play in normative human processes like aging?

Second, little is known about inborn errors in the genome. Are some areas more prone to such error?

Third, genes are at times highly conserved and at times not conserved at all. But is this an indication of disease? Or is it a way that the gene appropriately mutates and adapts to changes in the environment?

Fourth, is the low rate of human fecundity, 25 percent, a result of a continual rejection of alleles? What is the role of alleles in common and complex disease? What is the logical use of genetics and genomics in preventive medicine?

Anthropological Uses of Genomics

Two other uses of genomic information become immediately apparent. The first is the way that ancestry and identity can be constructed, and the second is the way that communities can be differentiated. Both present unique challenges to the Jewish community.

One interesting example of this anthropological use of genomics concerns the Kohen modal haplotype, a genetic marker that is found only on the Y chromosome and is thus passed from father to son. Since 1993 it has been known that this paternally inherited Y chromosomal marker is found in Ashkenazic and Sephardic Jews (the communities split into different regions of the world in the first century CE) in high frequencies, in contrast to the populations of central Europe. The particular microsatellite haplotype, within group J, occurs at a high frequency in men who share an oral tradition of lineage to the *Kohanim*, the priestly families of biblical times. The haplotype changes very slowly, and its emergence can be traced to a relatively small temporal moment in Jewish history between the First and Second Temple periods in the sixth century BCE.

Because Jewish communities were relatively isolated throughout history, and because Jewish lineage is determined by matrilineal descent (or by conversion, but that is still relatively rare), modern anthropologist Mark Thomas of University College London has traced the "founding mothers" of Jewish communities to very few women matriarchs.[1] Thomas's research rests on testing thousands of subjects at particular locations around the world for the haplotypes in question.

This research leads to other possibilities. Single nucleotide polymorphisms (SNPs) are not inherited individually but in larger chunks of DNA. SNP groups, or haplotype groups, allow a record of genetic ancestry based on statistical probabilities of such a combination occurring. In the early 2000s, a new project to map the haplotypes commonly found in the human species began, and this HapMap charts the difference and divergences in populations shaped by different environmental conditions. It is, in many ways, the continuation of the seventeenth-to-nineteenth-century project of dividing the human population by races, a project doomed by its assumptions that the entire species was based on a hierarchy that favored white European tribes and justified their colonial projects.

Tudor Parfitt, a social scientist who has used the genomic analysis in HapMap, described above, to study the reconstructed genetic ancestry of groups of people who claim to be Jews, describes two cases: the "Black Jews," or Lemba, of South Africa, a small African Judaizing tribe that has long claimed Jewish origins, and Jewish groups in India whose Jewish ancestry was long regarded as in doubt.[2] Parfitt became aware of the Lemba in the late 1980s. In the 1990s this group was tested for the Kohen modal haplotype and found to carry it in higher numbers than surrounding African tribes and indeed in the same frequency in men considered "priests" within the tribe as in European Jewish men considered *Kohanim*. This genetic link, of which Parfitt is still not convinced, is now largely celebrated by tribe members themselves, who see it as proof that their sense of Jewishness is warranted.

Applying Genetic Mapping to the Question, Who Is a Jew?

The essays in part 2 of this book explore the implications of these new methods of genetic mapping for Jewish identity. Two of them, the ones by Rebecca Alpert and Judith Neulander, ask whether these new mapping tools mark Jews as a diseased people, and, if so, what that means psychologically, politically, and socially for how Jews think of themselves. The essay by Laurie Zoloth explores the claims that the Lemba are Jews based on the Kohen modal haplotype, which is common among them.

The question of who is a Jew is a hotly debated one in our time, with the Jewish legal definition (a Jew is a person born to a Jewish mother or who converted to Judaism) vying with ethnic, cultural, and political definitions (the last as evidenced by the definition of a Jew in Israel's Law of Return, according to which one Jewish grandparent is enough to classify a person as a Jew for immediate citizenship in Israel). These three essays expand the scope of that question by invoking and applying to it the new science of genetic mapping.

NOTES

1. Mark Thomas et al., "Founding Mothers of Jewish Communities: Geographically Separated Jewish Groups Were Independently Founded by Very Few Female Ancestors," *American Journal of Human Genetics* 70 (2002): 1411-20.
2. Tudor Parfitt and Y. Egorova, *Genetics, Mass Media, and Identity: A Case Study of the Genetic Research of the Lemba and Bene Israel* (London: Routledge, 2005).

CHAPTER 7 *Folk Taxonomy, Prejudice, and the Human Genome*

JUDITH S. NEULANDER

The Ariadne thread in botany is classification, without which there is chaos.
—Carl Linnaeus (1707–78)

Efforts to map the human genome require modern taxonomists to define new, and sometimes unanticipated, categories of human variation. But human differences—biological or otherwise—have never been easy to define. Even Linnaeus, master taxonomist and father of modern scientific classification, dropped Ariadne's thread when he entered the maze of human distinctions. Drawing biological boundaries around the cultural characteristics of indigenous peoples, he constructed inaccurate, overly inclusive folk categories and, in so doing, reinforced a self-authenticating colonial worldview that justified savagery in the name of civilization. This is not because Linnaeus was an ill-intentioned or inferior scientist. Rather, despite good character and strong intellect, he was a product of his own cultural milieu, and his culturally determined knowledge—like our own—was so deeply entrenched as "the way things are" that he was unable to grasp its arbitrary nature. It follows that, faced with the task of defining newly discovered genetic populations—how to classify different groups at risk of heritable diseases, for example—some of our modern, analytical categories may be nothing more than folk taxonomies in disguise.[1]

To help avoid a modern recycling of colonial folk taxonomy, this essay will distinguish academic logic from folk logic in the context of the human genome. As an example of a modern, genomic folk taxonomy, we will examine New Mexican use of ailments labeled "Jewish" to classify Spanish Americans as descendants of Jews. As we are about to see, claims of Judeo-Spanish ancestry are used to assert an overvalued line of white ancestral descent in the American Southwest: a phenotypical line of ancestry regionally assumed for all Jews but not for all Spanish Americans. Hence, use of a "Jewish" label for ailments found, or imagined, among Spanish Americans of New Mexico both reflects and reinforces a regional scale of human valuation, striating into color-coded degrees of human degradation what was otherwise a socially cohesive community. We will therefore consider an alternative to the use of religious labels for heritable diseases, since it is precisely this academic practice that enables and legitimates the popular recycling of colonial folk taxonomies, like the one that persists in New Mexico, and the prejudice that accompanies them.

In defining categories of human variation, the distinction between academic logic and folk logic is straightforward. Academic logic discerns categories by differential, critical thinking; it strives for timeless, global accuracy, no matter what may be discovered. Folk logic, however, has a purpose different from accuracy. To paraphrase Hayden White, folk categories of human distinction are self-authenticating social devices involving not merely the clarity but also the self-esteem and perceived entitlements of the group doing the categorizing.[2] As folklorist Dan Ben-Amos puts it: "Analytical [academic] categories . . . have been developed in the context of scholarship and serve its varied research purposes. Native [folk] taxonomy, on the other hand, has no external objective.

The logical principles that underlie its categorization . . . are those which are meaningful to the members of the group and can guide them in their personal relationships."[3]

We overgeneralize, or create overinclusive folk categories, when we identify genetically linked populations the same way Linnaeus did according to nonheritable factors, namely, those particular traits and characteristics (like religious affiliation) that can only be acquired through learning, that may exist only in the eye of the beholder, or—as applies specifically to religion—that involve "volatile, inward states known subjectively, if at all."[4] Clearly, the extent, and even the reality, of a group's nonheritable traits and characteristics are at best unstable, if not entirely uncertain, making them essentially untestable. Thus, a heritable disease originating in a geographical area like West Africa will correlate more accurately to persons with ancestral origins in that area than to an arbitrary religious category like "Christians" or "Protestants." Similarly, populations carrying a heritable disease from a French Canadian founder will correlate more accurately to persons of French Canadian descent than to the category "Catholics," even though most descendants may be Catholic and may no longer live in Canada.

It follows that populations carrying a heritable disease from an eastern European founder will correlate more accurately to persons of eastern European descent than to the category "Jews." But we are conditioned to dispute the last instance, for the category "Jew" is by definition overinclusive in our culture, given the widespread European American failure to differentiate between heritable characteristics acquired only through DNA and the cultural characteristics widely attributed to Jews, some of which are wholly imagined and the rest of which are acquired only by learning.

Our societal propensity for overgeneralization of Cultural Others is hardly limited to Jews, but, as Gordon W. Allport writes: "The most clear of all is the case of the Jews. While they are primarily a religious group, they are likewise viewed as a race, a nation, a people, a culture. When religious distinctions are made to do double duty, the grounds for prejudice are laid, for prejudice means that inept, over-inclusive categories are employed in place of differential thinking."[5]

The "Discovery" of New Mexican Crypto-Jews

For a modern instance of genetic overgeneralization involving Jews as well as prejudice, we turn first to the state of New Mexico in the mid-1970s, when long-standing Spanish American claims of descent from Spanish conquistadors were fully discredited.[6] Regional claims of a prestigious aristocratic (ostensibly monogamous and white) lineage date back to at least the founding of modern New Mexico at the turn of the eighteenth century. By that time the Spanish American population included African and Native American admixture, a profile so undervalued by the founding fathers that they institutionalized what Chilean sociologist Alejandro Lipschutz calls a "pigmentocracy,"[7] a formal caste system based on skin color and phenotype, which in New Mexico was calculated in twenty-two degrees of increasing distance from the ideal of unbroken white European descent.[8] As Ramón Gutiérrez writes of Spain and New Mexico:

> The whiter one's skin, the greater one's claim to the honor and precedence Spaniards expected and received. The darker a person's skin, the closer one was presumed to be to the physical labor of slaves and tributary Indians, and the closer the visual association with the infamy of the conquered. In Spain families guarded their *limpieza de sangre* or blood purity through avoidance of Moors and Jews. In New Mexico, families of aristocratic pretension feared that their bloodlines might be metaphysically polluted by Indians, *mulatos*, and as one man put it, "castes which are held or reputed as despicable in this kingdom."[9]

Spanish Jews were undoubtedly among those ranked despicable on the Spanish colonial social scale, so much so that their exclusion from the brutally ostracizing New Mexican caste system is a powerful indicator that none were there to be brutally ostracized. Rather, reflecting the popular colonial conflation of indigenous peoples with "lost tribes" of Israel, Alejandro Mora, a resident of Bernalillo in 1751, gave what was then a socially acceptable explanation for beating an Indian slave: "'God has given me life,' said Mora, 'so that I might do to these Jews what they did to our Holiest Lord.'"[10]

By 1975, when Spanish American claims of aristocratic white descent were finally discredited, the notion of unbroken descent from (ostensibly monogamous and white) Jews, who introduced ostensibly telltale folkways into the region, became the best, and perhaps last, means of denying nonwhite admixture and restoring the local prestige lineage. With fortuitous timing, ambiguously Jewish artifacts introduced by early twentieth-century Adventist, Apostolic, or "messianic" Protestant experimentation (such as six-pointed stars on cemetery crosses, abstention from pork, Saturday observance of the Sabbath, giving Old Testament names to children, etc.) were touted as evidence of (white) Judeo-Spanish, or "Sephardi," descent by agenda-driven academics eager to discover local crypto-Jews.[11]

Upon careful investigation, there is no cultural evidence from the past, or anything visible in the present, to indicate crypto-Jewish descent for the founding fathers of Spanish New Mexico. Rather, the documented historical and cultural records refute any such claim.[12] Those seeking independent research that draws conclusions based on relevant genetic data should consult the work of Wesley Sutton and colleagues at Stanford and New York Universities.[13] It is generally accepted that maternal ancestry in New Mexico is predominantly Native American, but the Sutton study found that, for all relevant Y-chromosome markers, paternal ancestry of New Mexican Spanish Americans is *identical* (except for 2.2 percent Native American admixture) to that observed in modern, postexilic Spain. In both the Iberian parent population and in New Mexico, 10 percent of Spanish males have Middle Eastern origins, which must include populations like Phoenicians and Arabs as well as Jews, complicated by the fact that Jews cannot be distinguished in DNA from Arabs. Thus, Hispanics that descend from Jews comprise something under 10 percent of the male Hispanic population, nor is it clear that this group descends from a significant settlement of secretly professing eighteenth-century Spanish Jews, a population for which there is no cultural or historical evidence in either Spain or New Mexico.

Sutton found the genetic profile of the remaining 90 percent of males in Spain and New Mexico to be significantly different from all populations of Middle Eastern descent. Had claims of a significant crypto-Jewish settlement been accurate, he reasoned, there would be a higher rate of Middle Eastern ancestry among males in postexilic New Mexico than in postexilic Spain, reflecting a component of exiled Iberian Jews added to New Mexico's Spanish settlers. But the evidence from New Mexican Spanish American males is unequivocal: regional claims of a significant crypto-Jewish settlement are refuted by the genetic profile of this population.

Complicating the matter, there is historical and cultural precedent for falsely claiming Judeo-Spanish descent throughout the Spanish Americas. As Raphael Patai wrote on the phenomenon of Spanish American identity switching: "To trace one's ancestry to Spain has meant to establish a claim to high status, to a prestige lineage. It is a frequent phenomenon for an Indian to claim to be a mestizo [of mixed blood] and for a mestizo to claim pure Spanish [white] descent. . . . Spanish descent, even Spanish-Jewish descent, means a step up on the social scale."[14]

Hence, in 1980, New Mexico hired as its new state historian Stanley M. Hordes, who had written a doctoral dissertation on secretly professing Jews, or "crypto-Jews," in the Spanish Americas (although not in New Mexico). His arrival provoked a spate of rumors, gossip, and hearsay concerning ambiguously Jewish (and, therefore, ostensibly white) regional folkways. Hordes announced to the press and media that a significant component of "crypto-Jews" was indeed among the region's early Spanish settlers, the Jewish origin of their folkways having been ostensibly forgotten.[15] Shortly thereafter, a small but vocal group of Spanish Americans, their memories thus "restored," came forward to claim publicly their new prestige lineage. Undeterred by the indisputably Portuguese origin of colonial America's crypto-Jews,[16] or by the indisputably Spanish origin of modern New Mexico's founding fathers,[17] a similarly small and vocal number of academics—none of whom was a folklore specialist—also came forward to endorse primarily Protestant folkways found in modern New Mexico as both colonial and crypto-Jewish.

The embrace of crypto-Jewish claims, although at first made by very few Hispanics, is not unusual in this kind of circumstance. While Spanish

Americans had colonized New Mexico in the eighteenth century, they in turn were similarly colonized by hegemonic Anglo-American forces, beginning in the mid-1800s. Their own socially functional origin myth, stating direct descent from Spanish (white) conquistadors, had deflected the worst effects of the Spanish caste system by making it impossible for anyone in the mixed racial population to discriminate with certainty against anyone else. But in 1975 these claims were disconfirmed and by 1980 were being replaced by socially dysfunctional, academic reconstructions of a (white) "crypto-Jewish" ancestral past that many darker-skinned New Mexicans (who understand Jews to be white) feared they could not claim. Thus, the formerly cohesive society found itself once again threatened with social stratification along color-coded fault lines, but this time reinforced through the imperial eyes of an Anglo ruling class that classified Jews with overvalued whites and New Mexicans with people of color.

By 1980 the Anglo population finally outnumbered regional Hispanics, and local academics were variously citing 25 to 50 percent of the new minority as descendants of crypto-Jews. Most Spanish Americans chose to transcend this academic interference and move on, unburdened by old racialized caste issues. But a tiny, perhaps more threatened group embraced the purported crypto-Jewish past, a past that allowed them to redirect outrage at past Anglo abuses onto a "safe" Inquisitional villain, to ingratiate themselves to an ascendant Anglo ruling class, and to gain protective racial reclassification according to the new prestige lineage. At first, their accounts of the crypto-Jewish past were inconsistent not only with the crypto-Jewish past but also with each other, each vehemently accusing the other of lying. Only under the influence of constant media and academic "massaging" did accounts of the purported crypto-Jewish past gain stability over time.[18] This clearly demonstrates what Mary Louise Pratt called "instances in which colonized subjects undertake to represent themselves in ways that *engage with* the colonizer's own terms," hardly an unusual circumstance, she writes, but rather "a widespread phenomenon ... important in unraveling the histories of imperial subjugation and resistance as seen from the site of their occurrence."[19]

By the early 1990s, use of heritable disease as a Jewish ethnic marker was a topic of local gossip, but no purported instance was recorded in print. That would occur more than a decade later, after all cultural items cited as evidence of a New Mexican crypto-Jewish presence had been discredited by my detailed criticism of 1996.[20] In response, academic creators and promoters of the popular canon began to seek purportedly "scientific" means to strengthen their claims. To popular, pseudoethnographic constructions of New Mexican Sephardi folkways they added popular, pseudoscientific constructions of New Mexican Sephardi diseases. In this way, academic use of the label "Jewish" to determine who is at risk of heritable diseases paved the way for popular use of heritable diseases to determine who is a Jew.

Accordingly, criticism is due Kristine Bordenave, a physician who conducted a study meant to use pemphigus vulgaris (PV), a globally distributed, autoimmune blister rash, as a Jewish ethnic marker in New Mexico.[21] Bordenave concluded that the high incidence of PV among New Mexican Spanish Americans indicated their descent from Jews, since Jews also have a high incidence of PV.

The space allotted here will not suffice to criticize the agenda-driven, overinclusive logic required to support Bordenave's conclusion. Suffice it to say that a full year before a reprise of her study was published in Hordes's recent book,[22] Ron Loewenthal, director of the Tissue Typing Lab at Chaim Sheba Medical Center in Israel (cum laude 1985, MD 1987, Hebrew University; Ph.D. 1993, Cambridge University), a specialist and prolific author on medical biochemistry, molecular biology, and genetic tissue typing, found that disease haplotypes for PV are neither of ancient nor of Middle Eastern origin but are relatively recent and originate with a Mediterranean forebear.[23] Regarding Spaniards and Jews, the Loewenthal study found that the "distance between the two PV cohorts is relatively short, but the distance between Jewish patients and Jewish controls is greater compared to the distance between Spanish patients and Spanish controls."[24] Hence, the ancestral condition appears to have occurred first in Spaniards and then spread to Jewish populations. Moreover, as Sutton showed in 2006, the paternal profile of 90 percent of New Mexican Spanish Americans is highly significantly different from that of Middle Eastern populations and at the same time is indistinguishable from Mediterranean Spaniards (except for 2.2 percent Native American admixture).[25] Therefore, the more logical conclusion is that high incidence of PV among New Mexican Spanish Americans does not indicate descent from Jews but reflects instead descent from the same Mediterranean forebears who spread PV to Jews.

Constructing Legends

As Pratt puts it: "In any global classificatory project, the observing and cataloging of evidence itself becomes narratable, able to constitute a sequence of events, or even produce a plot . . . [to] form the main storyline of an entire account."[26] It is therefore revealing that the narrative genre most often employed for reporting disease-based discoveries of crypto-Jews is a traditional folk genre, one that sometimes employs academic formats (footnotes, endnotes, etc.) but never reflects academic substance. Such narratives are typically built on urban legends, that is, tales that circulate outward from modern urban centers, typically reflecting local preoccupations and social agendas.

In New Mexico, for example, typical reports of surviving crypto-Jewish customs are described as having been reported to, rather than witnessed by, the person telling us about them. This subtype of legend, from which we never get firsthand or eyewitness information, is classified as a FOAF tale in folkloristic scholarship. The acronym stands for "Friend-of-a-Friend," since the primary source of information in such tales (the friend of a friend, or the friend's friend of a friend, etc.) is always anonymous or too far removed to secure or verify the claims being made.

Unlike myths (which fill lacunae in time beyond recall when world and social order were still being formed), and unlike fairy tales (which take place in formulaic time, like "once upon a time" and "happily ever after"), legends always take place in historical time. Thus, a legend is a story involving a historical person, place, or event but is about something that cannot be secured or verified in history (e.g., a significant influx of crypto-Jews into colonial New Mexico, or sightings of Elvis along the nation's highways). Legends therefore gain credibility through a compelling combination of superficial plausibility and popular appeal. Relying on our propensity to lend credibility to any intrinsically satisfying or otherwise appealing plausibility, legends are always told as true stories.[27]

Thus it was in Santa Fe, in the summer of 1992, that an academic proponent of the crypto-Jewish canon mentioned to me that he was looking into a disease called "Niemann-Pick" as confirmation of Spanish American descent from crypto-Jews. This may not be the first attempt to use heritable disease as a Jewish ethnic marker in New Mexico, but it was the first time I had heard this proposition. The conversation would gain greater folkloristic significance six months later in a coffeehouse in Albuquerque, when a young man clearly convinced of his crypto-Jewish origin informed me that "a rabbi in Colorado" had recognized a Spanish American woman to be Jewish based on her mother's affliction with a "Jewish" disease. I could not determine whether the tale came first and had inspired academic interest in disease-based Judaism, or whether academic interest had come first, generating a self-authenticating tale of disease-based Judaism among would-be descendants of crypto-Jews. But, as we are about to see, the publication of this tale ten years later, as narrated by an anonymous source claiming to be the woman whose mother had the "Jewish" disease, confirms its ongoing circulation in local oral tradition. By the same token, the publication of this account by an academic, Janet Liebman Jacobs, constitutes yet another telling, one embellished by academic commentary and endorsed as "history" by virtue of the authority vested in a university press.[28]

However, had I not heard a variant of the same tale ten years earlier, the repetitions would still allow us to identify the story as a folk narrative according to Stith Thompson, grand old man of folktale classification. Thompson defines as a folk "motif" the smallest element of any tale that is striking or unusual;[29] thus, "mother" is not a folk motif, but "wicked stepmother" is. As Thompson notes, the narration of any striking or unusual element more than once not only classifies it as a motif but also indicates its "power to persist in tradition," adding that striking or unusual incidents comprise the vast majority of folk motifs that represent a category of tale, or tale-type.[30] When an unusual incident is (or can be) secured and verified, we automatically classify it as "history." But as long as an unusual incident remains unsecured and unverified, no one representing academe is at liberty to misrepresent it as a historical event (even if it reportedly occurred in a particular location and at a specified time). Thus, the unusual "rabbi in Colorado" incident, which has by now remained unsecured and unverified for more than twenty years, is by definition not history. It persists only as an unsecured, unverified narrative, therefore as a legend, and one that is destined to remain so by virtue of its FOAFtale format.

Although we could easily call this narrative the "rabbi in Colorado" story, there is nothing striking or unusual about a rabbi being in Colorado in and of itself. Therefore, folkloristic analysis requires parsing the motif—the striking or unusual incident itself—for productive study, in this case, for example: "individual's hidden identity revealed to an authenticating pundit through relative's heritable disease." Parsing the unusual incident in this minimalist way allows a given variant to be discerned from other possible variants within the tale-type, each according to its own regional or ethnic specificity. In this case, the southwestern Spanish American variant would be: "Spanish American daughter's hidden Jewish identity revealed to Colorado rabbi through mother's heritable Niemann-Pick disease."

For any folk narrative to persist, however, it must contain a number of negotiable traits: elements that can be modified without changing what the tale is about. This enables the tale to dodge later discrediting, to take on other ethnic identities, and/or to enhance its literary impact. Such negotiable traits include names and ages of protagonists, genders, national identities, religious affiliations, and—as would occur in the case of the Niemann-Pick tale—the name of the disease in question. The reason for this last modification is worth noting, since it does not reflect a change in the spirit or mentality of crypto-Jewish reporting in modern New Mexico.

Jewish by Disease

According to the medical literature, the term "Niemann-Pick" refers to a group of "storage" disorders in which waste materials build up in human tissue and cause it to deteriorate. Since storage disorders are found among Jews and also among Spanish Americans of New Mexico (as well as in other populations), Niemann-Pick allows for the superficial plausibility of a genetic link between Spanish Jews and Spanish Americans. But in overinclusive folk logic, the discovery of any supportive plausibility constitutes a full confirmation. Hence, the activity recognized as "research" by academics—the critical, differential logic required to confirm or deny a given plausibility—is never undertaken.

Not surprisingly, even the most rudimentary investigation reveals that Niemann-Pick types A and B alone are found among Jews, but only among Ashkenazi (Germanic and eastern European) Jews. Clearly, a disease not known to occur among Sephardi (Iberian) Jews cannot be used as an ethnic marker for Sephardi descent. With regard to New Mexico's Spanish American population, the tale fails to further differentiate (Ashkenazi) Niemann-Pick types A and B from (Hispanic) Niemann-Pick type C, which is the only type found among Spanish Americans of New Mexico and is a different disease at both the biochemical and genetic levels.[31] The Niemann-Pick investigation was apparently abandoned due to this discrepancy, but with selective inattention to its meaning: since the two populations manifest only biologically distinct types of Niemann-Pick, they cannot be one and the same people.

It was roughly six years after the folkloristic discrediting published in 1996 that the crypto-Jewish canon in general, and a variant of the Niemann-Pick legend in particular, first gained academic legitimacy through a university press.[32] In Liebman Jacobs's work on New Mexican folkways, we find the only printed version of the Niemann-Pick legend. Told as a true story, the text retains the daughter with the hidden identity, the rabbi in Colorado, and the afflicted mother, but it replaces Niemann-Pick with Creutzfeldt-Jakob disease (CJD). The substitute illness, known as heritable or "familial" CJD, is an extremely rare degenerative disease with a well-documented and unusually high incidence among Libyan Jews.[33]

The substitution of CJD for Niemann-Pick therefore indicates that those who tell the tale believe Libyan Jews to be of Spanish origin and also believe that heritable CJD has been found among southwestern Spanish Americans, although—as we are about to see—both beliefs are incorrect. The timing of the substitution indicates that Niemann-Pick lost credibility sometime after 1992, when it was first mentioned to me, and before 2002, when the CJD variant was first published. Alternatively, the substitution may simply represent a literary strategy inspired by the British "mad cow" epidemic of the late 1990s, which—having been identified as CJD by the media—would have lent CJD greater literary impact than the similarly heritable but less sensationalized Niemann-Pick disease.

The published narrative is important for understanding how easily an overinclusive folk category of human genetic identity can be constructed at the academic level. As can be seen in this particular case, the academic mission need only succumb to the thrall of discovering lost, hidden Jews, thus preserving the academic format while building exclusively on classic elements of identity-building folk logic: overgeneralization of superficially related parallels, facile placement of Jewish labels on persons who otherwise are not identifiable as Jews, use of pseudepigraphy (false ascription of authenticating statements to authorities who never made them), and—with regard to recognizing and interpreting folkloric material—a complete innocence (or wholesale rejection) of all theories, techniques, and methods of folkloristic inquiry, classification, and interpretation, in short, by wholesale violation of academic scholarship norms.

The FOAFtale and its commentary take up less than a single page in Liebman Jacobs's book, prefaced by a relatively long discussion of matrilineal descent as the traditional, Rabbinic means of establishing one's status as a Jew. Thus, the issue of matrilineal descent serves as a distraction from the primary factor that actually impedes Spanish American claims to Jewish descent: today's most elderly Spanish Americans (like their fathers before them) are professing Christians. This creates a problem for Spanish Americans seeking recognition as Jews, since Jewish status cannot be claimed by persons of non-Jewish descent who cannot document a personal Jewish history and who themselves profess Christianity or some religion other than Judaism. Hence, given the talmudic precept that descent from a Jewish mother automatically secures Jewish status, the solution for one non-Jewish individual (rendered anonymous by Liebman Jacobs) was to demonstrate the regionally coveted (white) bloodline through maternal affliction with a fancifully constructed "Sephardi" disease. Interestingly, this congenitally diseased construction of Judaism, emptied of all religious meaning, does indeed trace back to an ancestral Spanish legacy; it reflects the (obviously persistent) Spanish colonial belief that Jews do not constitute a faith community but instead belong to a biologically immutable "race" unaffected by religious affiliation.[34]

As we are about to see, all scientific and religious claims made in this tale are demonstrably unfounded, but, as a folktale, it has a different goal than that of accuracy: it exists to enable the Spanish American narrator to take a desired step up on the regional southwestern pigmentocracy by virtue of her mother's "Sephardi" disease, thereby compelling "Jewish" recognition according to the same Rabbinic law that otherwise impedes it. Liebman Jacobs, who frames the tale as a true story, upholds talmudic protocols regarding matrilineal descent but subsequently falls from grace by claiming CJD as one of them, informing her readers: "Only one descendant in this study was recognized as Jewish through blood ties to the mother" (as was purportedly "revealed" by heritable CJD).[35]

Notably, before the tale is even begun, the so-called descendant's crypto-Jewish lineage is given as rabbinically "recognized" (albeit by an anonymous character who himself cannot be recognized). The narrator's "Jewish" identity is additionally confirmed by academic fiat, since the author refers to the narrator as a "descendant," although the author herself never secured or verified the evidence given for that claim. This informant's Jewish-by-disease narrative is clearly a garden variety FOAFtale in which the author is our academic friend, who tells us what she heard from her Spanish American friend (the FOAF), while the key rabbinical figure—a friend of the author's friend (who could easily have given both women the credibility they so clearly desire)—is someone both anonymous and too far removed (living "somewhere" in Colorado) to be reached for verification. Failing to recognize the narrative as a biologically impossible FOAFtale (thus mistaking it for "history," and giving it as "medical history," at that) meets no standards set by any academic discipline. Not surprisingly, Liebman Jacobs's preamble mis- and disinforms her readers, biasing them instead of enlightening them: "In this case the descendant's mother suffered from Creutzfeldt-Jacob [*sic*] disease (CJD), a degenerative disease of the central nervous system that has been linked specifically to Sephardic ancestry."[36]

Her statement is inaccurate on two counts. First, CJD occurs in more than one form, only one of which is heritable, and, as we have seen, is found at much higher frequency among Libyan Jews than in other populations (in which it also occurs). But Libyan Jews have no Sephardi connection. By the fifth century BCE Libya had a substantial Jewish popu- lation, perhaps as many as thirty thousand, according to the somewhat impressionistic reporting of Pliny and Herodotus.[37] Nonetheless, the Jewish population was clearly large enough to suggest tens of thousands roughly seven hundred years before the first artifactual evidence of a Jewish presence in Spain in the third century CE.[38] Moreover, according to documented diasporadic patterns, Libya was not among the havens sought by Spanish Jews when they left Spain in 1492; the more attractive North African destinations were Fez and Tlemcen in the nearby Barbary states, as well as Algiers, Tunisia, and the city of Alexandria, then part of the Ottoman Empire.[39] Hence, while there may have been the odd unrecorded instance of Spanish settlement, Libyan Jews are not a Span- ish people; they have no such origin nor any subsequent history. On the other hand, while heritable CJD is well documented among them, there is no documentation of heritable CJD among descendants of Spanish Jews anywhere on the globe. Liebman Jacobs thus confirms, in a work published by a university press, that heritable CJD is a "Sephardic" dis- ease even though it has never been found in any Sephardi population. Furthermore, since there has also been no recorded instance of heritable CJD among Spanish Americans in the Southwest, the notion of CJD as a genetic link between Spanish Jews and southwestern Spanish Americans lacks any credibility at all.

Second, the source used by Liebman Jacobs to link CJD with "specifically Sephardic ancestry" is Richard M. Goodman, a recognized expert on genetic diseases found among different populations of Jews. But the book cited was published by Goodman in 1979, before any form of CJD was recognized as heritable. Hence, Liebman Jacobs resorts to pseudepigraphy, ascribing to Goodman an authenticating link—here between CJD and "specifically Sephardic ancestry"—that he never made. Thus, we should not be surprised that Goodman who addressed the question of heritable CJD under the chapter heading "Misconceptions," stated emphatically that CJD was *not* a heritable disease, exactly the opposite of what Liebman Jacobs attributed to him. Moreover, with regard to the one Libyan outbreak known at the time, he suggests that it may have been caused by eating infected animal parts; a link to "specifically Sephardic ancestry" was never mentioned and was clearly never considered by Goodman.[40]

Thus, following a medically untrained informant instead of any easily accessible medical encyclopedia—or even the outdated reference she cites, which also refutes her claims—Liebman Jacobs informs us that CJD is a recognized "Sephardic" disease and states that a Spanish American woman's search for Jewish roots led her to "approach a rabbi in Colorado." On hearing that the woman's mother had CJD (in the informant's own words),

> the rabbi said, "You're Jewish. It doesn't make any difference if you are an atheist, you are Jewish. You don't have to convert. You can just start practicing the laws if that's what you want to do." In recalling this aspect of her search for Jewish roots, she spoke of both the relief and sadness that accompanied her mother's diagnosis and the rabbi's affirmation of her Jewish lineage: "It was great to find out positively that I was Jewish. I had already been researching it, so I was pretty sure that we were Jewish. So I was relieved that I didn't have to research my mother's side anymore."[41]

Here, a modern Spanish American FOAFtale attempts to authenticate religiously empty, disease-based Judaism by appeals to the two most powerfully legitimating contexts of our time: science and religion. However, as we have already seen, both science and religion negate the claims being made. Genetic research has unequivocally refuted any significant influx of exiled professing Jews into colonial New Mexico, and there is no CJD-Sephardi link to indicate Jewish descent through an infected mother. Moreover, according to Rabbinic law, any individual professing a religion other than Judaism who seeks rabbinical recognition as a Jew must either convert or forgo that recognition. Hence, we can see that this tale has a more immediate purpose than mere accuracy. It is more clearly a self-authenticating social device meant to secure for the would-be descendant public recognition of a higher "rank" than she could otherwise claim on the regional Spanish American pigmentocracy, a rank she is determined to make unimpeachable in the face of any and all argument by means of a Rabbinic law on "blood ties" to a (CJD-infected = Jewish = white) mother.

Although CJD cannot be used as a marker for Sephardi descent, we might find a way to believe this narrator if we were to assume her mother suffered from an undocumented case of heritable CJD and if she was simply misinformed by a rabbi in Colorado. In bringing the narrative to an end, however, her own testimony eliminates this possibility; that is, if her mother had indeed had heritable CJD, the way in which this disease was diagnosed could not possibly have escaped her: "We knew my mother was ill. We knew she had a condition that wasn't getting any better. You know, it was hard knowing that my mother was ill in the first place. That was pretty tough. So the emotions were kind of mixed, because I was relieved that we were Jewish, but, you know, it's heartbreaking to see your mother go through that."[42] The informant never pays tribute to a mother who passed away before the encounter with the rabbi. Rather, she consistently refers to a living mother who wasn't getting any better at the time of the encounter. Hence, we learn that the rabbi's recognition evoked, on the one hand, relief that the heritable disease revealed the daughter's hidden identity and, on the other, heartbreak at the sight of the mother's ongoing suffering. Yet, at the time the tale was recorded, there was only one possible way to diagnose heritable CJD: after death, or postmortem, by brain autopsy.[43] Thus, no one directly or indirectly included in this narrative—not the physician, the mother, the rabbi, or the daughter—could possibly have known the patient had heritable CJD within the time frame of the tale, as told in the daughter's own words.

The Pitfalls and Prejudices of Folk Taxonomy

Following the tale's apparently fictitious rabbi, Liebman Jacobs recognized the narrator as a "descendant" based on her mother's "Sephardi" disease. What she did not recognize was the FOAF tale format, the tale's problematic time frame, the fact that Libyan Jews cannot be swept into an overinclusive Sephardi category, or the fact that heritable CJD could not at the time be diagnosed before death and, to this day, has never been diagnosed among Spanish Americans or among descendants of Spanish Jews.

Hence, it is important to stress, again, that when academics step out of their own areas of expertise into highly specialized fields in which they have no training, they cease to be experts and revert to being ordinary folks, enabled and disabled by the same strengths and weaknesses of the universal human condition. In the case we have just seen, the academic author had no apparent training in the highly specialized field she entered, and she clearly sought no professional consultation, which led her to discover crypto-Jewish "descendants" wherever she sought them and to slap "Jewish" labels on ambiguous cultural paradigms across an entirely non-Jewish landscape.

It is therefore significant that scientists charged with identifying patterns within the human genome are themselves not trained to recognize the negative scientific and social consequences of labeling diseases according to nonheritable, essentially untestable cultural characteristics or according to agenda-driven constructions of disease-based religious descent, as seen in the examples above. Moreover, from a scientific point of view, such folk taxonomies can seriously compromise fact gathering, as well as health-care delivery, since they inevitably result in wishful, impressionistic, or academically misled self-definition (as "Jewish") by whole populations selected for genetic screening (or simply seeking it) based on nonheritable (as well as non-Jewish) traits and characteristics.

In New Mexican social context, it is clear that use of the term "Jewish" for various and sundry heritable diseases serves only to strengthen a regional color-coded hierarchy, maintained in perpetuity by ongoing claims of descent that both reflect and reinforce a pseudoscientific social pathology, in this case, racialized colonial tenets aided and abetted by seemingly "innocent" academic authority. But the genetic labeling of faith communities can never be innocent, for faith communities are socially, not genetically, constructed, and there is no such thing as a culturally constructed, "congenital" line of religious affiliation that has no consequence in social stratification (for which modern New Mexico may be the best example).

As Pratt observes, it was colonial dominion, not genetic mapping, that first "exerted the power" to name, or place religious labels on, geographical and biological entities: "Indeed it was in naming that the religious and geographical projects came together, as emissaries claimed the world by baptizing landmarks and geographical formations with Euro-Christian names. But again, natural history's naming is more directly transformative. It extracts all the things of the world and redeploys them into a new knowledge."[44]

We can see how giving names to categories in the natural world is itself a powerful form of classification, one that brings into being a hegemonic world order: a typically overarching, self-authenticating notion of "the way things are." Hence, to name biological groups according to religious affiliation is effectively to follow Linnaeus not only by drawing genetic boundaries around cultural characteristics but also by transforming mutable religious affiliations into fixed congenital conditions. In this manner, all "despised" or demonized characteristics assigned to religious Others are cast as biologically immutable, a highly precarious position for any vulnerable minority. In particular, given the overgeneralized status of the category "Jew," and knowing that overinclusive folk logic is required to generate and sustain prejudice, it seems almost inevitable that academic classifications of heritable diseases as "Jewish" will fan the flames of anti-Semitism. Moreover, if the label "Jewish" is to be the only religious term associated with heritable disease, it can only bolster overinclusive folk constructions of a singularly contaminated, and therefore a singularly contaminating, faith community.

Following this concern, I conducted IRB-approved research in New Mexico in 2009 that was funded by the National Institutes of Health and the Center for Genetic Research Ethics and Law. Below is the first instance I found during that research of local, disease-based anti-Semitism assimilated to the crypto-Jewish canon, notable because it demonstrates the canon's support of anti-Semitic racism and because it occurred at the highest echelon of New Mexican social influence.

The Hispanic patient in this case was one of many siblings, several of whom suffered from schizophrenia. The mental health professional in charge took it upon himself to inform the family that they descended from crypto-Jews for the following reasons: first, because Jews have schizophrenia more often than non-Jews (an old anti-Semitic saw never scientifically substantiated); and second, because the family came from the general area of northern New Mexico and southern Colorado, a fictive crypto-Jewish demographic perpetrated by the crypto-Jewish canon, allowing him to further inform them that the first mental institution in New Mexico "had to be built there because that's where the Jews were."

Such claims are clearly based on a demonstrably unfounded, overinclusive category of *religion* as biologically inherited and identifiable by disease, as purported by the crypto-Jewish canon. But Gordon Allport never suggested that all overinclusive categories are necessarily prejudiced, only that they are undifferentiated and therefore pave the way for prejudice. To avoid paving that way, it may be helpful to distinguish between a mere misconception and a bona fide prejudice. As he explains:

> Not every overblown generalization is a prejudice. Some are merely misconceptions, wherein we organize wrong information. . . . Here we have the test to help us distinguish between ordinary errors of prejudgment and prejudice. If a person is capable of rectifying his erroneous

judgments in the light of new evidence he is not prejudiced. Prejudgments become prejudices only if they are not reversible when exposed to new knowledge. A prejudice, unlike a simple misconception, is actively resistant to all evidence that would unseat it. We tend to grow emotional when a prejudice is threatened with contradiction. Thus the difference between ordinary prejudgments and prejudice is that one can discuss and rectify a prejudgment without emotional resistance.[45]

Like Linnaeus in an earlier age of discovery, we are suddenly called upon to define new categories of human variation and, when we are mistaken, to correct our misconceptions without prejudice. Unlike Linnaeus, we have greater hindsight and voluminous academic enlightenment to guide us. But, as we have seen, in our haste to name heritable biological disorders according to nonheritable cultural characteristics, we are ourselves at risk of creating inaccurate, overinclusive folk categories, forgoing academic accuracy, and reinforcing our own self-authenticating social devices and/or those of historically prejudiced populations. As previously suggested, use of geographically based variance seems a more valid, reliable, and socially responsible way of categorizing heritable differences, since every heritable disease has a founder who can be located in time and place: a testable assumption in every case, which cannot be said for any founder's religious sensibility. As Joseph Graves puts it: "The practice of building whole theoretical constructs on essentially untestable assumptions is the hallmark of pseudo-science ... often associated with vested social agendas."[46] Assuming that one's social agenda is benevolent and entirely well-meaning, there is still no agenda more benevolent than accuracy in distinguishing human populations or in practicing health-related science. Therefore, to paraphrase Linnaeus, the Ariadne thread in human classification is disciplined, differentiating academic logic, without which only chaos can prevail.

Update: *BRCA*

Groundbreaking genetic discoveries began to proliferate with the completion of the Human Genome in 2004. But sophistry can also proliferate in the gap between groundbreaking discoveries and our ability to make good sense, let alone good use, of them. This is easily demonstrated using *BRCA1/2* mutations currently employed in New Mexico to show that local Hispanics descend from Spanish crypto-Jews. But as we are about to see, the same mutations given as evidence of Sephardi descent actually refute it. Moreover, scrutiny of New Mexican *BRCA1/2* claims will show that wherever Jews are a minority, there is no disease that can ever be used as a Jewish ethnic marker.

BRCA stands for breast cancer, and *1/2* stands for two genes that normally produce tumor-suppressing proteins in the human body. Because *BRCA1/2* mutations inhibit these proteins, individuals with these mutations are at greater risk of developing certain cancers. Among the many hundreds of mutations found in *BRCA1/2*, three have been associated with Ashkenazi (Germanic, eastern European) Jews. The discovery of one or more Ashkenazi mutations in the local Hispanic population is currently given as evidence of Sephardi descent by asserting that these mutations emerged before the Ashkenazi-Sephardi split.

But estimates for the origin of these mutations vary greatly among equally respected geneticists. Some lean toward emergence as late as the twelfth to fourteenth centuries, long after the Ashkenazi-Sephardi split.[47] Because the origin of these mutations is so uncertain, no one can assert anything conclusive at this time, and thus the New Mexican assertion is a conjecture, not a fact. Moreover, it ignores the well-established fact that Ashkenazi Jews and Hispanics have been living side by side in New Mexico for roughly 175 years, on which basis we should expect to find Ashkenazi admixture in the Hispanic gene pool, a modern admixture that does not indicate premodern Sephardi descent.

To show that Ashkenazi *BRCA1/2* mutations are useful Sephardi indicators, we would have to find them outside of New Mexico among Jewish populations whose Sephardi ancestry is not in question. But to date, no study has ever found Ashkenazi *BRCA* mutations among any non-Ashkenazi Jews. For example, a study in 2007 found a *BRCA1* mutation among Iranian Jews, but this mutation is different from any of the three mutations found in Ashkenazi Jews.[48] A study done in 2010 found no *BRCA1/2* mutations in a small sample of Sephardi women.[49] A more comprehensive study in Israel found two mutations (one in *BRCA1* and one in *BRCA2*) in "descendants of Jews from the Iberian Peninsula," but these mutations are also different from those found in Ashkenazi Jews.[50] To date, given the results of all studies on Middle Eastern Jews and Jews of Sephardi origin, it is clear that having one or more Ashkenazi mutations indicates Ashkenazi, not Sephardi, descent, eliminating crypto-Jews from the New Mexican *BRCA1/2* equation.

We now know that modern New Mexican males who self-identify as "Spanish American" or "Hispanic" are effectively identical to modern males in Spain and that 10 percent of both populations have Middle Eastern ancestry. But this 10 percent represents all Middle Eastern populations that settled in Spain, not just Jews. The issue is not simply that religious affiliation is not accessible through DNA, or even that New Mexicans with Sephardi ancestry comprise only a smaller fraction of an already small fraction of Hispanic New Mexicans. The real issue is that DNA cannot tell us if any instance of Sephardi descent in New Mexico is also crypto-Jewish. That determination requires historical and cultural evidence of a crypto-Jewish tradition, and, as we have seen, neither exists in New Mexico.

Moreover, even when Jews and non-Jews share mutations found in significantly higher numbers among Jews, the non-Jewish population is so gigantic compared to the tiny Jewish minority that the vast majority of affected people will *always* be non-Jews. For example, the frequency of Ashkenazi *BRCA1/2* mutations among Jews is roughly 2.5 percent; among non-Jews, roughly 0.5 percent. Nevertheless, there are so few American Jews compared to American non-Jews that only one in every dozen affected Americans is likely to be Jewish. Thus, as long as Jews remain a minority, no disease shared by Jews and non-Jews can ever be used as a Jewish ethnic marker.

The absurdity of New Mexico's agenda-driven logic may be most evident in non-Jewish contexts. That is, there are common *BRCA* mutations not found in Jews that occur with highest frequency in Norway and the Netherlands. If we apply New Mexican logic, that would make non-Jewish Hispanics who carry these mutations "crypto-Norwegian" or "crypto-Dutch."

As we have seen, New Mexico's Jewish-by-disease claims, endorsed as they are, without evidence and by sheer power of academic fiat, can only exist at a disturbing level of academic and social dysfunction. Thus, it should concern us that the new Sephardic / Crypto-Jewish Program at the University of Colorado at Colorado Springs (UCCS) has issued a mission statement that opens as shown below, emailed to me at my request in 2013 by the director of the program and available on its website. The mission statement defined Sephardic Jews in parentheses as "Jews of Spanish and Latino descent" and defined crypto-Jews in parentheses as "Sephardic Jews that retained their faith in secret." For better clarity, I have deleted the two parentheses so the statement can be read uninterrupted: "This endeavor aims to foster collaborative scholarly research on Sephardic Jews and *crypto-Jews of the southwest United States* and the world" (my italics).[51]

I hope I speak for the global academic community, including UCCS, in welcoming any new program that generates academic research in accordance with academic scholarship norms and fieldwork ethics. But a mission statement that explicitly validates a significant southwestern settlement of secretly professing Spanish Jews for which there is neither historical nor cultural evidence and that is refuted by the disease-based evidence given to support it promises only to perpetuate the pseudoethnography, bad science, and quack medicine of sinister interests in less enlightened times.

NOTES

1. Alan Dundes, "The Number Three in American Culture," in *Interpreting Folklore*, ed. Alan Dundes (Bloomington: Indiana University Press, 1980), 134–59, 155.
2. Hayden White, *The Tropics of Discourse: Essays in Cultural Criticism* (Baltimore MD: Johns Hopkins University Press, 1985), 151.
3. Dan Ben-Amos, "Analytic Categories and Ethnic Genres," in *Folklore Genres*, ed. Dan Ben-Amos (Austin: University of Texas Press, 1976), 215–42, 225.

4. Roy A. Rapapport, "Ritual," in *Folklore, Cultural Performances and Popular Entertainments: A Communications-Centered Handbook*, ed. Richard Bauman (New York: Oxford University Press, 1992), 253.

5. Gordon W. Allport, *The Nature of Prejudice, 25th Anniversary Edition* (Reading MA: Addison-Wesley Publishing Company, 1994), 446.

6. Nancie L. González, *The Spanish-Americans of New Mexico: A Heritage of Pride* (Albuquerque: University of New Mexico Press, 1969), 81; Fray Angélico Chávez, *Origins of New Mexican Families in the Spanish Colonial Period, in Two Parts: The Seventeenth (1598-1693) and the Eighteenth (1693-1821) Centuries* (Santa Fe NM: William Gannon, 1975), xiv, xvi.

7. Alejandro Lipschutz, *El indoamericanismo y el problema racial en las Americas* (Santiago: Nacimiento, 1944), 75.

8. Ramón A. Gutiérrez, *When Jesus Came the Corn Mothers Went Away: Marriage, Sexuality and Power in New Mexico, 1500-1846* (Stanford CA: Stanford University Press, 1991), 198; Pedro Alonso O'Crouley, *A Description of the Kingdom of New Spain* (1744), trans. from the Spanish by Seán Galvin (San Francisco: John Howell, 1972), 19; Adrian H. Bustamante, "Españoles, castas y labradores: Santa Fe Society in the Eighteenth Century," in *Santa Fe: History of an Ancient City*, ed. David Grant Noble (Santa Fe NM: School of American Research Press, 1989), 64-77, passim.

9. Gutiérrez, *When Jesus Came*, 198-99.

10. Gutiérrez, *When Jesus Came*, 195.

11. Judith S. Neulander, "The New Mexican Crypto-Jewish Canon: Choosing to Be 'Chosen' in Millennial Tradition," *Jewish Folklore and Ethnology Review* 18, nos. 1-2 (1996): 19-58; Neulander, "Cannibals, Castes and Crypto-Jews: Premillennial Cosmology in Post-colonial New Mexico" (Ph.D. diss., Indiana University, 2001); Neulander, "Jews, Crypto," in *Encyclopedia of American Folklife*, ed. Simon J. Bronner, 4 vols. (Armonk NY: M. E. Sharpe, 2006).

12. Neulander, "The New Mexican Crypto-Jewish Canon."

13. Wesley K. Sutton, Alec Knight, Peter A. Underhill, Judith S. Neulander, Todd R. Disotell, and Joanna L. Mountain, "Toward Resolution of the Debate Regarding Purported Crypto-Jews in a Spanish-American Population: Evidence from the Y-Chromosome," *Annals of Human Biology* 33, no. 1 (2006): 100-111.

14. Raphael Patai, "The Jewish Indians of Mexico," in *On Jewish Folklore*, by Raphael Patai (Detroit: Wayne State University Press, 1983), 447-75, 461.

15. Stanley M. Hordes, "The Sephardic Legacy in the Southwest: Crypto-Jews of New Mexico, Historical Research Project Sponsored by the Latin American Institute, University of New Mexico," *Jewish Folklore and Ethnology Review* 15, no. 2 (1993): 137-38.

16. Seymour B. Liebman, *The Inquisitors and the Jews in the New World: Summaries of Procesos, 1500–1810, and Bibliographic Guide* (Coral Gables FL: University of Miami Press, 1974).

17. Chávez, *Origins of New Mexican Families.*

18. Neulander, "The New Mexican Crypto-Jewish Canon," 24–26.

19. Mary Louise Pratt, *Imperial Eyes: Travel Writing and Transculturation* (London: Routledge, 1992), 7, 9.

20. Neulander, "The New Mexican Crypto-Jewish Canon."

21. Kristine K. Bordenave, Jeffrey Griffith, Stanley M. Hordes, Thomas M. Williams, and R. Steven Padilla, "The Historical and Geomedical Immunogenetics of Pemphigus among the Descendants of Sephardic Jews in New Mexico," *Archives of Dermatology* 137, no. 6 (2001): 825–26.

22. Kristine Bordenave and Stanley M. Hordes, "Pemphigus Vulgaris among Hispanos in New Mexico and Its Possible Connection with Crypto-Jewish Populations," in *To the End of the Earth: A History of the Crypto-Jews of New Mexico*, by Stanley M. Hordes (New York: Columbia University Press, 2005), 289–95. I was asked to review this book by the journals *Shofar* and the *Catholic Historical Review*; the reviews appeared in late 2006–7.

23. Ron Loewenthal et al., "Common Ancestral Origin of Pemphigus Vulgaris in Jews and Spaniards: A Study Using Microsatellite Markers," *Tissue Antigens* 63, no. 4 (2004): 326–34.

24. Loewenthal et al., "Common Ancestral Origin," 326.

25. Sutton et al., "Toward Resolution."

26. Pratt, *Imperial Eyes*, 27–28.

27. Jan Harold Brunvand, *The Vanishing Hitchhiker: American Urban Legends and Their Meanings* (New York: W. W. Norton, 1981), 1–4.

28. Janet Liebman Jacobs, *Hidden Heritage: The Legacy of the Crypto-Jews* (Berkeley: University of California Press, 2002).

29. Stith Thompson, *The Folktale* (Los Angeles: University of California Press, 1977), 114.

30. Thompson, *The Folktale*, 115.

31. Laith F. Gulli and Tanya Bivins, "Niemann-Pick Disease," in *The Gale Encyclopedia of Genetic Disorders*, ed. Stacey L. Blachford, 2 vols. (Detroit: Gale Group, 2002), 2:813–16.

32. Liebman Jacobs, *Hidden Heritage.*

33. Zeev Meiner, Ruth Gabizon, and Stanley B. Prusiner, "Familial Creutzfeldt-Jakob Disease: Codon 200 Prion Disease in Libyan Jews," *Medicine* (Baltimore) 76, no. 4 (1997): 227–37.

34. Jane Berger, *The Jews of Spain: A History of the Sephardic Experience* (New York: Free Press, 1994), 127.

35. Liebman Jacobs, *Hidden Heritage*, 103.

36. Liebman Jacobs, *Hidden Heritage*, 103.

37. John Wright, *Libya* (New York: Praeger, 1969), 45, 54.

38. Benjamin R. Gampel, "Jews, Christians, and Muslims in Medieval Iberia: Convivencia through the Eyes of Sephardic Jews," in *Convivencia: Jews, Muslims and Christians in Medieval Spain*, ed. Vivian B. Mann, Thomas F. Glick, and Jerrilyn D. Dodds (New York: George Braziller in association with the Jewish Museum, 1992), 2–27, 11.

39. Gérard Chaliand and Jean-Pierre Rageau, *The Penguin Atlas of Diasporas*, trans. A. M. Berrett (New York: Viking Press, 1995), 30–31.

40. Richard M. Goodman, *Genetic Disorders among the Jewish People* (Baltimore MD: Johns Hopkins University Press, 1979), 409.

41. Liebman Jacobs, *Hidden Heritage*, 103.

42. Liebman Jacobs, *Hidden Heritage*, 103.

43. Larry J. Lutwick, "Creutzfeldt-Jakob Disease," in *The Gale Encyclopedia of Medicine*, ed. Jacqueline L. Longe, 2nd ed., 5 vols. (Detroit: Gale Group, 2002), 2:950–54, 953.

44. Pratt, *Imperial Eyes*, 33.

45. Allport, *The Nature of Prejudice*, 9.

46. Joseph L. Graves, Jr., *The Emperor's New Clothes: Biological Theories of Race at the Millennium* (New Brunswick NJ: Rutgers University Press, 2001), 87.

47. Susan L. Neuhausen et al., "Haplotype and Phenotype Analysis of Nine *BRCA2* Mutations in 111 Families: Results of an International Study," *American Journal of Human Genetics* 58, no. 2 (1998): 271–80.

48. R. Ferla et al., "Founder Mutations in *BRCA1* and *BRCA2* Genes," *Annals of Oncology* 18, Supplement 6 (2007): vi93–vi98.

49. K. A. Metcalf et al., "Screening for Founder Mutations in *BRCA1* and *BRCA2* in Unselected Jewish Women," *Journal of Clinical Oncology* 28, no. 3 (January 2010): 387–91.

50. M. Sagi et al., "Two *BRCA1/2* Founder Mutations in Jews of Sephardic Origin," *Familial Cancer* 1, no. 10 (March 2011): 59–63.

51. http://www.uccs.edu/history/collaborations/sephardic-and-crypto-jewish-studies-program.html.

CHAPTER 8 *What Is a Jew?*

The Meaning of Genetic Disease for Jewish Identity

REBECCA ALPERT

My Jewish friends used to tell me that it was surely some Cossack who gave me my blue eyes and small nose, so comfortable were they with the idea that Jews carry a genetic imprint that makes Jewish eyes brown and noses large. Or perhaps it was their discomfort with the possibility that we wear our Jewishness on and in our bodies and genetic coding that caused them to joke. Jews have always experienced tension and lack of clarity around how we define ourselves as a group. We understand ourselves to be a people, religion, nation, ethnicity, or some combination.[1] This complex group definition has caused some confusion about how much who we are is about biology and how much it is about culture. And the relationship between our biological and cultural group identity raises some very interesting questions about how we see ourselves in relation to new scientific discoveries in the field of genetics. As science becomes more comfortable with the idea that "nature" and "nurture" interact to make us who we are, so we Jews are beginning to accommodate ourselves to understanding the ways in which our genetic and social identities interact to define who we are.

Defining Who Is a Jew

The standard halakhic definition allows for and seems to differentiate between the biological transmission of Jewishness (you are Jewish if you are born of a Jewish mother) and a religious transmission (you are Jewish if you convert to Judaism through a process including both accepting Jewish beliefs and performing Jewish actions). But the convert also has Jewish lineage bestowed upon him or her in the process, as he or she is expected to engage in rituals simulating rebirth and to take on the identity "son/daughter of Abraham and Sarah," becoming a Jew not only by practice but also by a fictive biology.

The dominant majority of Jewish texts and traditions assume it to be impossible to stop being a Jew, welcoming back even those who have converted to other religions if they wish to return, or accepting as Jewish the matrilineal great-grandchild of a Jewish woman whose family has not practiced Judaism in generations, which underscores the notion that Jewish identity is based on a combination of both inheritable characteristics and practice. But the Law of Return in Israel, which awards citizenship to all Jews and their relatives and spouses, does not include those with Jewish matrilineal lineage who practice other religions, as the Israeli Supreme Court ruled in the Brother Daniel case, and the tradition of placing someone in *herem* (excommunication, still practiced by some Orthodox Jews for what is considered unacceptable behavior or ideas) also denies Jewish identity to people who are considered to be outside the pale religiously.

The complicated nature of Jewish identity is also reflected in more recent efforts to augment the definition of who is to be included as a Jew in both Reform and Reconstructionist policies and in the Israeli Law of Return. Reform and Reconstructionist Jews now extend lineage to include those who have a Jewish father ("patrilineal Jews"). But to confirm their status, patrilineal Jews are also required to participate in Jewish rituals, like bar/t mitzvah and confirmation. And when the state of Israel created its Law of Return, the original version did not define "Jew" at all. Difficult cases related to the Law of Return and claims on

Jewish identity brought before the Israeli courts resulted in legislation that confirmed the halakhic concept of who is a Jew. Some of the magistrates and many of the commentators, however, questioned whether Jewish identity in the state ought to reflect nationality and commitment rather than religion or lineage. The question of who is included in the Jewish people and nation is very much alive today, in particular with the millions of Russian Jews who have immigrated to Israel and serve in its army but whose maternal lineage as Jews is not clear.

The Question of Election

The element of Jewish textual tradition that most poignantly reflects the confusion about identity is the interpretation of the doctrine of election, or what it has meant for the Jews to be identified as "the chosen people."[2] Election is one of the central categories that define Jewish identity. It is the predominant way of explaining why God cares about this people, calls Israel into being, and gives this group the inheritance of a land (Israel), a blueprint for living in that land (Torah), and a promise of future redemption. Yet traditional texts and commentaries differ significantly on the nature of that election. Did God choose this people because they were the descendants of Abraham? Does God continue to connect to them because of "the merit of the ancestors" or because the people assented to the covenant offered by God at Sinai? The answer given by Jewish traditional sources varies and reflects differing understandings of the nature of Jewish identity as primarily biological or religious.

This debate runs throughout Jewish history. The Babylonian Talmud seems to favor the idea that election was based on acceptance of the covenant and observance of the law (B. *Avodah Zarah* 2a–3b), and the Rabbis certainly favored passing the tradition down through their students over their biological sons.[3] Although becoming a rabbi was based on knowledge of Torah and reputation among the Torah scholars, Rabbinic Judaism still maintained the value of biological lineage through asserting that the status of *Kohanut*, the priesthood, passed from father to son and through honoring that lineage in liturgy and synagogue worship practices and also through defining Jewish identity through matrilineal lineage.

An examination of the traditional liturgy would suggest that Jewish difference from other peoples is based primarily on being chosen because we are the descendants of Abraham. Yehudah Halevi, in his medieval philosophical writings, supports this liturgical perspective, and it is also reflected in Lurianic Kabbalah and the writings of philosophers like Chaim Luzzato. Moses Maimonides takes the opposite stance, suggesting that God called Abraham not because of any inherent quality in him but because of his wisdom; Abraham was chosen not because of who he was but because of what he believed.

The argument over why the Jews were chosen has been carried on throughout modern times as well. Spinoza rejected the concept of election based on lineage. He argued that chosenness could only be reconceptualized when and if the Jews were reconstituted in a radically different kind of social organization in which Jews were self-governing, and therefore the concept is totally inapplicable in the absence of that. Cultural Zionists and the leaders of the Reform movement in the nineteenth and early twentieth centuries were also opposed to election based on Jewish lineage, supporting notions of chosenness based on culture and religion, respectively. Cultural Zionists like Ahad Ha'am suggested that election should be construed as national morality. The Reformers based their concept of election on the prophetic notion that the Jews have a mission to be "a light unto the nations," bringing the values of ethical monotheism to others. The cultural Zionists and Reformers both argued that election was a moral concept but differed about whether that morality was to be focused on building a nation or on spreading Jewish values among the host nations where Jews lived. But religious Zionists like Rav Kook and some other traditional thinkers like Michael Wyschogrod have continued to maintain a concept of election based on lineage.[4]

Other Jewish thinkers, like Mordecai Kaplan, reject any notion of chosenness and remove any mention of election from the liturgy, including the differentiation among the categories Kohen, Levi, and Israel, which support those who claim lineage from the ancient priests. Kaplan's concern about the latter was related not only to dispelling notions of hierarchy in the Jewish community but also to repudiating any sense of hereditary Jewish privilege. This argument has been echoed in contemporary times by Jewish geneticist Robert Pollack, who rejects any hereditary notion of Jewish identity primarily to avoid the inaccurate notion that there's a "Jewish gene."[5]

Many contemporary religious Jews also reject Jewish identity based on lineage, since they are uncomfortable with secular Jews claiming Jewish roots based either on nationality, as is the case of those living in the state of Israel, or on ethnicity, as is the case of those living in the United States and other countries of the Diaspora. Rather, they believe that Judaism is a religion that is predicated on being in a covenantal relationship with God. This perspective also includes a rejection of "Jewish culture," which they see as vacuous. Feeling Jewish because you eat falafel or love Woody Allen is woefully misunderstood by religious Jews. They do not comprehend the importance of Jewish culture for Jews who have no interest in the religious dimensions of Jewish life. The problem of election raises complex ethical questions about the hierarchy that is built into issues related to defining who is a member of the Jewish community.

The question of Jewish election parallels the issues raised above about defining who is a Jew. In both cases, much of the argument boils down to whether the author emphasizes lineage or religion when thinking about these issues. Rather than decide which is more important, it makes sense to assume that some combination of biology, culture, and religion is critical to our understanding of inclusion in the Jewish community. Although the vast majority of Jews are Jewish because they were born of Jewish mothers, with an increase in conversion, because of the acceptance by the Reform and Reconstructionist movements of patrilineal descent as sufficient to assert Jewish identity and an ongoing questioning of inclusive Jewish identity in the state of Israel, we are facing a much more complicated situation than ever before in defining who is a Jew.

This leads us to think more about the relationship between the biological, religious, national, and cultural dimensions of identity. It is therefore important to find resources to support the idea that Jewish identity is a combination of the genetic and the religious or cultural and not simply based on one dimension or the other. This perspective may be illustrated in two ways. First, it may be clarified through a reading of the biblical story of how God's covenant with Abraham relates to the Sinaitic covenant. Second, it is supported by current findings related to Jewish genetic disease.

The Burning Bush

The traditional Jewish model of defining identity is based primarily on the covenant with Abraham. Although there is much debate over whether Abraham was selected by God arbitrarily or because of his qualities of intellect and morality, we Jews understand Abraham as our common ancestor. We may be descended from Abraham by birth, or we may choose to identify as Abraham's descendants through conversion, but our lineage begins with him. When God reveals himself to Abraham (Genesis 15 and 17),[6] his promise is about the gift of the Land of Israel and the continuity of Abraham's descendants.

When God reveals himself to Moses in the desert at the Burning Bush (Exodus 3), he begins by announcing a connection through lineage. God identifies himself to Moses as the "god of your ancestors, the God of Abraham, Isaac and Jacob." But God also identifies himself as "I am what I am / I will be what I will be." This identity, known to mystical tradition as "the twelve-letter name" (an elaboration on YHVH, the four-letter name), implies that God is more than the God of Moses's ancestors. It is, as Martin Buber suggests, the revelation of a God of relationship, a God who will define a covenant not only based on ancestry but predicated on a set of beliefs and behaviors. The God of Abraham is a God of belonging. The God of Moses is a God of behaving and believing. But, of course, this God is one. And so the oneness of this God leads to the one God of two dimensions who revealed himself to Moses: the God of Abraham (the god of ancestors) and the God that is "what I am / what I will be." This twofold nature of God parallels the twofold nature of Jewish identity. Just as God's identity is described in terms of lineage (God of your fathers) and religion/culture (I will be what I will be), so the identity of the people of Israel is marked in this twofold way. In this way we Jews can indeed understand ourselves as made in the image of God.

In the episode of the Golden Calf (Exodus 32–34), we are given a further sense of what the biblical author had in mind in terms of portraying the twofold nature of God as it is expressed through Jewish identity. When the people make the calf, thus abandoning their identification with Moses, God makes Moses an offer. He will start this election process over again, making a new covenant with Moses and his progeny to replace the one made with Abraham. Moses talks God out of this idea, but the possibility itself is instructive. God's suggestion that the children of Moses could take on this covenant as well as (or better than) the children of Abraham reveals two key notions. First, it serves as an important reminder that from God's perspective the covenant is revocable not only from the perspective of the behavior of the Jewish people but also based on lineage. Second, and most important for our purposes, it reminds us that the covenant is indeed based on a combination of lineage and assent. God did not, for example, suggest that he might select righteous people for this covenant. He merely suggests a change in the point at which Jewish lineage would begin. There is no question that the biblical author understood a critical link between biological and cultural sources of Jewish identity. They would not exist apart from one another, no matter at which point in time the process is understood to start.[7]

So I would like to argue that Jewish identity through Abraham, through the ancestors, is as important to Jewish self-understanding as is identity based on a connection to the religious and cultural tradition of the Jewish people and that these two dimensions are inextricable. This understanding of "who is a Jew" provides a way of making the Jewish community inclusive but not unbounded. It incorporates the halakhic definition, based on the idea that Jewishness is handed down through generations or received through the fictive adoption by converts of Jewish lineage, but it is not limited by it. It also recognizes that birth is not enough; there must be in addition some connection to Jewishness, as is understood in the Reform and Reconstructionist adoption of patrilineal descent, where living a Jewish life is required in addition to biological heritage.

Because this definition requires assent and engagement in the religious or cultural life of the Jewish people, it excludes Jews by birth who do not want to be associated with the group, but it would welcome them back in case they do.[8] By this definition, the assent can be either religious or cultural, therefore including both the secular Jews in the United States and elsewhere who understand their connection to Judaism as ethnic, as well as those secular Israelis who see their Jewishness as based on nationality. It would also make room for Brother Daniel, the priest who was born a Jew, fought to save Jews during the Holocaust, and wanted to claim the right of return because he identified culturally as a Jew in addition to being a priest but who was denied Israeli citizenship under the Law of Return by Israeli courts. And it would include Jews who see themselves as Buddhists or members of other religious groups but who also want to remain faithful to their understanding of their ethnic heritage.

This definition does not solve the problem of those who want to join the Jewish people but who are not of Jewish lineage and are interested in becoming Jewish because of a connection to a Jewish culture but not Jewish religion and so are uncomfortable with conversion rituals. These cases would require some kind of acknowledgment or ceremony different from conversion. There is not much support for this position in the official ranks of the Jewish community, but there are indeed many people, at least in the United States and Canada, who would find this possibility to be important to them in choosing a connection to the Jewish people, especially when that connection stems from something they've learned in college or the influence of a particular Jewish person to whom they wish to join their lives. The definition also does not solve the problem of connections with other groups who identify as Jews, like some black Hebrews or the Messianic Jews, who reject the claim of the organized Jewish community's exclusive authority to define Jewish lineage or to limit the boundaries of Jewish culture and heritage to the antecedents of Rabbinic Judaism only. However, as recent findings in genetics are now providing new information to enable us to ascertain lineage, we may have to reckon with the questions of Jewish identity that are raised by these other groups in new ways.

The Chosen People and Ethical Dilemmas

I am not, however, trying to suggest that acknowledging Jewish difference based on lineage as well as assent is always, or even often, a good thing for the Jews or for any group that claims that their difference has some meaning attached to it. It not only fails to remove the ethical problems faced by an exclusive community but actually serves to underscore them. Many of the rabbis and teachers who throughout Jewish history have downplayed the importance of lineage have done so to avoid the problems of Jewish claims of special status based on heredity. What I would argue, however, is that removing the idea that our status is based in part on lineage, no matter how it is explained, has failed to remove the problems encountered by being different. Claims of special status based on observing the law or even on the idea that Jews have a vocation (Mordecai Kaplan's answer to chosenness) or that our claim to chosenness is based on our social organization (as in Spinoza) or the moral mission suggested by Reformers or the national mission supported by Zionists do not remove the essential quality of the claim of chosenness. It makes no difference if chosenness means belonging to a tribe or being the bearers of a mission, for in any case the Jews are marked as different. We cannot get away from the fact that viewing a group as different in any way is the very quality that defines a bounded community that includes and excludes. Experiencing Jewish identity as both hereditary and by assent simultaneously does not alter this dimension of the problem.

Understanding Jewish identity as both hereditary and voluntary also does nothing to diminish the other problematic dimension of chosenness—the fact that our difference has also led to our being stigmatized and placed in danger. Surely, the biological dimension of our difference has led to heinous racism against us, from the horrendous eugenic programs that halted immigration of Jews and other "undesirables" to the United States in the early twentieth century to the ultimate degradation perpetrated through Nazi racial policies.[9] But we should not forget that Jewish exclusiveness and claims to be chosen by God to bring ethics to the world have also been the source of much hatred against us, and counterclaims by Christianity and Islam throughout history have also led to animosity, competition, and bloodshed. Claiming that our difference has some meaning, whatever that meaning is, subjects us to the same problems that linking our identity to our lineage does. It is not for us to surrender our difference, no matter what its basis. The goal is to work toward a society that no longer sees difference as a mark of superiority or inferiority but accepts difference as a normal part of what it means to be human. Then it will not matter on what our difference as Jews is based.

The other ethical dilemma that we face when we accept our status as a separate group is a corollary to being labeled inferior or superior because of our difference. Seeing ourselves as a tribal community can also lead us to favor our own group over other groups and make ethical decisions based on what is good for the Jews before we consider what is good for humanity or the planet. This has often been described as "concentric circle" ethics or the ethics of care, and many ethicists have argued that it makes perfect sense that we should take care of those who are close to us before we attend to those who wish us harm or merely are outside our circle.

But it is important to question that ethical position. For example, what differentiates concentric circle ethics from the argument that might also be used by politicians who favor special interests? Those politicians are, after all, only taking care of those who are close to them. We need to be alert to the possibility that this is indeed a danger that often befalls bounded communities and an ethic we might wish to reconsider.

Despite the problems it raises, this twofold identity based on biology and culture is the situation in which we Jews have lived in the past and currently find ourselves. Thinking about the possible ethical problems we face when acknowledging our status as a community linked by lineage as well as religion and culture is critical to the discussion about Jews and new developments in genetics that follows.

The Case of "Jewish Genes"

The conversation about identity is important for many reasons, but one of the most crucial aspects is helping us understand what the findings from the Human Genome Project (HGP) mean to us. As Robert Pollack points out, new discoveries in genetics do not in any way suggest a single "Jewish gene." But they do open up new insights about the biological dimension of our identity to which we must attend. The findings of the HGP explain the genetic resemblances of Ashkenazi Jews, and that should not surprise us. But it also has provided evidence to suggest that the people who have considered themselves inheritors of the status of Cohen (and many who have not thought of themselves in that way) have to a great extent passed on that heritage genetically for several thousand years, while more recent findings about descendants of the Levites suggest genetic connections that began only about one thousand years ago in central Asia. This genetic mapping provides links to groups like the Lemba in Africa and the Khazars of central Asia who also claim Jewish ancestry, and it raises new questions about what we really do mean by lineage. It also raises questions about limiting our notions of Jewish culture and religion to those passed down exclusively by Rabbinic Judaism.[10] It is important to note that current thinking about the nature of the connection between biology and culture in general, or the so-called nature/nurture debate, has led to conclusions similar to that to which the discussion of the covenant draws us. As we know now, it is the complex interaction between the biological and the cultural that creates the social phenomena of our world as we know them. We no longer need to ask whether nature or nurture is the root cause of depression, sexuality, laughter, aggression, or spirituality or even of many diseases. We have come to know that while there are genetic components to each of these, it is the interaction of the complex biological processes within individuals and with the environment in which they find themselves that creates individual selves and their social world. So, too, we must recognize that as Jews we inherit and pass on a number of genes that, in interaction with each other and the environments in which we live, produce many different possibilities for Jewish bodies, including physical characteristics, dimensions of personality, and propensity toward certain diseases. As it says in the Talmud, "All is foreordained and free will is given."[11] In other words, we live with the understanding that while much of what makes us human (and Jewish) resides already in our genes, the choices we make, the situations in which we find ourselves, and random chance interact to create the outcomes that define our lives.

Tay-Sachs Disease

As the above discussion about lineage suggests, we Jews have always been conscious of a biological dimension to our identity. Developments in genetic science over the past couple of centuries have supported the idea that we Jews share a genetic heritage that we pass on through our children. In the 1950s Tay-Sachs disease (TSD), a genetic disease that generally causes death by age four, was found to be prevalent among Jews of Ashkenazi descent.[12] Jews were very involved in research (both as doctors and subjects) that resulted in the discovery of a genetic test for this disease in the 1970s. When genetic testing for carriers became available, the Jewish community was quick to respond, setting up foundations and volunteer organizations to educate the Jewish community and provide testing and counseling. The parts of the community that were open to genetic testing that could result in abortion immediately began to encourage Jewish couples to get tested prior to planning a family. Major information campaigns took place in synagogues and community centers. In more Orthodox segments of the community, where abortion was not a religiously acceptable option, an organization, Dor Yeshorim, was created to provide confidential services to test individuals and discretely discourage two carriers from marrying.[13] Over the years, TSD has virtually been eradicated from the Ashkenazi Jewish population. As Paul Edelson points out, Jewish communal response and distribution of information through Jewish sources were central but not the only factors in the eradication of this disease among the Jewish population.[14] The fatal nature of the disease, the high correlation of predictability of outcome, the fact that Jewish medical and philanthropic institutions and voluntary organizations of affected families framed the disease as a personal family tragedy rather than blaming carrier parents, and the fact that middle-class men and women were the target testing population were also important factors.

The example of TSD screening clearly indicates the interrelationship between Jewish genetic and cultural identity. Scientists identified a gene carried by a particular group of Jews,[15] thus emphasizing the genetic component of Jewish identity. But the TSD allele in and of itself would have no meaning for Jewish identity apart from the cultural and religious significance attached to the genetic test and the various Jewish community reactions to it. The fact that so many data are now available as a result of TSD testing has led to new knowledge about other genetic diseases and disease susceptibilities found primarily among Ashkenazi Jews. Ashkenazi Jews are not alone (either among Jews or among other human populations) in having genetic diseases and susceptibilities, but a close connection between genetic and cultural identities has made Ashkenazi Jews excellent subjects for genetic studies.

Ashkenazi Jews and the Founder Effect

Genetically, Ashkenazi Jews are a good population to study for several reasons. First, like other small, bounded communities (e.g., Finns), Jews tend to be endogamous and therefore more likely to pass on a smaller variety of genes to the next generation. From the time of the ancient Rabbis until the past two hundred years, to be a Jew meant to be born of a Jewish mother almost without exception. To be included in the community, Jews had to marry one another. The Jewish population migrated, sometimes willingly and sometimes because they were expelled from their homes, from the area around the Mediterranean by various routes through North Africa, Spain, France, and Italy over the course of several centuries in the early Middle Ages and then farther into central and eastern Europe.[16]

But because of repressive governmental policies, Jews were forced to remain in the "Pale of Settlement" in isolated areas of what are now Poland, Lithuania, Belarus, Ukraine, and Russia for a period of four hundred years, from the fifteenth to the nineteenth century. Scientists suggest that the population in this sector started out quite small, numbering between ten and fifteen thousand. In the mid-seventeenth century the Chmielnicki massacres killed off about a quarter of that population.

By 1800, however, the population had increased to close to eight million. Population scientists again speculate that the population increase derived from a very small percentage of wealthy individuals who had large families. According to this theory, very few families (perhaps no more than a thousand individuals) were responsible for producing very large numbers of Ashkenazi Jews over this time period. Scientists suggest that this "founder effect" is the main reason why there are high frequencies of individual genes based on single mutations in the gene pool of Ashkenazi Jews.[17] The founder effect hypothesis is not the only explanation given for the prevalence of diseases like TSD in this population. Others have suggested that the propensity for such diseases could also be based on selective advantage. In a way similar to the effects of sickle cell anemia, which is prevalent in African populations and is presumed to have provided some protection against malaria, TSD and other Ashkenazi-linked diseases are thought to have protected these populations from tuberculosis.

Ashkenazi Jews are not the only population to bear the putative results of the founder effect. If this story were only about genetic population groups, we would most likely not know any more about Jews and genetic diseases than we know about French Canadians and genetic diseases. This result has come about precisely because Jewish identity is based on both lineage and religion/culture. Genes do not tell the whole story; we need also to understand Jewish religious and cultural values. Without Jewish eagerness to participate in genetic testing in the TSD case it is unlikely that scientists would have learned about the founder effect. The Jewish religious/cultural tradition, which places a high value on knowledge, including and especially scientific knowledge that advances healing, played a major role in this development. So does the fact that the Jewish community has developed sophisticated mechanisms of communication both through communal structures and through the media.

The Dilemma of Difference and Gaucher Disease

Medical science has uncovered a long list of single gene mutations, often similar in genesis to TSD, that affect the Ashkenazi population and some that affect the Sephardi Jewish population as well. Jews have responded to each of these diseases with great interest. They have established foundations that create educational campaigns to alert Jews to the potential threat, finance further research, encourage (often Jewish) hospitals to create sites for genetic testing, and are involved in legislative campaigns to champion laws to provide resources for those with the disease. They have also developed networks to provide care and support for people with these diseases. What differentiates some of these diseases from TSD is that they are not fatal at an early age and do not have a clear trajectory. In certain of these diseases, the symptoms may be mild or severe, and they may also have a late onset.

Caring for and providing social support for people with these diseases is a more complicated social issue, and the need to do this raises the ethical dilemma regarding the concentric circles of care. Should Jews be more concerned with Jewish genetic diseases than with other social problems? Should the burden of care for people with these diseases fall only on the affected families and their extended networks, or do all Jews have responsibility to be involved in the complex issues related to these diseases?

This issue became important to me over two decades ago when close friends of mine gave birth to a daughter who was soon diagnosed with familial dysautonomia (FD), an autosomal recessive disorder similar to TSD. Had Sam been born a decade earlier, this disease would most likely have been fatal. FD is a disease of the autonomic nervous system. Those born with it cannot regulate breathing, swallowing, body temperature, blood pressure—the things we mostly take for granted. Before doctors understood the disease, FD babies often died from choking. But Sam was diagnosed early. Fed through a tube for many years and with careful monitoring, she has grown into a delightful, witty, charming young woman whose bat mitzvah in the summer of 2002 was an event that made family and friends weep with joy. Sam continues with schooling, goes to a Jewish summer camp for kids with FD, and works on behalf of the Familial Dysautonomia Foundation.

Some years ago, researchers developed a genetic test for carriers of

FD with the support of the FD Foundation. In the past few years they have dedicated themselves to working for legislation to provide accommodations so that kids with FD can lead their lives with better support from social networks. Because of Sam I both am aware of this disease and have given *tzedakah* over the past years to the FD Foundation. It is clear that being close to Sam and her parents has changed my life and my intuitions about working for the rights of people with disabilities. I do not have the distance from this issue to discern the extent to which the fact that FD is a "Jewish disease" is part of my decision as an individual to be committed to and deeply concerned about the work of the FD Foundation, but this situation did raise my awareness of how close relationships serve as a factor in ordering ethical concerns.

Another genetic disease prevalent among Jews of Ashkenazi descent raises similar questions about what it means to make ethical decisions based on connections of lineage and cultural/religious bonds. During the course of the project that led to this volume, I came across an advertisement in a magazine called the *Jewish Woman*. The ad covered a full page, mostly in black and white but with some yellow background used in a very interesting way. Half the page depicted a woman with a sad expression and downcast eyes. In a magazine targeted to a Jewish audience, she was clearly meant to have a Jewish appearance. Her Jewishness was indicated by her dark hair and high cheekbones but also, ironically, by her tapered nose, clearly the result of a nose job, which while meant to eradicate a Jewish visage can also stand in for it. The copy on the top of the ad read, on the left in white on black, "If you often feel tired, it could be anemia." But next to this was a yellow block with black print that stated: "If you're Jewish, it could be Gaucher Disease." The ad copy goes on to explain that Gaucher disease is the "most common genetic disease affecting Jews of Eastern and Central European descent," "far more prevalent than Tay-Sachs," and "not gender or age specific." A list of symptoms followed, along with information on how to contact the National Gaucher Foundation. The foundation's symbol bears a logo of the planet with silhouettes of a mother, a father, and two children playing happily. It is unremarkable except for the fact that it is in the shape of a yellow triangle, a subtle reminder of the yellow stars that the Nazis forced Jews to wear and thus of the specter of what happens to Jews when they do not attend to caring for their own needs.

Gaucher disease and the Jewish response to it raise the ethical dilemma of concentric circle ethics in a dramatic way. Those with Gaucher disease may lead symptom-free lives or suffer from anemia, enlarged liver, and brittle bones, which can cause anything from minor disruption to serious debilitation. Gaucher can be detected through either genetic testing or a simple blood test. Unlike TSD, Gaucher can be treated now with enzyme replacement therapy. However, the treatment is expensive and ongoing throughout the course of a patient's life. Annual costs can reach $400,000 per patient,[18] and because the disease is rare, drug companies have little incentive to lower costs, because profit is based on volume. This presents an acute difficulty for countries with national health insurance, as they must decide whether or not to purchase the drugs. It is a particular problem in Israel, where there is a large number of cases of this disease, which disproportionately affects Ashkenazi Jews.

Israel did decide to purchase the enzyme replacement drugs, treating approximately 350 patients in 1999. Michael Gross questions the ethical basis of this decision when weighed against the cost to the overall medical system. He suggests that Israel adopted this course of action "from some vague sense of sympathy and solidarity for identifiable individuals who are often close to us. Perhaps noble, these sentiments," he argues, "do little to advance the cause of equity in health care."[19] He raises concerns about the possibility that in a society where a single ethnic group dominates, the willingness to fund treatment for this disease will encourage treatment for diseases that affect the majority's welfare while ignoring the needs of minorities.[20] Although Israel does fund treatment for diseases like thalassemia, a genetic disorder that mostly affects Arabs,[21] Gross raises another concern based on Israel's status as a Jewish state. It is possible, he argues, that Gaucher disease represents not the result of the founder effect but rather a selective advantage that protected Jews in eastern Europe against TB. If that is true, those who have Gaucher disease as a result represent a special population, namely, those who have suffered to protect the Jewish people in times of crisis and, like veterans of war, on that basis deserve special consideration.[22] Gross summarizes: "This makes Gaucher a 'Jewish' disease and confers a special obligation on the Jewish people to rescue Gaucher patients without creating the same obligation toward others suffering from similar diseases."[23] This consideration is troubling if the goal is a health-care policy based on absolute and abstract considerations of equity. Gross goes on to argue that Jewish consumption of health-care resources in Israel is higher than Arab consumption by one-third and that the Arab population has higher infant mortality rates and shorter life spans. He puts the question succinctly, asking "whether a health care system should provide any specialized high-cost treatment until it ensures basic care for all citizens,"[24] and he goes on to answer the question in the negative. Of course, Israel is not alone in this dilemma, but the case of Gaucher disease highlights the problem sharply. It is here that we may need to question our special concern for those close to us or those who are part of our genealogical, religious, and cultural community. It is my hope that this concern may lead us to look at our process of decision making when our ethic of care is in direct conflict with our sense of justice.

Breast Cancer, Colon Cancer, and the Stigma of the Jewish Gene

When studies came to light in the mid-1990s that indicated a greater susceptibility of Ashkenazi women for inheritable breast and ovarian cancers, Jewish women responded in a similar pattern to that which had developed around single gene mutation diseases like those discussed above: they gathered information, considered questions about the value of testing, and participated in further research. But after researchers discovered a higher prevalence for Ashkenazi Jews in a gene linked to the inheritable type of colorectal cancer, the alarm was raised in the Jewish community. The media carried scare headlines about "mutant Jewish genes," and communal leaders began raising concerns about stigmatization. They also used their resources to organize a national meeting with the leading genetics researchers to publicize and clarify the issues involved. They discussed possibilities for legislation against discrimination in insurance and employment based on genetic susceptibilities.

What makes this interesting, of course, as Nancy Press and Wylie Burke have stated so clearly, is that "none of these findings indicate that a Jewish woman is necessarily more likely to get breast cancer than a non-Jewish woman."[25] A Jewish woman or man is not more likely to get colorectal cancer, either. To further elucidate:

> It must be remembered that for the majority of individuals, cancer is a sporadic event, occurring in the absence of an inherited predisposition to malignancy. However, a small proportion of adults and children carry an inherited, germ line mutation in a cancer predisposition gene. This genetic alteration places them at an increased risk of developing cancer over their lifetimes. Such genetic susceptibility does not mean that all mutation carriers will develop cancer. In fact, the causative genes for most family cancer syndromes, including breast cancer, are not fully penetrant. The expression of the cancer, therefore, depends upon a complex chain of events where mutations in a major predisposition gene interact with mutations in other modifier genes and with other environmental factors.[26]

The discoveries of the breast and colorectal cancer genes simply indicate that scientific research has had the opportunity to locate certain single genes that are related to increased susceptibility for certain types of inherited cancers. No doubt as research continues and new discoveries are made about the makeup of the human genome, we will learn more about the susceptibilities of other communities and about other "Jewish genes" as well. At this point, Ashkenazi Jews are more like the canary in the mine than a targeted group, and this has happened precisely because we are a small, endogamous community of a shared gene pool with a history of an interest in being educated about and involved in scientific research.

Once again, the link between being a group based on lineage and religion/culture has placed Jews in an interesting position. Because the findings are about cancer, a particularly stigmatized disease in our culture, the Jewish community is concerned about its own stigmatization. As a group that has suffered horrifically from the results of being labeled an inferior race, we Jews are particularly concerned when we find ourselves associated with a stigmatized disease, especially when the press uses this stigma to create sensational headlines.[27]

Now that the panic over the discovery of certain inherited cancers has calmed down and people have been educated about the true nature of the connection between these inherited cancers and "Jewish genes," it is possible to reflect on the larger ethical questions related to this dilemma. Is it worth maintaining our status as different if it means that we will be stigmatized? Of course, the answer to that question is simple; it is unimaginable to me that we would give up being Jewish because it has led to stigmatization and mistreatment. If the Crusades and the Holocaust did not change our minds, why would predispositions to certain cancers? But this new genetic knowledge also gives us an opportunity to reflect on the meaning of difference and perhaps to make some suggestions for an ethic of justice that respects rather than ignores difference.

When Jews are being stigmatized, we do everything in our power to fight discrimination. It is my hope that increasing knowledge about genetic diseases particularly affecting Ashkenazi Jews will lead us to a place where we do everything in our power to fight not only for ourselves but also for others who are similarly stigmatized because they are also thought to be different. This notion could be translated to think about ways we need to respect the rights and humanity of Palestinian Arabs who live in Israel. Or it could remind us that the experience of being African American in the United States has led to stigmatization based on physical and genetic characteristics similar to the oppression we Jews have experienced in other times and places. Or it could lead us to work for the rights of people with disabilities to equal access to the legal and social supports that will allow them to live fulfilling lives free of the stigmatization they experience because they, too, are different.

We find truths in often-repeated aphorisms; that is why they are repeated, even if the lesson is not easily learned. So I close with Hillel's questions:[28]

"If I am not for myself, who will be for me?" We must accept ourselves as a people with a genetic heritage, both good and bad, and a cultural and religious tradition, also good and bad, but one that is ours.

"But if I am only for myself, what am I?" The heritage that makes us different must also make us conscious that difference is still not respected in the societies in which we live and that we have an obligation to work not only on our own behalf but on behalf of any group's right to be different. We must question whether or not it makes sense to give preferential treatment to our own causes or to work for an end to the absence of preferential treatment at all.

"And if not now, when?" Although we come from a tradition that cares deeply about the past and takes building for the future seriously, we must live in the present and work for change now, even though we know that everything will continue to change.

NOTES

This chapter originally appeared in the Reconstructionist *71, no. 2 (Spring 2007): 69–83. Reprinted with permission.*

1. I have excluded the term "race" from this description because of the complex and confusing ways in which the term has been used. When race was thought to be a biological category, Jews were thought to be part of the Semites, a category presumed to encompass the populations of the Middle East. Given the evils that have been perpetrated in the name of racial theory, we should be pleased that that notion of race has been thoroughly discredited, although it is important to note that it remains part of our consciousness because of the misuse of the term "anti-Semitism" to refer to all hatred of Jews, whether biologically, culturally, or politically based. Race as a social construction, on the other hand, is a valuable tool in understanding social issues, but its application to the Jewish people is extremely complex as well. In the United States, Jews have come to be seen as part of the socially constructed "white" race, complicating their role both in the racial politics of America and in understanding who is to be defined as Jewish. The assumption that Jews are white and European has serious ramifications for internal Jewish "racial" politics,

as Jews of African and Middle Eastern descent (the original Semites, if you will) are often discounted or oppressed.

2. I base much of the discussion on the question of election on traditional sources culled from Michael Walzer et al., eds., *The Jewish Political Tradition*, vol. 2, *Membership* (New Haven CT: Yale University Press, 2003). Additionally, I thank Noam Zohar, coeditor of that volume, for his thoughtful comments on this section.

3. Daniel Boyarin, *Carnal Israel* (Berkeley: University of California Press, 1993), 197–225.

4. Michael Wyschogrod, *The Body of Faith: Judaism as Corporeal Election* (New York: Seabury Press, 1983).

5. Robert Pollack, "DNA and Neshamah: Locating the Soul in an Age of Molecular Medicine," *Cross Currents* 53, no. 2 (Summer 2003): 231–47.

6. I refer to God with masculine pronouns when discussing texts that use them to portray God as an active character in anthropomorphic terms.

7. Maimonides understood this exchange to suggest the opposite, that beginning with Abraham would instead imply transcending ethnic confines of the relationship with God (see his letter to Ovadya the Convert).

8. Dena Davis reminds us in her work that while a community can decide whom to include and whom not to include, the individual also has the right to decline participation or identification if he or she chooses to do so ("Groups, Communities, and Contested Identities in Genetic Research," *Hastings Center Report* 30, no.6 [2000]: 38–45).

9. Daniel Kevles, *In the Name of Eugenics: Genetics and the Uses of Human Heredity* (New York: Knopf, 1985); and Sandra Gilman, *The Jew's Body* (New York: Routledge, 1991).

10. The implications of these findings are discussed by Laurie Zoloth in her chapter "Yearning for the Long-Lost Home: The Lemba and the Jewish Narrative of Genetic Return" in this volume.

11. M. *Avot (Ethics of the Fathers)* 3:19.

12. TSD is a recessive single gene disorder; to produce a baby with TSD, both parents need to be carriers of the recessive gene and would then have a 25 percent chance of having a child with TSD and a 50 percent chance of producing another carrier. TSD only affects Jews who are defined as Ashkenazi, meaning of German or eastern European descent. While Ashkenazim make up 80 percent of the current world Jewish population, most studies done on North American Jewry in particular have a tendency to see this majority as the only Jewish population, ignoring Jews of Sephardic descent entirely. I do not mean to add to or support that perception, and I hope that readers remember

that the Judaism that derived from northern Europe is only part of the larger Jewish cultural heritage.

13. Etty Broide et al., "Screening for Carriers of Tay-Sachs Disease in the Ultraorthodox Ashkenazi Jewish Community in Israel," *American Journal of Medical Genetics* 47, no. 2 (1993): 213–15.

14. Paul Edelson, "The Tay-Sachs Disease Screening Program in the U.S. as a Model for the Control of Genetic Disease: An Historical Overview," *Health Matrix: Journal of Law and Medicine* 7, no. 1 (1997): 125–33.

15. While Jews think of TSD as a Jewish genetic disease, it is important to remember that TSD carriers are also prevalent in other endogamous populations.

16. Recent genetic research suggests that the female founders of the Ashkenazi gene pool originated from the Roman Empire, converts who married men from the Near East. See Martin B. Richards et al., "A Substantial Prehistoric European Ancestry amongst Ashkenazi Maternal Lineages," *Nature Communications* 4 (2013): 2543.

17. Arno G. Motulsky, "Jewish Diseases and Origins," *Nature Genetics* 9 (February 1995): 99–101.

18. Michael Green and Jeffrey Botkin, "Genetic Exceptionalism in Medicine: Clarifying the Differences between Genetic and Nongenetic Tests," *Annals of Internal Medicine* 138, no.7 (2003): 571–75, 151.

19. Michael L. Gross, "Ethics, Policy, and Rare Genetic Disorders: The Case of Gaucher Disease in Israel," *Theoretical Medicine and Bioethics* 23, no. 2 (2002): 151–70, 154.

20. Gross, "Ethics, Policy," 155.

21. Gross, "Ethics, Policy," 156.

22. Gross, "Ethics, Policy," 156.

23. Gross, "Ethics, Policy," 156.

24. Gross, "Ethics, Policy," 158.

25. Nancy Press and Wylie Burke, "How Are Jewish Women Different from All Other Women?," *Health Matrix: Journal of Law Medicine* 7, no.1 (1997): 135–62.

26. G. L. Weisner, "Clinical Implications of *BRCA1* on Genetic Testing for Ashkenazi-Jewish Women," *Health Matrix: Journal of Law Medicine* 7, no. 1 (1997): 3–30, quote at 14.

27. See also K. Egan et al., "Jewish Religion and Risk of Breast Cancer," *Lancet* 347, no. 9,016 (1996): 1645–46; M. Waldman, "Jewish Leaders Meet NIH Chief on Stigmatization Fears," *Nature*, 30 April 1998; and Laurie Zoloth, "Mapping the Normal Human Self: The Jew and the Mark of Otherness," in *Genetics: Issues of Social Justice,* ed. Ted Peters (Cleveland: Pilgrim Press, 1998), 181–202.

28. M. *Avot (Ethics of the Fathers)* 1:14.

CHAPTER 9   *Yearning for the Long-Lost Home*

The Lemba and the Jewish
Narrative of Genetic Return

LAURIE ZOLOTH

"How can we mingle? We look different. Smell different. Are different."
—Sonia Levitin, *The Return*

After the Human Genome Project completed the first drafts of the
genomic map, in which it was persuasively argued that human beings
are largely genetically identical, attention turned to the ways humans
are genetically different: What variations in the genetic code occur, and
how does such different coding affect biology and, in turn, traits? In
2002 the National Institutes of Health (NIH) entered into an international cooperation to map the haplotype variation of the human species.
This project was able to be conceived because the human genome contains patterns of single nucleotide polymorphisms (SNPs), or variants
of common genetic regions. The project was intended to reveal, as had
been suggested by preliminary data, that such SNPs are inherited with
distinction in particular ethnic communities that have their origins in
historically continuous geographic regions, suggesting common ancestry.

The ethical issues then brought into being are the questions of whether seeking biologically essential difference is a morally permissible task, the meaning of such findings, and the use of the data. There are obvious issues here, for some patterns of alleles may produce differences in behaviors or traits that certain cultures consider advantages and highly reward—such as intelligence, physical prowess, empathy, and the abundance of secondary sexual characteristics—or disadvantages—such as poor coordination, shyness, and depression, which some Americans consider an illness and for which medical intervention is warranted. Is the haplotype map a moral enterprise, and if so, why?

In this chapter I argue at the outset that the power and authority of such a project are based on two important goals. First, like all Linnaean systems, genomic classification, while introducing inherent bias in the choice of ordering, can always be useful in understanding data in a visible order. Simply knowing as much as we can know, being always attentive to new ideas, data, and observations, allows us to better understand ourselves and the world. We are curious; much is to be learned. Second, the use of the SNP mapping processes may allow extraordinary advances in how drugs are designed and delivered.

A New Exodus

Some of the early uses of SNP mapping are already in play, and one example of this is the use of a particular SNP pattern, the *Kohen* modal haplotype, which is found on the Y chromosome and in significantly higher proportions in the Jewish community. What are the ethical implications and what might be the ethical norms as society confronts such technology? As a Jewish scholar, I ask, what texts, history, and norms can be brought to bear as we reflect on the implications of the pursuit of this knowledge? This chapter explores these questions using several sorts of texts. Specifically, we look at the "text" of the code of DNA through the texts of *halakhah*, of medieval literature, of early modern literature, and of contemporary popular culture.

As I was preparing a commentary on the meaning of ethnicity after the Human Genome Project allowed scientists to chart similarities in genetic heritage, thus calling into being complex relationships among peoples, my son was carefully reading one of the books assigned to him in sixth grade called *The Return*. It is, he told me, "a true story" based on the accounts from the Ethiopian refugees, "Falashan Jews," who were brought into Israel in a secret rescue operation called, lyrically enough, Operation Moses. It is a harrowing account, involving dead parents, starving children, painful and brutal beatings, plucky eleven-year-olds, innocent natives who think a toilet is a special water bowl and a zipper a big deal, pious grandparents, evil Communists, and white Jewish saviors from America who smell sweet and wear "fine, expensive" (but tasteful) linen.

This award-winning book has gained an iconographic status in American Jewish Life, and my older son sensed a Moral Lesson and told this to his brother. "This," he said, "is a cliché." He is correct, and yet not. The story of the Ethiopian Jews, still ongoing, is a true story, and it is told frequently in the Jewish community to rally the marginal into affiliation. Between November 1984 and January 1985, eight thousand Ethiopian refugees were airlifted from refugee camps in Sudan to Israel and relocated there. Many who claimed Jewish membership were left behind, more were allowed to immigrate later, and now, in my local Jewish paper, there appears a desperate new account—more and more Ethiopians now claim "Jewish ancestry," and the state of Israel is deciding whether to accept them. The latest group is Christian, of whom it is said, "They are Jews inside" or "They are Jews by blood."

The Return is the thinly fictionalized account (some real names are used) of one family who leaves Ethiopia, flees to a refugee camp in the Sudan desert, and is rescued weeks before the border is closed. It is, for a small book, an enormously complex text, and I too carefully read it, noting the construction and the reification of the social construct that was "the Falasha" (*falasha* is the word for "stranger" in the language of the surrounding tribes) and now "the Ethiopian Jewish" story. This account is the master narrative, variants of which can be found in much of Jewish popular culture.

Years after the events in the novel, a second story of lost black Jews made the news, the South African Bantu tribe of the Lemba. In this story, the Lemba also had a narrative of Jewish continuity dating back not to the eroticized tale of Sheba and Solomon, as did the Ethiopians, but to a pre–Second Temple, more contemporary story of traders who found refuge and a comfortable new home in southern Africa.[1] Although it did not have the biblical resonance of the Ethiopian narrative, the story of the Lemba does contain classic elements of both Jewish text and history. Further, the "discovery" of the Lemba was amplified by new genetic tools that can match the genetic patterns of males in the tribe. The Lemba were first found and then confirmed by genomic tests to be "authentically Jewish," meaning that they carried the trace of the *Kohen* modal haplotype on the Y chromosome, just as do Ashkenazi Jewish men. These patterns of genetic coding, these haplotypes, occur on the Y chromosome, as well as on other chromosomes. The Y chromosome, however, is stable in that it changes slowly, and mutations occur at a predictable rate. Haplotypes can be found in human group "clusters" that closely followed other descriptors such as language grouping and ethnogeographic origins. One of the early patterns noted was that of men who claimed they were *Kohanim*, hereditary priests descended from the family of Aaron within the tribe of Levi.[2] Among Ashkenazi and Sephardic Jews who claim such a lineage, over 80 percent carry the *Kohen* modal haplotype. Among Ashkenazi Jewish men in general, the haplotype is likely to occur in a high frequency of cases, far greater than in the non-Jewish population. This finding correlated to similar patterns among the Lemba.[3]

It is not the first time that nearly identical narratives have shaped Jewish culture. The story of the Lemba is a genetic version of a classic yearning for Jewish history. The narrative, moreover, of lost Jews in Africa, or in America, or in the Islamic world appears with frequency in the Jewish literary and epistolary tradition. In these tales, the lost Jews are then found, and found to be powerful, wealthy, and still vibrantly Jewish; they reflect a deeper yearning, that of the repair of Exile. In a post-Holocaust world, in which so many Jews have been irrevocably lost, finding new Jews is both a potent symbol and an assertion of continuity.

The Lemba tribe has come to the foreground of this classic narrative in part because of the very incongruity of their tale and in part because of the very familiarity. Like the Ethiopian rescue story a decade earlier, the entire construction of the narrative is both truth and reification of a dream, and it calls forth both a response of real compassion and an instrumental use of the narrative. That both are true simultaneously sets the stage for the ongoing story of how genetic marking and identification amplify and deconstruct the oral history of lost and found groups of African Jews. What are the key elements in this response?

The Theme of the "Lost Tribe"

Lost tribes have a persistent resonance for Jews. In the account of the Hebrew Scriptures, the Jewish people trace their historical origins to Abraham, patriarch of one Mesopotamian family, one line of whose descendants were enslaved in Egypt but win liberation through the intervention of a God concerned with both history and justice; they then organize themselves into a political/legal/religious state. This people and state journey to a land, divinely promised to them in the sacred text, and live there amid other Mediterranean peoples, building first one and then a second temple as its religious center and a state-based monarchy. The core of the nation is repeatedly separated after battles and taken into exile, regaining and losing land, population, kingship, and statehood, until the Roman Empire finally occupies Palestine, leaving behind a small remnant of Jews and taking the remaining into exile or captivity. The idea of a people split and dispersed dates from these exiles during which first the Assyrians, then the Babylonians, and then the Romans seized parts of the historical Davidic Kingdom and marched its inhabitants into captivity. Tribes that are lost, then, begin to appear as a concern in the biblical canon, and they find their way into medieval Jewish and Christian literature. Lost tribes turn up as native peoples in both North and South America and in travelers' tales in India.

Literary accounts of what happens to lost tribes and what the original community's responsibility is to them vary greatly, and they stand largely outside of the elaborate system of interpretive discourse that constitutes the Talmud and responsa literature, which are the record of how religious authorities led their communities to understand and act on moral and religious queries. Much of the tradition of lost tribes is carried on outside this literature, in folk accounts or in travelers' tales written by non-Jews, often Christian missionaries with a deep confidence in biblical history. In their accounts, missionaries have sometimes openly speculated that various groups of peoples encountered by the missionaries are lost tribes of Jews. The tradition of belief in these tales continues in the modern period, with contemporary Jewish travelers, often social scientists or professional journalists, stumbling across evidence of such lost peoples. The story told by the tribal members is the same: a vague sense of connection to rituals, a strong sense of identification with biblical narratives, little or no understanding or practice of Rabbinic Judaism, and a persistent tale that a lost ancestor was forced to pretend to convert to Christianity (or in some cases, another faith). This congruence in the stories told by the tribe and its foreign visitors is seen as confirmation of the truth claim, not a challenge to it. Hence, it is not new to discover lost Jews.

However, the use of genetic knowledge to confirm narrative accounts *is* new. What is at stake is how, after the charting of the Human Genome Project and the search for genetic origins and variants that the haplotype map represents, our sense of identity, citizenship, membership, and kinship will change to make the "true story" the genetic one. The tension in the tales of the Lemba and the Ethiopians illustrates this point of divergence, for we now can and do look for genetic confirmation of their stories. Let me argue, though, that the story of these "lost" Jews is not based on genetics alone but rather was so seamlessly accepted because it so closely followed central type scenes that were already powerfully embedded in the Jewish textual and historical tradition.

Bearing the Mark of Israel

For the Ethiopian Jews, whose ritual activity was outside mainstream practice (the invention of new holidays, the reading of the Torah at night, etc.), the one defining thing that marked them as Jews was their yearning to go to Israel. For the Jew, the classic sense of not being at home, being always estranged and in exile, is intrinsic to the notion of "otherness." That the Ethiopian Jews, and now, perhaps, others such as the Lemba, are not "at home" is enough to make them ours.

This yearning to go home is the ultimate mark of the Jewish condition. In this sense, even Israeli Jews, who are home in the land, await liberation to "Jerusalem" of the perfected, redeemed world. This is true liturgically, and it is part of the structure of normative, Rabbinic Jewish thought after the destruction of the Second Temple in 70 CE.[4] In the nineteenth century, after several failed attempts to enter into European nationalisms, Jewish intellectuals made common cause with religious traditionalists in calling for a physical return to the land of the biblical narrative, Palestine. The events of the Shoah and the later expulsions/ flights from the Middle East, northern Africa, and Russia repeated this theme—that the Jew was not really a citizen of any other place, that identity of place was a fiction, and that only the reality of exile, only the yearning to return was permanent. The actual Jews who return from African tribal lands and whose story my son can retell is a parallel to the story of Moses himself, whose journey to Israel from northern Africa marks the central ritual activity of Passover, a story mentioned several times in each of the three daily prayer rituals. It could be argued that the entire linguistic and physical ritual praxis of the Jew (e.g., turning east to pray toward Jerusalem) is the reenactment of the Return itself, in this way marking both what the Jew does (yearn for return) and who the Jew is (the always exiled).

For the Lemba, merely telling the oral narrative was not enough to lend credence to their story. It remained one of several questionable claims of lost tribal identity that exist in Africa, Latin America, and a few places in India and China, marking in many places the trace of an eighteenth-century missionary cadre particularly fond of the book of Exodus.[5] The existence of a marker that is both hidden, "in the blood," and embodied redefines the notion of the "in the blood" sense of those who have no physical evidence to substantiate their claim. We are then moved to ask: In what sense are the Lemba (or anyone, for that matter) "really" Jews? The Lemba claim Jewishness through genetics, not through practice (which is Christian, in large part), and this claim reorganizes what claims need to be valid. On some level, the story that is the oral tradition is "authenticated" by the neutral frame that genetics provides.

The Lemba story did not have the biblical resonance of the Ethiopian narrative, with its reference to the wisdom, complex technology, and power of the Israelite king and his sensuous female African queen. Much of this story is rehearsed in the book my son read innocently enough, with wise, white, and technologically able Israelis welcoming the heroine of the novel, who, like Sheba herself, is beautiful, clever, fecund, and, like Sheba, not inclined to actually marry. The Lemba story, though, does contain classic elements of both Jewish text and history.

At the same time that the sixth grade was reading this book and that the narrative of the Ethiopian Jews was routinely invoked to solidify Jewish unity and pride, communities around the world were learning of several other groups that are also claiming Jewish genetic identity and asking to be allowed to settle in Israel as Jews. These included people from India, China, and countries that made up the Soviet Union and its allies. The Lemba tribe, though, is remarkable for its claimed genetic link. It is a group that tells the story of migration from Israel, to Yemen, and then to southern Africa. In this story, the oral history is amplified by new genetic tools that can match the genetic patterns of males in the tribe, called haplotypes, to patterns on the Y chromosome that correlate to similar patterns among Ashkenazi Jews. Reflection on this earlier tale of the lost-and-found black tribe offers insights into how we now understand the new tale of the tribe called the Lemba. The entire construction of the narrative, which is both a construction of a genetic story and an amplification of a classic and long-standing "tale," is a narrative of a colonialist vision (Africa as the pathetic "Dark Continent" in serious need of rescue). In this, it creates both a real tale of rescue and redemption and a fictive tale of triumphalism. That both are true simultaneously sets the stage for the ongoing story of how genetic marking and identification amplify and deconstruct the oral history of groups.

The Texts of the Story

Let me turn to some exemplary texts. An early reference to the idea of lost tribes living in a specific area, identified in some texts as Africa, occurs in midrash:

> In the time to come the Holy One, blessed be He, will bring them together; this may be inferred from the text, "Behold, these shall come from far; and lo, these from the north and from the west, and these from the land of Sinim" [Isaiah 49:12]. The other exiles shall come with them, and the tribes who are living beyond the River Sambatyon and beyond the Mountains of Darkness shall gather together and come to Jerusalem. Isaiah stated: "Saying to the prisoners: 'Go forth'" [Isaiah 49:9]. He was referring to those exiles who were living beyond Sambatyon. "To those who are in darkness: 'Show yourselves'" [Isaiah 49:9]. By this he was referring to those who were living beyond the cloud of darkness. "They shall feed in the ways, and in all high hills shall be their pasture" [Isaiah 49:9]. By this he was referring to those who were living in Daphne Antiochene. When the time comes they will be redeemed and will come to Zion in gladness; as it says, "And the ransomed of the Lord shall return, and come with singing unto Zion, etc." [Isaiah 51:11]. (*Lamentations Rabbah* 11:9)
>
> R. Judah b. R. Simon said: The tribes of Judah and Benjamin were not exiled to the same place as were the other ten tribes. The ten tribes were exiled beyond the River Sambatyon whereas the tribes of Judah and Benjamin are dispersed in all countries. (*Genesis Rabbah* 25)

The Sambatyon River, a magic river that responds to the Sabbath cycle by becoming impassable and violent, in some cases on the Sabbath, in some cases during the rest of the week, is identified in various places as being located in Africa (most likely), Turkey, or Iran. In this text, we see some themes that will be repeated—exiled Jews hear of Jews in even deeper or primal exile, they are both in darkness and unseen, and yet they live in a place of high hills, or in an advantaged setting, but, finally, they yearn to return.[6]

The medieval period is a rich source of such epistolary accounts. In this literature, Jewish travelers appear with persistent accounts of lost tribes. In many settings, the tribe is not only mysterious but valorized, a genre of the *Übermensch*, a sort of Jew unlike the community from which the traveler comes. These Jews are fighters, successful, wealthy, and, most consistently of all, far more pious than the community that hears the tale.[7]

Consider the following excerpt from a letter from Elad the Danite to the Jews of Spain in 883 CE:

And this was my going forth from the other side of the river of Ethiopia. . . . A Jew, a merchant of the tribe of Issachar, found me and bought me for 32 pieces of gold, and brought me back to his country. They live in the mountains of the seacoast and belong to the land of the Medes and the Persians. They fulfill the command "The book of this Law shall not depart from your mouth." The yoke of sovereignty is not upon them, but only the yoke of the Law. Among them are leaders of hosts, but they fight with no man. They only dispute as to the Law, and they live in peace and comfort, and there is no disturber and no evil chance. They dwell in a country ten days' journey, and they have great flocks and camels and asses and slaves. . . . And the sons of Zebulon are encamped in the hills of Paran . . . , and they practice business and they observe the four death penalties inflicted by the courts. . . . The tribe of Rueben is over behind them, and there is peace and brotherhood between them, and they go together to war, and make roads and divide the spoils and they go on the highroads of Media and Persia, and they speak Hebrew and Persian, and they possess [know] Scripture and Mishnah and Gemara and Agaddah. . . . And the tribe of Ephraim and the half tribe of Manesseh are there in the mountains against the city of Mecca, the stumbling block of the Ishmaelites. They are strong of body and of iron heart. They are horsemen and take the road, and they have no pity on their enemies, and their only livelihood comes from spoil. They are mighty men of war. One is a match for a thousand. . . . And the tribe of Simeon . . .

take tribute from five and twenty kingdoms, and some Ishmaelites pay them tribute. [Others] live in the land of gold, and every year they make war with the seven kingdoms and seven countries. . . . They have gold, silver, precious stones, and many sheep. . . . Every three months a different tribe goes out to war. The tribe of Moses . . . dwells in . . . castles. No impure [i.e., forbidden] thing is found with them, and from one comes forth a hundred. They are of perfect faith, and their Talmud is all in Hebrew. But they know not the Rabbis, and they are all Levites.

Benjamin of Tudela, a twelfth-century traveler, wrote this account in 1165:

Here dwell the Jews called Kheibar. . . . It is a great city. . . . The Jews own many large fortified cities. The yoke of the Gentiles is not upon them. They go forth to pillage and to capture booty from distant lands in conjunction with the Arabs, their neighbors and allies. All the neighbors of these Jews go in fear of them. . . . Their land is extensive, and they have in their midst learned and wise men.

In 1448 Obadiah Jare da Bertinoro wrote to his family about his discovery of his version of lost tribes beyond the Sambatyon: "There are high mountains and valleys . . . that are certainly inhabited by descendants of Israel. . . . They have carried on great wars against the Ethiopians." In another letter, he wrote:

Jews have come here from Aden; said to be the site of the Garden, it lies southwest of Ethiopia. . . . On the other side of the Sambatyon the Children of Israel are as numerous as the sands of the sea, and there are many kings and priests among them . . . but they are not so pure and holy as these surrounded by the stream. The Jews of Aden relate all this with a certain confidence, as if it was well known, and no one ever doubted the truth of their assertions.

David Reubeni, in a diary in 1520, relates a similar tale. He is a lost Jew, a king descended from King Solomon, and he comes to Rome "on a white horse" from beyond the Sambatyon. In 1624 Yiddls of Prague returns from travels in Africa and tells of a crossing of the river Sambatyon, meeting a charismatic Christian (and a recurring figure in the later medieval tales) called "Prester John" who lives near the lost tribe of Jews.

> It is said that beyond the river live the Jews, and they have four and twenty kings . . . and large and fortified cities and countless villages. [The king] is a terribly brave warrior, and when he rides on his donkey, one hundred and fifty thousand armed warriors follow behind with armor and spears, and their horses are large and terrible and always look forward and back, and they are trained in war. They bite and trample and kill anyone who comes near. . . . All the men and women know a trade so as to support themselves with ease, and the land lacks nothing in crops and anything you might mention.

Menassah ben Israel in 1649 relates to John Dury, chaplain to Mary, princess of Orange, the finding of a lost tribe in America:

> The first inhabitants of America, I believe, were of the Ten Tribes; moreover that they were scattered also in other countries. They keep their true religion, as they are hoping to return into the Holy Land in due time. . . . I declare that our Israelites were the first finders of our America.

Bearing the Body of Memory

"You told them you are a Jew?" "No, I said nothing. They looked at me. Somehow they knew." "How?" He sighed, turning his head to hide the bruised eye. "How do they ever know? They just do."[8]

A central anxiety for the Western Jew is precisely both this very disappearance, this very ability to vanish into the population that the Ethiopians cannot seem to achieve, and the anxiety of nondisappearance and continued anti-Semitism. Both recent popular books (such as *The Vanishing American Jew* by Alan Dershowitz) and scholarly ones (such as *Imagining Russian Jewry* by Steven Zipperstein) deal with themes of memory, a world lost, a people vanished, and so on. The African Jews come, then, into the present bearing the mark not only of a stranger but of the past, history and memory. That they are "an ancient people" is emphasized in two ways: first, the narrative repeats this theme, and second, it persistently has the characters act confused and astonished by modernity. This last point is made in virtually every story written about the Africans. In one scene, they do not know what a zipper is; in another, the refugees encounter a toilet but do not know how to use it: "Then we lifted up a lid and found a deep and shining silver bowl, but no place for . . . comfort."[9]

What is being told by this detail is the charm of the lost, yet inferior, native past. Here the African Jews are simple, devout, and not terribly smart. They do not even know what an eleven-year-old, the American child reading about them, knows about the world of things. Yet they are far better—they know they are Jews who are destined, marked, and made for Israel, no matter how African they are, and they do not give up. They need us to "bring them into the present" from a primitive "then" to the "real" time of our now. It is the basic premise upon which anthropological field studies are based, that one can travel both geographically and temporally when one enters into a "primitive people's physical space." Earlier appears as more natural, more authentic, less artificial. The Africans who inhabit this world with the Jews are indeed different—they are portrayed not only as looking different ("we did not trust noses like that") but as wielding, if somewhat pathetically, the tools of modernity, primarily because they are always bad guys with guns, with which they callously slaughter the Jews.

Europeans appear in the tale as salvific and as white. In this passage, Desta, the narrator in *The Return,* describes the American Jewish women who come to the village to tell the Ethiopians they are really Jews: "More than anything else in the world I wanted to smile at her, to touch her skin. . . . [W]hat is that gold in your teeth? A treasure? . . . I could only stare, seeing the faint light hairs on her arms. . . . She was rich, you could see it . . . such softness. . . . I also wanted that touch."[10] Importantly, it is the Westerners who are the ones who, with sophisticated knowledge and finally with linguistic "evidence," both rescue (for the Ethiopians) and corroborate the reality of the oral story, which is identity itself. Here again, the narrative is both true and heartbreaking. It is also deeply colonialist in its existence, if not by intent, then by design. The Africans personify "pastness" in their very primitive way in the world. They are even more authentic than the Europeans—they are ancestral, they are of the First Temple period itself. (And what could be more essential, more Israel, than that?)

Desta is sure she is Jewish because she "looks like a Jew." In the novel, she and her aunt and grandmother are described as looking just like Sheba. In all cases they are pious, even about ritual practice that bears no relationship to normative rabbinate practice, such as sitting in menstrual huts during the days of separation.

The Jews cannot "mingle" and are marked. They are persecuted in the novel with the precise medieval libels that mark classic Western anti-Semitism: "One day there was cholera in the town. Some people died, and everyone was scared, plenty. People asked each other—how did this cholera come to us? Who did it? Someone remembered seeing a Jew in the town; Jews bring the evil eye, don't you know? They can turn themselves into hyenas at night and suck away people's breath. . . . Everything that happens bad in Ethiopia they blame on us—sickness, bad crops, drought and famine."[11] The novel includes the classic tropes: The Jew carries death and disease (compare the medieval Christian canard of Jews poisoning wells). The Jew's blood or body is different, and the real body of the Jew is that of an animal with cloven feet and horns and a tail, that is, satanic, and hence the Jew must drink the blood of mortal children to look human. The blood carries the self from which it was sucked into the body of the Jew. The Jew is wandering, homeless, and likely disloyal to the country in which he lives—all elements that fit precisely into the narratives of the lost African Jews. The link to the language and narrative tropes of Christian persecution draws us toward the validity of the story.

Bearing the Belief in the Inner Thing One Cannot See

In the novel about the Ethiopians, the inner code that is critical is Hebrew—like DNA, it is inner, secret, and linked to their identity. (The only Hebrew the people know is a few phrases from the Bible, but that is enough for them to prove that they are really-real Jews to the Westerners, and that is but one of the oddities of the plot.) The Lemba need no such test, for what we now understand as really-real is the code of the body. For that which needs to be proven (such as crime or parentage), what has come to matter is the code of DNA and not the story of self.

This too fits into a Jewish sense of what it is to be Jewish, for to believe in God as a Jew is centrally about a rejection of the need for physical objects as holy. One cannot have a physical object toward which to pray—in fact, it is a theological error to consider any object itself a god. (A Torah is regarded as important, the text honored and described as holy, but the object is not supernatural.) To believe in the inner life that one cannot see, the theology that the very name of God is a transient verb form, all potential, is the particular frame of reality that has animated Jewish texts for two thousand years. One of the problems of the classic biblical tradition is how to ensure that a people recently freed, slaves tempted to pray to idols and large Golden Calves, will understand a God they cannot see, who enacts history through legally binding social justice. Of course, such an idea, that one mirrors and carries the idea that each person possesses a rich inner life that cannot be seen but that is what truly matters, is part of a strong folk tradition as well as the religious one.

Genetic knowledge is hidden knowledge. The language of the genetic code is difficult, based on a limited set of base pairs and expressed as code. Like Hebrew, it is both the sign, meaning the actual language of the people, and the symbol of their difficult, ancient, unknowableness. Hebrew is *lashon kodesh*, the holy tongue, God's language, and these very things are what is said about DNA base pairs. Both are ambiguous (in biologic terms, they have redundancy) and need interpretive settings to do the work of creation of the self. The theme of a hidden soul, a hidden self, and a codal key are familiar ones for a Jewish community hearing of the Lemba's story. Although it clearly matters to the rabbinate if the Lemba decide to be Jewish in practice, for the lay community, it clearly does not seem to: one can find hundreds of web citations, announcements of lectures, and other specialized references for the Lemba. That the lost Jews are profoundly Found just as American Jews are feeling the greatest loss of Jews in the Holocaust is a paradox not lost on anyone who reflects carefully on the response to the Lemba's discovery.

The Cousins Club

When I was my son's age, we lived in Los Angeles, and so did all of our cousins, all of whom looked like us, with speech patterns and hand gestures that so marked us that years later I will meet new people who immediately ask me if I am "from New York." This explains why I am marked as a Jew in these precise ways. We had an actual cousins club, consisting of our extended family and everyone they knew from their home in Russia, cousins or not. The club also functioned as a mutual aid society, with dues that were collected and sent abroad to the relatives who had been scattered by the Holocaust and needed resettlement in Israel. The resonant story of return for our generation was the Shoah. The imprint of memory is of the lost, shattered world of European Jews, Russian Jews in particular.

When one meets a new Jew, one plays a game called "Jewish geography," which is to say, you name people until you find that you know a person in common. This is the geography for a people with no nation, no land, based only on the ties of remembrance, memory, and friendship. In the postwar years, "Jewish geography" represented a way linguistically to reconstitute a community broken and scattered by immigration to America and by the Shoah. But for my own children, the geography has location, and the cousins who are the most far-flung are the ones of greatest value, acting as a stand-in for our own return.

In the multinational world of my children, the idea that Jews are from all over, need to return, and look different yet are authentic is pivotal at a point where the worry is that assimilated Jews will disappear, intermarry, and fail to continue as a community. For our family, after 450 years living in close geographic proximity in the Pale of Russia along with hundreds of thousands of other Jewish families, the 1950s represented a fundamental globalization and dispersal. World War II left the members of our family who survived in four countries: Israel, Mexico, America (three thousand miles apart), and Australia. That cousins could turn up everywhere did not seem farfetched.

What is the Jewish community's connection to the Lemba, the cousins they seem to turn out genetically to be? Kinship implies duty; hence, what is our new duty toward the other who is our new kin? And in what sense are Ashkenazi Jews more "like" the Lemba than they are "like" the Celts? How is it that their fates are (to put it figuratively) intertwined? This question turns out to be harder to unpack in a world in which social grouping around what is understood as race turns out to be so intensely connected to identity, standing, and health. For the cousins, what matters was our shared family story. Our cousins who migrated to Mexico, for example, and who intermarried, and who spoke Spanish seemed entirely Mexican to me as a child, yet they had been told the same narrative about our family as I had, and we shared the same grandparents and the same intense set of values.

Blacks and Jews

That the Jewish community is linked as kin across such racial lines is highly significant to the discourse. Jews are attracted to the story of the Lemba at a time in American history when the enmity runs high between blacks and Jews. The Lemba serve as a concrete example of this highly vexed relationship suddenly working out, oddly, but possibly. A final point of convergence in the story of the African Jew is not as clearly made in the children's novel, but it is strongly made in contemporary accounts and uses of the black Jews. In a time of tension between black and Jewish Americans, the Ethiopian case is nearly always put forward as concrete evidence of how committed Jews and Israelis are to defeating racism. This is a complexity. As Karen Brodkin notes, even American Jews of European ancestry were ambivalently white until after World War II. Along with Greeks and Italians and, in an earlier period, the Irish, Jews were understood, "read" culturally, as ethnics of (some) color. Jews and blacks between 1945 and 1968 were further bound by common social and political commitments, in particular, the civil rights and trade union struggles. Later tensions and the emergence of separatist and nationalist movements created ongoing unease between the groups, which, fanned by an eager press, erupted into contentious debates—"the black-Jewish conflict."

On every poster about Israel that features Israelis in their dress of origin, the point is made graphically that Jews are of all colors, from all over the world, and are still Jews. Jews are black, and Jews are African. In the 1990s on my campus, San Francisco State University, one historically riven by just this contentiousness, the Jewish agencies funded groups of Ethiopian teenagers to come to speak to students on campus about how, since Jews were black, it was absurd to have a conflict at all. As Tudor Parfitt, an expert on the Lemba, told the group whose essays are included in this volume, this political positioning was, at least in part, one of the uses made of the Lemba.[12]

Ethics and Ethnicity: Identity, Coloration, and the Margins

In popular accounts of genetics, a person is best known and best understood medically by his or her DNA. DNA is an inner, secret code linking identity to creation, for what we now understand as really-real is the code of the body. For that which *needs to be proven* (crime, parentage), the DNA identity is the definitive one, the one that can exonerate you, and the one by which you claim family, kinship, and finally, perhaps, self. The way to know identity in the twenty-first century is to know the genetic self—the inner helix that names us as unique, each one by one. Hence, the Lemba can be "really Jews" because they "really" have the correct genes of identity. A feature of this genetic self that has come to matter is the code of DNA.

This way of thinking, however, establishes a persistent tension between the idea that one can choose identity and membership (e.g., a premise of conversion) and the idea that one's blood or genes are normative. The idea that behavior, membership, affiliation, and hence participation in a series of moral and ritual activities is a part of identity and a part of what is genetic and heritable creates a vexing problem. It arises when identity becomes predicative, prognosticative. It implies the future choices that self will make—that is, that ontology implies morality in that it defines commitments and relationships and thus morality as well. The stage becomes set for assumptions about group behavior, one step removed from classic tropes of marginalization and discrimination.

The case of the Lemba suggests the first step in this process, and to the extent it does, it alerts us to the caution that should accompany the interest in the altogether interesting finding of genetic similarities. It implies that such identity has become temporally descriptive—in other words, that the self implies the past, that ontology implies history as well as morality.

Legal Implications

Judaism has long been a religion of complicated entrance. There is, as it were, no exit, because Judaism exists both as a religion to which one can convert and also as an ethnicity from which, once in, one cannot be discarded entirely, as the Jews of the Third Reich came to understand and as the Lemba seem to personify. The research on the *Kohen* modal haplotype raises immediate questions about citizenship and the obligation of the Israeli state to support any Jew anywhere. In a more profound way, it rehearses the oldest of Jewish dilemmas—namely, what does a covenantal community mean? Who am I, as a person, standing within it, and who are you, standing there next to me?

This research raises and restimulates what in some sense—the Jewish legal or halakhic sense—is a settled issue, that of genetic determinism. A community that allows for the full legal transformation of the self by conversion, such as the Jewish community, has a carefully developed methodology for ascertaining this aspect of the self.

There are several discussions about how kinship and heredity determine identity and destiny in Jewish law, but they are contradictory in nature and based on differing sources of case law. For certain specific diseases, hemophilia being the outstanding example, there seems to be a recognition of familial, not just maternal, transmission. In the case in which a boy bleeds to death following a circumcision, and then a second boy also dies in this manner, it is understood that all males born to that set of parents and to the sisters of that family are exempt from the requirement of circumcision, since this bleeding disorder is understood as what we would now see as a genetic disease.[13] But in the matter of behavior, of possibilities, of character, and of achievements, the law is complex. Each person must shape his or her own fate, and this construction of the self is achieved only with study, decent community activity, and prayer. Can even the greatest of scholars be assured that their children will be scholars?[14] Not at all: the Talmud is clear. Each scholar must study himself; wisdom is not acquired genetically. The language of the Gemara is that "*yerushah,* an inheritance, cannot confer status (for knowledge or wisdom)."[15]

Further, traditional rules of conversion place obligations on the community that transcend genetic ties, yet ambivalence about identity persists. A sense of self as acting "as a Jew" or not "as a Jew" is linguistic testimony to this. Such notions of identity as inherently Jewish, or "in the blood," persist across cultures and historical periods despite the law that fully recognizes converts as Jews. In part, this occurs in reaction to externalized anti-Semitism, and in part, it is an explanation and defense of behaviors that communities see as advantages, such as education and later marriage. A legal system such as Judaism also must normalize the concept that some activities are permitted to Jews (abortion in some circumstances, loaning money at interest to non-Jews, etc.) but not to Christians, and some activities are permitted to non-Jews (eating nonkosher meat, cooking on the Sabbath) but not to Jews, and hence the legal system must decide if acceptance of norms or family origin is the relevant category to be described as a Jew.

Further, traditional rules of conversion place obligations on the community and on the new member of the community that endorse the new, nongenetic identity as the actual one. It is prohibited to identify a convert as such, as a *ger,* and it is considered impolite to refer to the convert as distinctive, or less than fully Jewish. Converts are required to rename themselves, taking on the name of son/daughter of "Abraham, our father, and Sarah, our mother," for all ritual purposes. Some rabbinical authorities allow consanguineous marriage of two former siblings once they are converts, since they are no longer the son/daughter of those non-Jewish parents, an extreme form of identity swapping.

The problem of identification of Jews based on the use of the *Kohen* modal haplotype marker raises several other intriguing questions. Would the presence or absence of such an SNP ever impact the oral tradition of who is a *Kohen* in the first place? Because traditionally this denotation is entirely self-identified—a man knows he is a *Kohen* only when told so by his father, and the community knows he is a *Kohen* only when he asserts that status—and because in the American immigrant context family names may or may not give accurate claim to *Kohen* status, is the deconstruction of the tradition to the genetic evidence yet another slow shift that might in fact be vulnerable to this sort of knowledge?

The African Jewish stories raise fulcrum issues for the Jewish community specifically and for the wider community as a whole about narrative, description, and obligation. The Lemba are only the beginning of this new way of knowing, for as our ability to make visible the invisible inner world expands—a process that begins with the microscope and leads us to digital computational biology—the way we see—and see morally—is also affected.

In the story of the African Jews, we learn that the Falasha, the quintessential stranger, is really (inside) ourselves but carries a different story. Then, in the recovery of the Lemba, that undoubtedly noble tale is destabilized by another level of complexity: the story the stranger tells may no longer be "true" in the historical sense unless it is also "true" at the molecular level. It is the hope, expressed by many, that haplotype linkages might provide a sort of tethering to a shared past, and behind this, there is the hope for repair of estrangement itself.

The idea of haplotype mapping is destabilizing, and it is so for both of the reasons I have noted above, identity and authenticity, and for the signal ethical problem that is discrimination based on newly reified constructs of race. Is this a reasonable fear? Much of the history of genetic science argues that it is. Yet can such dangers be avoided? Here is the junction at which ethics can imply a response and in which Jewish ethics might best be employed. What protects us from genetic abuse or genetic fantasy tales? The answer of religion might be to ask the question of allegiance, of love, and of justice. Our problem is not one of "too much knowledge" but one of too little agreement on issues of its just use.

The Kuzari: In Defense of a Despised Faith

A final text is *The Kuzari: In Defense of a Despised Faith*, written in Judeo-Arabic in 1140 by Yehudah Halevi, a Spanish Jew, and widely and intensely studied in the fourteenth century. In a sense, the narrative is a reflection on the problem of identity and a discussion of reason and free will, yet it is also a text that puts forward the idea that Jews bear a special inherited quality, passed on through direct lineage, that makes them chosen for a particular receptivity to the moral law.

As scholars of text with an obscure and dense codex such as the haplotype map, we are faced with "a difficult text." In considering *Kuzari* as a source for reflection on the problem of the special nature of Jews, we have two difficult texts: *Kuzari*, and Jewish genetic identity. We have the text of *Kuzari*, in which Halevi put forward a tale of conversion based on his understanding of the actual historical account of what happened in the ninth century with the king of the Khazars. As Halevi tells it, the king had an existential crisis marked by a dream in which he is told to find truth, and so he sets out to explore the systems of veracity available to him: classic Hellenistic philosophy, Christianity, Islam, and Judaism. The arguments of the Jewish rabbi make the most sense to him in the narrative, and the rational philosopher-king decides that he will convert himself and his people to Judaism (there is some genetic trace that this actually happened). He converts because it seems that Jews have the superior idea about God and the human task in relationship to God. The king wants the truth about the world. The text about conversion turns out to be a text about chosenness. Halevi's text about reason is also a text about "essences"; his text about language and argument also makes claims about blood and "sparks" that Jews uniquely possess. Yet in the description of this account, the text is clear in its assertion that Jewish identity is fundamentally genetic and heritable. The rabbi describes the history of the people thus:

For me it is sufficient that God chose them as His people from all nations of the world, and allowed His influence to rest on all of them, and that they nearly approached being addressed by Him. It even descended on their women, among whom were prophetesses, while since Adam only isolated individuals had been inspired till then. Adam was perfection itself, because no flaw could be found in a work of a wise and Almighty Creator, wrought from a substance chosen by Him, and fashioned according to His own design.... The soul with which he was endowed was perfect; his intellect was the loftiest that it is possible for a human being to possess, and beyond this he was gifted with the divine power of such high rank that it brought him into connection with beings divine and spiritual, and enabled him, with slight reflection, to comprehend the great truths without instruction.... It [the essence of Adam] passed to Seth, who also was like Adam, being [as it were] his essence and heart, while the others were like husks and rotten fruit. The essence of Seth, then, passed to Enoch, and in this way the divine influence was inherited by isolated individuals down to Noah. They are compared to the heart; they resembled Adam, and were styled sons of God. They were perfect outwardly and inwardly, their lives, knowledge, and ability being likewise faultless. Their lives fix the chronology from Adam to Noah, as well as from Noah to Abraham.... Thus the divine spirit descended from the grandfather to the grandchildren. Abraham represented the essence of Eber, being his disciple, and for this reason he was called *Ibri* .... The essence of Abraham passed over to Isaac ... the special inheritance of Isaac. The prerogative of Isaac descended on Jacob, whilst Esau was sent from the land, which belonged to Jacob. The sons of the latter were all worthy of the divine influence, as well as of the country distinguished by the divine spirit. This is the first instance of the divine influence descending on a number of people, whereas it had previously only been vouchsafed to isolated individuals. Then God tended them in Egypt, multiplied and aggrandized them, as a tree with a sound root grows until it produces perfect fruit, resembling the first fruit from which it was planted, viz. Abraham, Isaac, Jacob, Joseph, and his brethren. The seed further produced Moses, Aaron, and Miriam, Bezaleel, Oholiab, and the chiefs of the tribes, the seventy Elders, who were all endowed with the spirit of prophecy; then Joshua, Kaleb, Hur, and many others. Then they became worthy of having the divine light and providence made visible to them.[16]

What is the claim in this text? First, that there is a physiomystical property that is the marker of chosenness that predates Sinai and is in fact transmitted genetically from Adam to Moses through a complex line of descent. That this can even skip generations yet remain intact is a point Halevi makes at the end of his book:

After this [conversation] the Khazar king, as is related in the history of the Khazars, was anxious to reveal to his vizier in the mountains of Warsan [Varshan, near Balanjar] the secret of his dream and its repetition, in which he was urged to seek the God-pleasing deed. The king and his vizier travelled to the deserted mountains on the seashore, and arrived one night at the cave in which some Jews used to celebrate the Sabbath. They disclosed their identity to them, embraced their religion, were circumcised in the cave, and then returned to their country, eager to learn the Jewish law. They kept their conversion secret, however, until they found an opportunity of disclosing the fact gradually to a few of their special friends. When the number had increased, they made the affair public and induced the rest of the Khazars to embrace the Jewish faith. They sent to various countries for scholars and books, and they studied the Torah. Their chronicles also tell of their prosperity, how they beat their foes, conquered their lands, secured great treasures, how their army swelled to hundreds of thousands, how they loved their faith, and fostered such love for the Holy House that they erected a tabernacle in the shape of that built by Moses. They also honored and cherished the Israelites who lived among them. While the king studied the Torah and the books of the prophets, he employed the rabbi as his teacher, and asked him many questions on Hebrew matters.[17]

For Halevi, it is this "order" that allows the Jewish people to survive, even across the centuries of error and failures that have led to the exile as "the despised faith." Yet when the Khazars take on the Torah for themselves, they become a sort of superhero Jew, even erecting a fake Temple, presumably in the Caucasus Mountains, thus creating another of the mythic kingdoms so reminiscent of travelers' tales.

Mutability and Incompleteness in Research

A critical point in all genomic research is the constancy of the mutable knowledge base. The mutability and incompleteness of most of the research in genetics and genomics means that most of the chapters in this book, including mine, were rewritten in response to new science many times over the last ten years. Why is it so important to stress this point? It is important because so many of the discussions about the "essences" of the Jew and the lost tribes miss a core moral and historical truth about Jews and history, namely, the centrality of the construction of the Talmud and the moral architecture at the core of normative Judaism. Halevi's narrative of the Kuzar king is more subtle. Jews are both made and becoming, chosen and not exclusively chosen. The long, difficult, and divisive discussion about Jews, race, and identity is now largely reread as "DNA" in modernity. Yet religion, race, and identity are three of the most vexed and burdened concepts in our American public conversation. There is no scholarship on these topics without the context, and the social framing, material conditions, and human suffering that have attended these discussions.

How do we talk about human difference? It is a fact that we are different, and it is a fact that we are embodied creatures. That these two facts are linked is not a surprise. Of course, anyone in an actual family knows this, despite the caprice of birth, or the Bible would not be so resonant with brothers who are rivals. Actually, it is the worst sort of politeness to avoid the problems of embodiment, for when we do, we flee from our duty to attend to the actual needs of the other. It is a fact that transcription factors, proteins and how they are folded matter to the shaping of every human body, and those sensorial stimuli play defining roles in our beings, including our social beings.

Here is the difficult problem we face in thinking about the genetic basis for Jewishness: people look different and can do different things well or poorly. We, meaning particular societies, tend to prefer and then value some aesthetics, capacities, and talents over others. The trouble begins when the desired, pretty, lucky, agile people get more social goods, when the word "undesirables" becomes linked to populations, and when the power of the state is wielded to marginalize or oppress them. This is not a Marxist fantasy; it is the terrible legacy of much of our history. This sort of distinction only is made more dangerous when distinction is allied with the idea that God is involved in the fray and when the language slips from acknowledgment of human difference to thinking that the other is barely human at all.

Human differences, if read as "diversity," are political assets, for without full moral and intellectual attention to differences, how can justice be maintained? Because we are not alike, despite our very nearly identical endowments noted in the Constitution, we need to make the world we share just, which requires of us a system of judgment. Part of this is because of Hobbesian self-interest: we are all potentially befallen, all vulnerable, all living in fragile bodies. Part of this is because of commitments of faith, which insist on turning us from the world we see to the world that might be. If we are not allowed to know how we differ, then difference looks like personal blundering or lack of will. Human differences can read as justification for the powerlessness of the state to do anything very well or as evidence of the triumph of the market. The problem that DNA presents us is to find a coherent balance between freedom and responsibility, between the American idea that we can be anything we want if we try and the constraints of biology.

Why is this balance difficult to achieve? To some extent, it is difficult because of the history of medicine and the care of bodies. It was the great, democratizing insight of Enlightenment science that while the status and power of the person varied, disease did not. The normalization and standardization of human bodies as essentially alike allowed the study of disease to be rationalized and cures found for infectious diseases. Because the treatment for smallpox was preventative vaccination, and because the disease itself made no distinctions, it made sense to provide treatment to entire populations. The victory was for depersonalized medicine, which stresses our similarities. This was important in a world in which strangers are feared and resources are scarce.

But research about physical difference is quite different, for it is held in tension with ideas about the universality of the body. Thus the conversation about research in difference has historically turned to hatred, plain and simple, in particular, historically and scientifically endorsed hatred of Jews and of blacks so fierce and persistent that it now shapes nearly every conversation about the duty of the state internationally and domestically not just in the European past but in our time as well.

What was it that made the Jews "smart"? Was it the genes in my family that also made one aunt have epilepsy and another obsessive and fearful? Or was it that my grandfather, despite being desperately poor, his hands always speckled with paint from his job, made it his work to also start a shul and a study house down the block in a tiny storefront? Was it his genes or the fact that when he fled the tsar's army, he fled with a prayer book in his pocket, and one morning, in the slanting winter light of a barn window in Kovner, he knelt to pray and was found there by my grandmother? He never told me the names of his family, but he took my hand in his and showed me the letter aleph, and he listened with pride when I wrote my first poem in the English he had struggled to learn. What of the fact that the *Chicago Jewish News* this week, as it does every week, included an envelope telling me that I needed to give money to the superfund for education, linking it to my very survival? That the entire community is told over and over that we are utterly responsible for one another, so deeply in debt to one another that we do feel stricken when any Jew anywhere fails or is shamed and proud when any Jew anywhere does well? Of course, we are tangles of love, duty, history, and transcription factors that trigger proteins. This is why the politically incorrectly named *bubbameises* (grandmother's tales) get it right: everything you do, say, and yearn for shapes your dopamine level.

And here is where the Talmud, so curiously absent from the account of Jewish genetic history, is so important. In part, the notion of the Talmud, claims Jonathon Schofer, professor of Jewish ethics, is the question of being: how to be a being who understood the terror and the gift of chosenness; how to be a moral being in a vulnerable human body; how to be a Jew in a world of violence and death.[18] Reading the Talmud is a complex matter. The Talmud is large, like the genome, and full of serious discussion about the puzzle of a number of things, including the great puzzle of chosenness. The Talmud has many such discussions, like the one in my family last night: What is the meaning of illness in families? Does physical difference make us Jews, or is it what we know and what we do? What are the duties and rights of a proselyte? How are children shaped for good? How are they shaped to be scholars? If we are so chosen, why are we so despised, hunted, and killed? How ought we to live with courage and humility? To think about such questions as a matter of daily life does indeed create a certain sort of place into which to be born.

The project on which this book is based began when the *Kohen* modal haplotype was just being published, and since then, work on genetic difference and diversity has continually expanded. For many of us, the problem was most quietly explained by our colleague Mark Thomas, the University of London geneticist who, at the end of a long meeting with the authors of this book's chapters, reflected on our work. Thomas was quiet that night, and thoughtful. "We are going to learn things about how we are made," he said, "and learn that we are made differently, and that some of these differences will be in our brains, in how they are made. And then it will get hard." He looked at us seriously. "I hope by then, you bioethicists will have figured out how we talk about this, because we will need to talk about this."

Thomas had a point, too. We lack the vocabulary to discuss differences. We know we are deeply committed to democracy, and that democracy requires equal opportunity, and that we really do believe in liberty and justice for all. To say that we are genetically better, smarter, or more socially adept says at the same time that you are not, and that is that.

Everyone would agree that what I will call ethnogeographic stories of origin (EGSO) are critical; "race" is not. But there is little that is fate: there are exceptions to norms. Take the example of Karl Gauss, ironically, who described the Gaussian curves that are the way statisticians plot characteristics to describe means and averages. Gauss arose out of a poor family to become perhaps the greatest mathematician of his generation.

The puzzle of chosenness requires a toughness of mind and a willingness to reflect on our moral duty and consider where our science might lead. The question of moral philosophy in Jewish thought is ultimately based on the fact that to be chosen means to ask these questions: What are our duties to God and to one another? How do we, with what we have, become who we yearn to be? Was Einstein smart because his brain was dense in specific regions? Or was he smart because six different women devoted their lives to his care and feeding and career?

It is a smart thing to develop a tradition that valorizes charity and education and gives award dinners to honor them but not, for example, beauty contests for preteens, and sorting out chosenness from law is complex and perhaps futile. But if the claim for chosenness is to be based in genetic science, it must be supported by experiments and evidence, and it will be only verifiable if it is falsifiable. If it is moral philosophy, it must be defensible; and if it is theology, it must be supported by authoritative texts or thoughtful teachers.

We have different jobs and a lot of work to do to develop the language needed to learn about ourselves; a moral language is as necessary as the need to figure out algebraic number systems. As a bioethicist, I have to reflect on the gaze of the research, insist on an ironic historical analysis, and be alert to the use and misuse of human subjects and data. As a moral philosopher, I need to construct an argument for justice in an unjust world and be wary of ideas that make the injustice worse. As a Jewish theologian, I need to speak to the trouble and the possibilities of the world. For a Jew, the point of chosenness is to inhabit not only a land, which we never quite enter in the Torah, but a state of being, an illumination, participation in a nation that is the light to the world. It is to teach children the story of an unredeemed, broken, and unfinished world, one where love is made durable by duty, and duty is sweetened by love. To be a Jew is to carry that combination of love and duty, as surely as our genotype—both things of defining importance—into the world we share.

NOTES

The epigraph is from Sonia Levitin, The Return *(New York: Fawcett Juniper, 1983), 36–37. This book is in its eighth printing, and it has won the following awards: the Association of Jewish Libraries' Sidney Taylor Award; the 1988 National Jewish Book Award in the category of children's literature; the* PEN *Los Angeles Annual Award for Young Adult Fiction.*

1. Tudor Parfitt, *Journey to the Vanished City: The Search for a Lost Tribe of Israel* (New York: St. Martin's Press, 1993).

2. In traditional Jewish communities, *Kohanim* are prohibited from contact with the dead and from marrying a divorced woman, and they are given several public ritual tasks, such as the first turn at blessing the reading of the weekly Scriptures and blessing the congregation in a specialized ritual on holidays.

3. Mark Thomas et al., "Y Chromosomes Traveling South: The Cohen Modal Haplotype and the Origins of the Lemba—the 'Black Jews of Southern Africa,'" *American Journal of Human Genetics* 66 (2000): 674–78.

4. In the linguistic and ritual universe, the Orthodox Jew prays for the restoration of the Temple and the Temple services three times a day.

5. This is in addition to trader communities whose origin is well documented and was never seen as in doubt.

6. The Lemba do not claim a need to return, but it is early in their new relationship with the world's Jewish community, and we will see how the yearning to return becomes what is salient about their Jewishness. *Midrash Rabbah,* Lamentations 11:9.

7. Of interest is that the golem tales and midrashic narratives about the golem contain many of the same characteristic of a pious, powerful, "big" Jew.

8. Levitin, *The Return,* 144.

9. Levitin, *The Return,* 159.

10. Levitin, *The Return,* 22.

11. Levitin, *The Return,* 11, 27.

12. Tudor Parfitt is emeritus professor of modern Jewish studies at the School of Oriental and African Studies (SOAS), where he was the founding director of the Centre for Jewish Studies. He is now distinguished professor at the School of International and Public Affairs and a lecturer to the group that wrote this volume.

13. S.A. *Yoreh De'ah* 263:2–3; see B. *Yevamot* 64b.

14. The rabbinical system constantly undertook to transform the priestly hierarchy of the Temple into a world where what mattered was one's moral and academic achievement. In the Rabbis' world, power, wealth, and class were of little importance before a persuasive argument about interpreting and applying the Torah. Because the Torah includes many laws that require Jews to attend to the needs of the widow, the orphan, and the poor, Pharisaic Judaism—which is what we today call simply "Judaism"—sought to repair and heal the ruptured world after the Roman occupation and through the centuries of exile by a steady attention to both distributive and procedural justice and community solidarity, made manifest by rational discourse, not hereditary claims. In a world of medieval feudalism, the Jewish system of the democracy of study negated claims of genetic inheritance.

15. B. *Nedarim* 81a.

16. Judah Halevi, *The Kuzari: An Argument for the Faith of Israel*, trans. Hartwig Hirschfeld (New York: Schocken, 1964), pt. 1, para. 95, pp. 64–66.

17. Halevi, *The Kuzari*, pt. 2, para. 1, pp. 82–83.

18. Jonathan Wyn Schofer, *Confronting Vulnerability: The Body and the Divine in Rabbinic Ethics* (Chicago: University of Chicago Press, 2010).

PART 3 *Genetic Testing*

CHAPTER 10 *Summary of the Science of Genetic Testing*

ELLIOT N. DORFF AND LAURIE ZOLOTH

Jews have had a curious and complex relationship with the use of genetic testing, a rapidly growing part of the medical armamentarium in the modern era ever since hereditary disease was first understood in the mid-nineteenth century. The concept of heritable traits was a fascination of the emerging science of genetics. The measuring of a Jew's body was a nineteenth-century project in which biological "difference"—for example, in the "gaze" or gait of the Jew—was asserted and then used to mark the body of the Jew as not only disparate and non-Aryan but a danger to the larger body politic.[1] As Sander Gilman noted, it was the nineteenth-century testing and marking of the Jewish body as non-Aryan that provided the "scientific" rationale for Hitler's Final Solution in the twentieth century. Thus, for many Jews this linkage between the testing and marking of the body and the capacity for discrimination and oppression is unbreakable. Genetic testing enters the stage of Jewish history as the threatening first step of genocide. This link to the Holocaust has affected not only Jews but also non-Jewish European bioethicists, who have raised concerns that to test any population genetically is to close off an open future or to allow the first temptations toward the return of state authority over vulnerable bodies.

Yet Jews, particularly American Jews, are simultaneously devoted to the idea that medicine is a great, indeed an imperative, social good. There has been a strong internal cultural trust of medicine and the medical profession among American Jews, including genetic medicine in its first iteration, that of genetic testing.

It is the tension between these two very strong pulls in opposite directions that has led to some uncertainty about how to weigh the concern about the possibility of abuse against the promise that genetic information might provide for Jews both as a group and as individuals. The articles in this section discuss various aspects of the deep ambiguity that genetic testing presents for Jews.

Genetic testing entered the stage of Jewish history for the second time in the 1970s, when it became known that the clearly defined single allele mutation that caused the fatal childhood disease of Tay-Sachs occurs much more often in the genes of Jewish families than in those of the general population. When a clear genetic cause for the devastating, universally fatal, neurodegenerative disease was found to be a single allele mutation that could be detected through a blood test, the entire Jewish community, from Orthodox matchmakers to Jewish communal youth leaders in the more liberal movements, joined in a massive effort to assure that Tay-Sachs testing would be widespread. It was a message proclaimed, quite literally, by rabbis in weekly sermons. The testing for the Tay-Sachs allele was so effective in both premarital and prenatal screening that Tay-Sachs is now less likely to occur in the Jewish community than in the community at large. This experience with Tay-Sachs opened the way for widespread acceptance of testing to avoid a series of diseases and for the conviction that testing prior to conception was firmly acceptable within Jewish law and tradition.

The Origins of Ashkenazi Jewish Diseases

Tay-Sachs is not the only gene that naturally occurs in higher frequency in Ashkenazi Jewish populations—that is, those descended from northern and eastern European Jews. (Sephardic Jews, those descended from Jews living in the Mediterranean basin, are carriers of other genetic mutations that cause other genetic diseases.) In fact, the number of genetic conditions affecting Ashkenazi Jews—far more than the general population—for which tests now exist are so numerous that they are part of a Jewish genetic screening program offered to pregnant Jewish women and to couples planning to begin a family. In some Orthodox circles, these tests are part of an elaborate and confidential matchmaking process. Scientists' interest in genetic testing of Jews may be largely due to the results of the widespread testing done during the Tay-Sachs era, which produced thousands of available samples for further research, but it is also linked to the specific historical reality that Jews are a relatively small community with strong religious and cultural prohibitions that encourage marriage only within the Jewish community (endogamy). This historically produced the founder effect, that is, the continuing appearance in succeeding generations of the mutation in the genetics of one ancestor, for if one person with the recessive mutation married another person with that same recessive mutation, they had a one-in-four chance of producing a child with the disease.

In his book *The Genetics of the Jews*, A. E. Mourant, one of the founders of population genetics, pointed out that generally each major Jewish community has the tendency genetically to mirror the local population of the region from which that community arose.[2] In reviewing that book, geneticist Arno Motulsky and his colleagues instead asserted the "genetic distinctiveness" of the Jewish populations, one example of which are the many diseases unique to Jewish communities. They thus claimed that Mourant's book is "somewhat selective and less complete" in respect to these diseases compared to the literature on polymorphisms. Mourant responded that the high occurrence of "recessively inherited sphingolipidoses" such as those for Gaucher and Tay-Sachs diseases among Ashkenazi Jews is purely a matter of chance. But Motulsky, in response, contended that a rare genetic mutation influencing the appearance of these diseases may have taken place in a reduced Ashkenazi population following Jewish "persecution and extermination." Motulsky does admit that the argument over chance occurrence versus natural selection is quite heated in most recent scientific literature, with advocates of natural selection "asking how chance can explain why three different genes affecting sphingolipid metabolism occur in the Ashkenazi population." This debate began in 1954 and continued for the next several decades.[3] At stake was how deeply the biological traces of oppression could be seen in the genes of Jews and the existence and nature of Jewish genetic distinctiveness. It is now largely understood that the disease burden of Ashkenazi Jews is linked to specific histories and not to chance alone.

BRCA1/2

In the late 1990s, several more mutations, some more likely to be found in Ashkenazi Jews, were linked to a much higher incidence of breast cancer, and they were labeled *BRCA1* and *BRCA2*. As genetic testing expanded, bioethicists became interested in understanding the evolving attitudes toward that testing. In her article "A Population-Based Study of Ashkenazi Jewish Women's Attitudes toward Genetic Discrimination and BRCA1/2 Testing," Lisa Lehmann expressed her desire to analyze the female Jewish community at large to determine how the majority of Jewish women feel about genetic testing. "Ashkenazi Jews have an increased frequency of founder mutations in *BRCA1/2* and therefore have an increased rate of inherited breast and ovarian cancers," Lehmann states. "They have also been found to be at high risk for the *APC I1307K* allele associated with the development of colon cancer." Lehmann theorizes that while genetic testing for disorders common in Jews is beneficial, resulting in successful treatment, it is a major concern of some that this information could be misused by third parties.[4]

Lehmann conducted a population survey to determine the depth and types of concern Jewish women have over the possibility that this genetic information could be used to discriminate against Jews attempting to obtain insurance and employment, thereby inhibiting individuals from getting tested and, at the same time, impeding the progress of genetic research. In this specific study, the majority of the population-based sample of Jewish women was not concerned that genetic testing would result in group discrimination. In fact, most women (71 percent) in the group thought that there were scientific reasons for testing Jews, and a large majority of the respondents (95 percent) responded that research focused on Jews was either neutral or good.

In 2013 two events converged to make the need for testing for *BRCA1/2* and its acceptability more public. One was a very well publicized letter about the need for testing for these mutations from a prominent American actress, Angelina Jolie. The second was that the test itself, which had been a proprietary test patented by Myriad Genetics and prohibitively expensive, became more widely available and more reasonable in price.

Testing for Jewish Genetic Diseases Today

The *BRCA1* and *BRCA2* mutations are certainly not the only genetic predispositions prevalent in the Jewish community, and Jewish communities throughout the United States and Canada have created programs to educate young Jewish adults about the genetic diseases especially common among Jews, to test for these diseases through a simple blood test, and to inform Jews who find themselves to be carriers of one or more of the diseases about what they can do to avoid having children with those diseases. In 2001, for example, the Jewish United Fund (JUF) and the Jewish Federation of Metropolitan Chicago established a national center to serve and educate the community about Jewish genetic diseases. Called the Chicago Center for Jewish Genetic Disorders, it has expanded its genetic disorder screening panel from four disorders to eighteen, and this number grows annually. This expanded prenatal screening panel now includes Bloom's syndrome; Canavan disease; cystic fibrosis; dihydro-lipoamide dehydrogenase deficiency; familial dysautonomia; familial hyperinsulinism; Fanconi anemia, group C; Gaucher disease, type I; glycogen storage disease, type 1A; Joubert syndrome 2; maple syrup urine disease; mucolipidosis IV; nemaline myopathy; Niemann-Pick disease, type A; spinal muscular atrophy; Usher syndrome (type 1F and type III); Walker-Warburg syndrome; and, of course, Tay-Sachs.

"Within the Ashkenazi Jewish community, approximately one in five individuals carries the gene mutation for a Jewish genetic disorder," the

JUF's website explains. The Chicago Center's testing began in 2002 with only three diseases, and now screening for the eighteen disorders listed above requires only a simple blood test. Although this testing involves a "prohibitive cost and lack of insurance coverage" for much of it (it can cost over a thousand dollars), the center operates a subsidized screening program to assist the Jewish community. In addition to these eighteen diseases, which are included in "the panel of Ashkenazi Jewish diseases" for which American and Canadian hospitals now test, Israeli hospitals also test for fragile X syndrome, which, although not more prevalent among Ashkenazi Jews than among the general population, is nevertheless a commonly occurring genetic disease.

Thus, on the one hand, the history of Jews as a persecuted people makes Jewish individuals wary of tests that single them out as different and diseased. When such tests were rare and unfamiliar, they were all too easily understood as akin to eugenic tests in the early twentieth century that served as the basis for Jews being thought of as tainted and diseased in the minds of non-Jews and of Jews themselves. On the other hand, the enormous benefit of genetic testing is now quite clear, for we have seen that the powerful predictive value of tests and the certainty of carrier status can do several things to transform medicine and benefit patients. First, genetic testing can inform couples about whether either or both of them carry the recessive gene for one of the Ashkenazi or Sephardi genetic diseases. If both members of the couple are carriers of the same disease, thus giving them a one-in-four chance of having a child affected by the disease, they can avoid passing on that genetic disease in many single allele illnesses by using preimplantation genetic diagnosis (PGD), a technology first successfully used in October 1989. In this process, the couple provides sperm and eggs that are joined together in a petri dish and cultured for several days. Then one cell is taken from each embryo to determine whether it carries the genetic disease, and only those embryos free of the disease altogether or that are only carriers but not diseased themselves are implanted in the woman's womb. In previous decades, the only way couples could avoid passing on a genetic disease to their offspring was to become pregnant, wait until the fifth month, undergo an amniocentesis, and then abort the fetus if it was shown to have the disease; now, through PGD, couples can produce children without the disease (and possibly even without being carriers) without having to wait through five months of pregnancy and without the need for an abortion.

Second, people can be tested early in life, allowing specific monitoring, lifestyle modification, or even, in the case of *BRCA1* and *BRCA2*, prophylactic surgery to avoid breast or ovarian cancer. Thus, as a way both to identify genetic diseases in their early stages so that people can take steps to avoid them and to prevent passing on genetic diseases to one's offspring, genetic testing is rapidly moving toward a far wider acceptance among Jews than in the past, together with measures to make the testing financially more accessible.

Over the years that we have worked on this manuscript, the search for genetic causes in catastrophic diseases has accelerated, such that in the last few years the number of known diseases to be caused by specific genes has actually doubled. As whole genome sequencing becomes more commonplace, separate testing for these diseases will no longer be necessary. If a whole genome sequence becomes part of the baseline for medical testing in general, the idea that particular populations carry a special risk may become an historical veracity but not a medical reality. These essays consider the issue as we stand at the threshold of this future.

In the last decade, genetic testing has moved out of the medical establishment and has become a direct-to-consumer industry to illuminate any person's genetic heritage. This increased availability of genetic testing makes the issues discussed in the previous section of this book on group and personal identity all the more acute. In this section, however, we address the medical uses of genetic testing.

NOTES

1. Sander Gilman, *The Jew's Body* (New York: Routledge, 1991), introduction.
2. A. E. Mourant, *The Genetics of the Jews* (Oxford: Clarendon Press [Oxford University Press], 1978).
3. British Medical Association 0157118: Eugenics Society Symposia, Volume 1: Biological Aspects of Social Problems, 1965; Arno Motulsky, "Jewish Genetic Diseases and Origins," *Nature Genetics* 9 (1995): 99–100; see also David B. Goldstein, *Jacob's Legacy: A Genetic View of Jewish History* (New Haven CT: Yale University Press, 2008).
4. L. S. Lehmann, J. C. Weeks, N. Klar, and J. E. Garber, "A Population-Based Study of Ashkenazi Jewish Women's Attitudes toward Genetic Discrimination and BRCA1/2 Testing," *Genetic Medicine* 4, no. 5 (2002): 346–52.

CHAPTER 11 *Genetic Testing in the Jewish Community*

PAUL ROOT WOLPE

Genetic tests analyze bodily fluids or tissues to determine the donor's genetic information. The information may include whether the donor has a genetic disease or syndrome, has a predisposition to develop a genetic disease or syndrome, or carries the genes that would pass such a condition on to future generations. Genetic tests can be performed throughout the entire span of human development, from the embryonic stage to postmortem. Genetic tests have other uses as well, such as the identification of workers who may have particular susceptibilities to toxins in work environments, DNA fingerprinting of criminals, and archaeological testing of human samples to determine their origins and relatedness. For the purposes of this chapter, however, we will focus on genetic testing for diagnosing or predicting genetic susceptibility for disease.

Although genetic tests are not fundamentally different from other kinds of medical testing, they have characteristics that raise ethical concerns. First, genetic tests are often predictive; that is, they suggest a likelihood (or in some cases, a certainty) that the patient will develop a condition. Such prognoses raise the question of medical management after testing, which can be particularly thorny when there are no possible preventive actions. Second, genetic tests often reveal information about the patient's relatives, as we will see below, who may be unaware that such information exists or who do not want the information. Third, genetic information can influence reproductive decision making, or it can prompt decisions about terminating pregnancies. Finally, genetic testing of embryos raises the specter of eugenics and selection of desirable traits in offspring.

Jewish tradition and law can offer guidance on these matters and suggest whether and how Jews might best make use of these technologies. "To be religious is to identify with a community," writes the Protestant religious scholar Ronald Cole-Turner,[1] and Jews who affiliate with a religious community often look to it for guidance when making difficult ethical decisions. To varying degrees, the four major denominations of Judaism have offered guidance to their adherents about how to make use of genetic testing in a way that conforms to the Jewish standards each denomination follows. Jews who are not observant—that is, do not follow communally defined standards of religious behavior—often turn to more secular cultural values to help them make such decisions. Even so, when confronted with thorny bioethical problems, many secular or unaffiliated Jews do turn to Jewish tradition and halakhic perspectives even if these have little influence over other kinds of moral decisions these Jews make in their lives.[2] The ambiguities and pain of making such morally salient and emotionally fraught decisions can drive people back toward their faith traditions, or at least to the moral frameworks that inform those traditions. Rabbis often report getting calls from congregants or strangers with whom they have had little or no contact when the person is involved in end-of-life decision making or a similar bioethical dilemma.

In this chapter, I will discuss the ethical challenges of three areas of genetic testing: testing for clinical purposes on adults, testing for preventive purposes on embryos, and research into genetic testing. We will look at each of these areas below.

Clinical Testing on Adults

Genetic testing has advanced from a mere hope to an established practice within medical care in a relatively short time. The first medical genetic test was for Huntington's disease, developed in 1983. A decade later, genetic testing was available for just over one hundred diseases; by 2011, testing was available for upward of sixteen hundred diseases.[3] Between 2000 and 2006 alone, the genetic testing market rose from about $320 million to about $877 million and is expected to reach $10 billion by 2015.[4] So the concerns that ethicists and religious scholars have over aspects of genetic testing are not simply theoretical; these concerns are already appearing in doctors' offices around the world.

Diagnostic Testing

The primary purpose of clinical testing on adults today is either to diagnose a disease or to identify genetic susceptibilities to a disease. With the rapid development of pharmacogenomics, however, genetic testing will also soon be used to identify genetic subtypes in order to target personalized pharmaceutical treatment. For a person already suffering symptoms for which a genetic test will determine the cause, there is little ethical or religious concern; virtually everybody agrees that such genetic testing is permissible and even desirable. All the basic bioethical standards need to be met, of course: the patient's confidentiality must be respected; informed consent must be obtained during which the psychological, familial, and physiological risks and benefits are fully explained; and access to sophisticated genetic counseling must be assured, among other concerns. If these important ethical standards are met, clinical testing as a diagnostic tool to explain a patient's symptoms remains fairly uncontroversial.

Susceptibility Testing

Testing to determine susceptibility to genetic disease is far more problematic. Such testing may be conducted to explain disease frequency in families; genetic testing on members of families with a history of breast cancer, for example, is now common. In such cases, the justification for testing may be to determine who needs increased surveillance or to provide information for preventive medical management, such as prophylactic mastectomies. However, testing may also be conducted in the absence of disease or any history of it; in many Orthodox communities, for example, premarital testing is performed to determine whether potential mates carry genetic diseases that would put them at risk to have offspring with genetic disease. If they do, as we will see below, the couple is often advised not to marry.

Genetic testing for disease susceptibility raises a number of concerns. First, unlike most other tests, genetic tests often give a patient's family information about their own genetic status that is sometimes unwanted. While confidentiality can often be maintained, a woman who opts for a prophylactic mastectomy after discovering that she is positive for the breast cancer (BRCA) mutation will not be able to keep that test result from family members. Furthermore, many patients feel an ethical obligation to inform family members who may themselves be at risk for disease. Testing can also raise anxiety. In an Israeli study, Ashkenazi women reported that one reason for declining genetic testing was "genetic anxiety," or a fear of known and unknown dangers in one's genetic makeup.[5]

Some Jewish concerns have also been raised. Rabbi Josef Eckstein, founder of the Tay-Sachs screening program at the nonprofit organization Dor Yeshorim in 1983, did not offer the same religious sanction for early testing for breast cancer susceptibility as he did for testing for Tay-Sachs. His concern was that without an adequate treatment (except prophylactic mastectomy, which at the time was not proven to be preventive), testing would create a class of stigmatized women, unmarriageable and socially ostracized.[6] Rabbi Moshe Tendler, a professor of biology at Yeshiva University and a respected authority within the Orthodox Jewish community, has also expressed his aversion to testing for diseases that have no treatment or cure, such as Huntington's disease, seeing no advantage to knowledge that he believes may harm more than protect the patient.[7] Another, broader concern is the general stigmatizing effect of identifying the Jewish people as a whole through the use of genetic disease susceptibilities.[8] A Google search of the term "Jewish genetic diseases" finds thousands of websites that use the term (278,000 in a search on June 5, 2014). Perhaps more ominous, the media have often seized on the testing of Jews to suggest that Jews have higher rates of genetic disease than other groups. In fact, Jews have no higher rates of genetic diseases than most other ethnic groups, though the willingness of Jews to participate in genetic research may make it seem so, as I will discuss below. Still, media reports have, in the past, spoken of the genetic "flaws" of Jews or portrayed them as a class as prey to genetic disease with headlines like "Ashkenazi Jewish Women Stalked by Second Mutant Breast Cancer Gene."[9] Such reports have been used by anti-Semitic websites and periodicals to uphold the notion of Jewish genetic inferiority.[10]

Despite concerns about stigmatizing the Jewish population, Jewish spokespersons are generally in favor of genetic screening. Because Ashkenzic and *Mizrahi* Jews are relatively genetically homogeneous compared with the general population, there are specific diseases that do affect Jews at a higher rate than the population as a whole. Therefore, genetic testing is a desirable strategy for identifying such diseases among people with Ashkenzic or *Mizrahi* heritage (whether they identify as Jews or not). The Center for Jewish Genetics in Chicago, for example, partners with the Chicago Board of Rabbis to provide guidance for Jews about genetic testing.[11] Using basic halakhic principles, such as *pikku'ah nefesh* (preservation of life), the dictum *v'nishmartem m'od l'nafshoteikhem* ("And you shall diligently preserve your souls," which in traditional interpretation includes the physical body), the duty to heal, and the Jewish acceptance of technology to further healing, rabbis fully support the right, and in some cases the obligation, of Jews to undergo genetic screening for therapeutic purposes.

Moreover, many religious Jewish communities have not only accepted therapeutic genetic testing in individual cases but also encouraged community screening programs. The Dor Yeshorim program in the Orthodox community tests potential mates for a variety of genetic diseases. If the partners are both carriers, and thus their marriage increases the risk of having a child with a serious genetic disorder, they are advised of their genetic incompatibility; however, their exact carrier status is not revealed to them. Dor Yeshorim has tested tens of thousands of people, and this effort is partially responsible for the virtual eradication of Tay-Sachs from the Jewish community. Still, Dor Yeshorim has not escaped criticism. Some have argued that the program violates civil liberties by failing to reveal people's carrier status even when they ask; that by not revealing test results the program maintains a monopoly on testing in a community; that matchmakers and genetic testing laboratories have an unhealthy relationship with Dor Yeshorim; that the program at times refuses to accept results from other laboratories; and that Dor Yeshorim requires screening of children when screening the parents alone will reveal if the children even need testing.[12] Despite these complaints, Dor Yeshorim is generally considered a model program for community screening of genetic disease.

Not all community screening programs have enjoyed the success of Dor Yeshorim. An attempt to institute a screening program in the Jewish community in Sydney, Australia, for example, failed due to insufficient support from community leaders and lack of access to Jewish institutions.[13] In Melbourne, Australia, in contrast, the community split: those in the ultra-Orthodox community declined to participate in the local effort, preferring to register their results with the international Dor Yeshorim database; other Jewish groups were more welcoming of the screening program and less concerned with the Dor Yeshorim policy of anonymity and marital screening, preferring simply to be informed of their individual testing results.[14] Both examples show that access to synagogues, day schools, and other community institutions is key in developing programs with community participation.

Recently, efforts have been made to encourage testing of people of Ashkenazi Jewish descent before they make reproductive decisions. For example, the Atlanta Jewish Gene Screen is a program providing education and testing to youths to facilitate testing before reproduction, with the explicit aim of reducing the incidence of genetic disease in that population.[15] There are plans to expand these types of testing programs nationally.

Despite the success of screening programs like Dor Yeshorim, it is not clear who should be counted as part of the Jewish community for purposes of screening. In fact, such programs are not actually aimed at Jews as members of a religious group but at people of Ashkenazi or *Mizrahi* ancestry, a large proportion of whom are Jews. Through conversion, assimilation, mixed marriages, and other kinds of genetic mixing, many non-Jews, or people who no longer affiliate with the Jewish community, have Ashkenazic heritage. Attitudes toward testing are also influenced by the type and degree of affiliation. For example, there is a difference between Jewish religious affiliation and Jewish cultural affiliation; one study of 220 Ashkenazic Jewish women found that those who identified more culturally as Jews were more likely to express interest in being tested for *BRCA*, while those with a stronger religious identity were more likely to express an intention to adhere to mammography recommendations and decline testing.[16] So "Jewish identity" must be further clarified in order to understand attitudes toward genetic testing.

Despite these issues, genetic testing of adult populations is common and generally accepted. However, many of the issues mentioned so far, along with a host of new issues, arise when we consider using genetic testing to screen embryos for disease.

Embryo Testing

The most controversial aspect of genetic testing is its use in preselecting embryos for implantation. In preimplantation genetic diagnosis, or PGD, embryos are created in a dish and allowed to divide until they reach eight or sixteen cells. A single cell is then removed from each embryo and tested for genetic mutations that might lead to disease. Only those embryos that are free of the genetic trait are implanted, and the rest are discarded. Sometimes the tests for a disease marker are very specific. In other cases, more general criteria can be used. Some genetic disorders are sex-linked, for example, and it is possible to sex-type the embryos and only implant the unaffected sex. Prenatal genetic testing usually results in selective implantation, meaning that the undesirable embryos are discarded. For the Catholic Church, as well as other religious groups, an embryo should be accorded all the rights of any human being. Discarding an embryo, even one that has been determined to carry significant genetic risk or disease, is unacceptable. Judaism, in contrast, does not consider embryos in a petri dish (in vitro) to be equivalent to embryos in a woman's uterus (in vivo). Most halakhic authorities hold that untransplanted embryos have little standing and may be discarded. Although embryos should be accorded respect and not used for trivial purposes, they have no "right" to be implanted, nor is it our duty to do so.

On the other hand, the act of selecting is itself fraught with moral implications, no matter the status of the embryo. Selection, after all, is a eugenic act—the selection of one (potential) person over another to improve genetic fitness. Although Judaism sees no intrinsic problem with such a process, it is sensitive to its implications both for halakhic reasons and because of the unique history of the Jewish people as victims of eugenic ideology, most graphically, but not exclusively, in the Holocaust.

Most Jewish authorities allow at least some PGD and selective implantation for therapeutic purposes. However, as new genetic tests arise, the question of where to draw the line becomes an important consideration. Clearly, selecting against Tay-Sachs is allowed. Cystic fibrosis is a bit more difficult, as people with the disease now often live into their thirties, although debilitated and in pain, but most authorities would allow such selection. As more and more genes are identified for conditions that are not immediately life threatening, however, we need to begin to think about what is or is not an ethical criterion for selection. Predisposition to disease is not a diagnosis and not a disease in itself; in fact, everyone reading these words undoubtedly has some genetic predisposition to certain diseases. Which ones should we select against? Should we select against a genetic propensity for Alzheimer's disease, which will not show up for decades (by which time there may be a cure)? Should we select against predisposition to developing heart disease, or against embryos with *BRCA*? If genes for susceptibility to alcoholism or obesity were found, should we select against them even though the effects could be modified by lifestyle choices? What of a gene for homosexuality, were it found? Finally, should we allow trait selection, or sex selection, simply because parents want their children to have certain attributes?

Most halakhic authorities do not allow screening for nontherapeutic reasons, such as sex selection or desired physical traits. Sex selection, however, is a special case; although one may not select for gender simply out of preference, other reasons may justify sex selection. For example, one may screen for gender to avoid a gender-linked genetic disease. The policy in Israel also allows a degree of sex selection for family balancing. A family with four children of the same sex may apply to the health ministry for permission to use preimplantation selection to choose an embryo of the other sex. In an interesting case in Israel a few years ago, a man petitioned a court to allow sex selection in order to keep his infertility private. In that case, a *Kohen*, a member of the priestly caste, wanted to select for a female child, for he and his wife had to use artificial insemination to achieve pregnancy. His offspring would therefore not be considered a *Kohen*. The father argued to the court that, should their child be male, the son could not be called to the Torah as a *Kohen* at his bar mitzvah, thus publicly revealing that he was not the biological son of the father. The court ruled that the man could use sex selection in that case.[17]

While PGD has been permitted in these circumstances, abortion remains more controversial. The vast majority of rabbis in the more liberal streams of Judaism would allow the abortion of a fetus afflicted with Tay-Sachs and many other genetic diseases as well.[18] Aborting a fetus found to have Tay-Sachs has been allowed in the Orthodox world as well, as in the decision by Rabbi Eliezer Yehudah Waldenberg to permit such an abortion; other rabbis, however, have been critical of his decision. Premarital screening has become important in the Orthodox community precisely because it preempts the possibility of couples deciding to remain childless or to undergo abortion when genetic disease susceptibilities are discovered after marriage.[19]

Interestingly, the distress and psychological pain of the parents is the argument generally used to permit screening and disposal of embryos in vitro rather than any negative qualities the child might suffer or the poor quality of life it would lead were it born. This is generally the attitude taken for abortion as well; when permitted, it is usually justified on the basis of the suffering of the parents. As Rabbi Yitzhak Zliberstein ruled, "One cannot close the door in the face of despondent people who suffer mental anguish in the fear of giving birth to sick children, pressure which can drive the mother mad. Therefore, in the case of a serious genetic disease which affects the couple, it is difficult to forbid the suggestion [for genetic screening through IVF]."[20] Even so, although the stricter halakhic approaches are less tolerant of aborting a fetus that shows less than life-threatening genetic abnormalities, the more liberal branches of Judaism may place greater weight on the suffering of the fetus or the parents as a reason to permit abortion.

Israeli women report feeling pressured to engage in embryo selection.[21] A study by Professor Larissa Remennick of Bar Ilan University done in 2006 found that elective genetic testing was more acceptable, if not expected, among educated middle-class Ashkenazic women, while it was more often questioned and refused by lower-class *Mizrahi* women and religious women of any ethnic origin. Women who rejected genetic testing often reported that they did so because of objections to abortion or the eugenic aspects of prenatal screening. The key forces that drove women's choice of prenatal genetic diagnosis in the study included the fear of having a sick and/or socially unfit child in an unsupportive environment; strong endorsement of testing by gynecologists; popular and professional discourse about the common Ashkenazi mutations causing genetic anxiety in the group; and the emerging social pressure that considered comprehensive prenatal screening an indispensable part of good motherhood. Obviously, Israeli society is manifesting the tension between the desirability of using genetic testing for screening debilitating diseases and the inherent ethical dilemmas involved in any eugenic selection.

The complexity surrounding ethical issues of genetic testing for embryo selection will undoubtedly increase as more tests are discovered and as testing becomes more routine. Already, fertility clinics are offering sex selection and limited trait selection in certain cases. The consideration of ethical standards for PGD will be an ongoing part of the Jewish bio-ethical agenda for many years to come.

Research on Genetic Tests

Developing genetic tests often requires comparison of the genomes of a number of individuals.

Ashkenazic Jews are considered a desirable population for genetic research. Aside from the fact that American Ashkenazim are well educated, medically sophisticated, and often willing to volunteer for genetic research, Ashkenazic Jews also represent a unique genetic profile. Ashkenazic Jews are believed to have undergone a population "bottleneck": a small population—as low as 15,000 to 30,000 people—in the fourteenth century were the ancestors of all Ashkenazic Jews alive today. Thus the genomes of Ashkenazim are relatively similar to one another compared to those within other groups, which makes genetic analysis easier.

Unfortunately, however, the claim that Ashkenazic Jews represent a genetically unique population is historically problematic. Many scholars find the demographic data on which that claim is based suspect, and some historians question whether the Jewish population underwent the extreme expansions and contractions that the theory requires. Large percentages of the mutations found in Spanish, Dutch, German, Czech, Hungarian, Polish, and Greek women are the same mutations found in Ashkenazic Jewish women, which suggests that there is nothing uniquely "Jewish" about the mutations.[22] The use of mutations to identify ethnic groups leads researchers to label certain mutations as belonging to certain ethnic groups and then, when they are found in other ethnic groups, to try to find explanations as to how the favored group transferred those mutations to the others. This ethnocentric way of discussing mutations could, of course, be reversed. In a "mutation-centric" formulation, the mutation is the focus of attention, shared among many ethnic groups; Jews would be considered one population with a mutation, for example, rather than the mutation being "Jewish" (or Amish or Icelandic).

The importance of this issue was demonstrated in the resistance of many Jewish communities to genetic research in the 1990s. Tension arose between the Jewish tradition of valuing knowledge acquisition and medical advances, on the one hand, and the fear of stigmatization by having mutations labeled as "Jewish," on the other. Participating in early genetic research, it was feared, could have very practical adverse consequences. If the disease mutations that Ashkenazic Jews have in greater numbers were identified before the disease mutations of other groups, and if insurance companies were to consider those mutations as grounds for denying coverage or raising premiums, then Jews might be disproportionately penalized for their willingness to participate in research.

For these reasons and others, some Jewish lay and religious leaders suggested that Jews not volunteer for genetic research. The Boston Rabbinic leadership, for example, dissuaded populations from participating in a Dana-Farber Cancer Institute study in 1996.[23] Rabbi Moshe Tendler, a prominent modern Orthodox biologist and bioethicist, discouraged participation until legal protection against genetic discrimination was passed.

This concern culminated in a meeting organized by Hadassah in 1998, when Jewish leaders met with Francis Collins, head of the Human Genome Project, and Richard Klausner, then the director of the National Cancer Institute. In that meeting (which I attended), representatives of the Jewish community expressed their concerns, and Collins and Klausner promised to support and lobby for anti–genetic discrimination laws. Some representatives left the meeting still planning to encourage coreligionists to decline participation in genetic clinical trials.

The concern of Jewish leaders over discrimination has abated as genetic testing has become more common and as states have passed anti–genetic discrimination laws. The general public also seems less concerned. In a 2006 study of concerns about genetic discrimination in Jewish women, only 17 percent reported that they had significant worries about the issue, and only 13 percent believed it might lead to an increase in anti-Semitism. Women who were highly educated reported more concern than those who were less educated. The overwhelming majority (95 percent) felt that, all in all, research that focused on Jews was either neutral or positive.[24] Similar findings were reported in Israel.[25]

Genetic testing is a promising and important part of medical treatment. It will become increasingly important as new genes are found for susceptibility to a host of maladies that plague human beings. Ethnic groups in general, and Jews in particular, should not fear genetic testing or oppose it but should remain vigilant about the possibilities for discrimination or misuse of the tests. Jews have both a unique history and a set of unique religious tools to use to illuminate modern genetic challenges.[26] The lessons of Jewish history and the long tradition of Jewish thought about medical issues can provide valuable guidance as genetic tests become a common feature of medical treatment.

NOTES

1. R. Cole-Turner, "Genes and Genesis: Religion and Genetic Testing," *Bull Park Ridge Center* 13 (2000): 5-7.
2. G. Goldsand and Z. R. S. Rosenberg, "Bioethics for Clinicians 22: Jewish Bioethics," *Canadian Medical Association Journal*, 23 January 2001, 219-22.
3. J. Couzin-Frankel, "NIH Wants to Hear about Genetic Tests," AAAS *Science Insider*, 18 March 2010, http://news.sciencemag.org/scienceinsider/2010/03/nih-wants-to-hear-about-genetic-.html.
4. R. Park, "Genetic Testing Market on the Rise," IVD *Technology* 7, no. 9 (2001), http://www.devicelink.com/ivdt/archive/01/11/009.html (15 June 2007); Renub Research, www.renub.com Genetic Testing Market and Forecast—Global Analysis 2010-2015.
5. L. Remennick, "The Quest for the Perfect Baby: Why Do Israeli Women Seek Prenatal Genetic Testing?," *Sociology of Health & Illness* 28, no. 1 (2006): 21-53.
6. S. G. Stolberg, "Concern among Jews Is Heightened as Scientists Deepen Gene Studies," *New York Times*, 22 April 1998, 24A.
7. Elliot N. Dorff, "Jewish Theological and Moral Reflections on Genetic Screening: The Case of *BRCA1*," *Health Matrix Cleveland* 7, no. 1 (1997): 65-96.
8. Paul R. Wolpe, "Bioethics, the Genome, and the Jewish Body," *Conservative Judaism* 54, no. 3 (2002): 14-25.
9. Daniel Kurtzman, "Geneticist Conference Sheds Light on Jewish Cancer Genes," JWeekly.com, 1 May 1998, http://www.jweekly.com/article/full/8199/geneticist-conference-sheds-light-on-jewish-cancer-genes/.

10. K. H. Rothenberg and A. B. Rutkin, "Toward a Framework of Mutualism: The Jewish Community in Genetics Research," *Community Genetics* 1, no. 3 (1998): 148–53.

11. James M. Gordon, "Overview of Halakhic Issues," Center for Jewish Genetics (no date), http://www.jewishgenetics.org/overview-halakhic-issues.

12. Netty C. Gross, "When the Genes Don't Match," *Jerusalem Report* 17 (2005): 20.

13. Leslie Burnett et al., "The Tay-Sachs Disease Prevention Program in Australia: Sydney Pilot Study," *Medical Journal of Australia* 163, no. 6 (1995): 298–300; M. Levin, "Screening Jews and Genes: A Consideration of the Ethics of Genetic Screening within the Jewish Community: Challenges and Responses," *Genetic Testing* 3 (1999): 207–13.

14. Alexandra A. Gason et al., "Evaluation of a Tay-Sachs Disease Screening Program," *Clinical Genetics* 63 (2003): 386–92.

15. http://www.atlantajewishgenescreen.org.

16. D. J. Bowen, R. Singal, E. Eng, S. Crystal, and W. Burke, "Jewish Identity and Intentions to Obtain Breast Cancer Screening," *Cultural Diversity & Ethnic Minority Psychology* 9, no. 1 (2003): 79–87.

17. Judy Siegel-Itzkovich, "Israel Allows Sex Selection of Embryos for Non-medical Reasons," *British Medical Journal* 330 (2005): 1228.

18. Dorff, "Jewish Theological."

19. Elliot N. Dorff, *Matters of Life and Death* (Philadelphia: Jewish Publication Society, 1998).

20. Richard V. Grazi and Joel B. Wolowelsky, "Preimplantation Sex Selection and Genetic Screening in Contemporary Jewish Law and Ethics," *Journal of Assisted Reproductive Genetics* 9, no. 4 (1992): 318–22.

21. Yael Hashiloni-Dolev, "Between Mothers, Fetuses and Society: Reproductive Genetics in the Israeli-Jewish Content," *Nashim: A Journal of Jewish Women's Studies and Gender Issues* 12 (2006): 129–50.

22. Zoe H. Rosser et al., "Y-Chromosomal Diversity in Europe Is Clinical and Influenced Primarily by Geography, rather than by Language," *American Journal of Human Genetics* 67 (2000): 1526–43.

23. Stolberg, "Concern among Jews"; Richard Saltus, "Jewish Women's Group Warns of Risks of Cancer-Gene Testing," *Boston Globe*, 17 January 1997, B2.

24. L. S. Lehmann, Jane C. Weeks, N. Klar, and J. E. Garber, "A Population-Based Study of Ashkenazi Jewish Women's Attitudes toward Genetic Discrimination and *BRCA1/2* Testing," *Genetic Medicine* 4, no. 5 (2002): 346–52.

25. Remennick, "The Quest."

26. Paul R. Wolpe, "If I Am Only My Genes, What Am I? Genetic Essentialism and a Jewish Response," *Kennedy Institute of Ethics Journal* 7, no. 3 (1997): 213–30.

CHAPTER 12 *Jewish Genetic Decision Making and an Ethic of Care*

TOBY L. SCHONFELD

Consider the following case: Stephanie T. is thirteen weeks into her first pregnancy when she learns that both she and her partner are carriers of Tay-Sachs disease (TSD). This means that there is a 25 percent chance that the child she is carrying will have this disorder. Her obstetrician and a genetic counselor have discussed several options with her: (1) proceeding with the pregnancy with no changes; (2) terminating the pregnancy; (3) testing the fetus for TSD. Stephanie is considering where to go from here.

Cases like Stephanie's are increasingly common occurrences in health care. As advancing medical technology enables providers to be better predictors of future health concerns, patients find themselves in new territory when it comes to decision making. This information confers both benefits and burdens on the individual patient: benefits in terms of the preparation, planning, and (sometimes) peace of mind that increased information confers, but burdens with regard to the pressure both to obtain increased information and to make decisions based on these data.

Genetics researchers isolated the mechanism of TSD quite early in the history of the field. This development was particularly significant for the

Jews who trace their ancestry back to eastern Europe (Ashkenazic Jews), as TSD disproportionately affects this population. Unfortunately, the prognosis for children born with TSD has not advanced much in the forty-plus years that the genetic test has been available. As a result, patients like Stephanie who discover their carrier status after getting pregnant still have a very difficult choice to make: proceed with the pregnancy, terminate the pregnancy, or test the fetus to get more information (and then either continue or terminate the pregnancy).

Patients often consult a number of sources when making such important decisions. Discussions with clergy are frequently among such sources, both for assistance with understanding the religious precepts involved in such a decision as well as emotional and spiritual support in the decision-making process. And while texts on Jewish bioethics routinely include a statement recognizing the importance of this individualized contact, scholars nevertheless focus on the rules, principles, and precedent cases in the determination of an appropriate course of action.

Yet it is the individual patient who has a decision to make. When Stephanie consults with her rabbi about what to do, she needs more than simply a recitation of the relevant principles of Jewish law that should guide her decision. She also needs to understand how those features of her tradition coincide with the other values, goals, and priorities she has set for her life.

In this chapter, I will argue that using an ethic of care will help Stephanie understand how to order the various values, goals, and priorities in her life by incorporating both the tenets of her religious tradition and her other moral commitments. The primary vehicle for such incorporation will be the conversation she has with her rabbi, which will not only illuminate the important religious precepts relevant to the decision but also demonstrate how an ethic of care describes the pastoral process of counseling regarding medical decision making.

Tay-Sachs Disease 101

TSD was first described in 1881 by Warren Tay, an ophthalmologist who reported a "cherry-red spot" in the macula of the eye in a child who suffered from degeneration of the nervous system. Six years later, Bernard Sachs, a neurologist, described the clinical nature of the disease and its pathology. Sachs noted that the disease tended to run in families, and it was characterized by rapid neurological degeneration. He initially called this disorder amaurotic familial idiocy.[1] It was later renamed Tay-Sachs disease in honor of its discoverers.

TSD is an autosomal recessive disorder, which means that an individual must possess two copies of the affected gene to develop the disease.[2] Since we receive half of our genes from each parent, both parents must pass on a mutated gene in order for the offspring to develop a recessive disorder like TSD. Because the disease requires two copies of the faulty gene, carriers—those who have only one copy of the faulty gene—are not clinically affected.

Children with TSD appear normal for the first four to six months of their lives, after which time the neurologic degeneration becomes apparent. The specific cause of this disorder is an absence of the enzyme hexosa-minidase A (Hex A). The function of this enzyme is to break down *GM2* gangliosides, a specific lipid, and in the enzyme's absence, this fatty material accumulates in the neurons to such a degree that life's functions are impaired. Specifically, children develop "progressive mental and motor deterioration, blindness, paralysis, dementia, seizures and death," usually by three to five years of age.[3]

What makes this disorder particularly interesting for the Jewish community is the frequency of its occurrence. One in 3,600 Jewish children born has TSD, compared to 1 in 360,000 non-Jewish births.[4] While less than 0.3 percent of the general population are carriers for TSD, 3 percent of the Ashkenazic Jewish population are carriers.[5] Hence, the disorder is one hundred times more common in this population than in the international society at large.

Jewish Bioethics 101

The most common approach to Jewish bioethics uses a principle- or rule-based approach, and as such it is an example of what is sometimes called an ethic of justice. These deontological rules are based on certain biblical imperatives. In this case, four fundamental religious principles are relevant to the issue of genetic decision making. (1) The sanctity of human life and (2) the preservation of life and health require individuals to do anything possible to save a human life (with limited exceptions). This mandate to heal derives from three biblical sources where permission is given to individuals to heal illness. Some have interpreted that failing to act to save a life when one has the ability to do so results in a transgression. (3) The principle that the body belongs to God implies that individuals are required to take reasonable care and avoid any unnecessary harm. (4) Finally, the principle of mental anguish recognizes the connection between physical and mental health and requires the consideration of psychological as well as physiological outcomes of a proposed action.[6]

Typically, scholars apply these principles to the dilemmas involved in a case like Stephanie's by looking at the issues surrounding the decisions. Some of the questions traditionally discussed in relation to TSD include the following: Is abortion permissible (at all) in Jewish law? If so, does this situation qualify? How should we approach questions involving having and raising special needs children in the context of a tradition that mandates procreation? What is the value of genetic information in such a culture? Applying these four principles enables rabbis to give guidance about the (im)permissibility of various alternatives.

These principles serve as rough frameworks for guiding our action. As Elliot Dorff discusses, the principles of Jewish law help to establish a "moral bottom line."[7] Their use must still be specified according to a particular context, which is often accomplished by looking at previous cases that are similar in nature to the one under consideration, a process known as casuistry. In determining how we apply principles to a particular context, casuistic reasoning allows us to compare this situation to those that have come before. The wealth of Judaism lies in its rich heritage of oral and written traditions. It is by appealing to what has been historically decided and elucidating its similarities and differences to current dilemmas that this heritage remains relevant for contemporary thinkers.

Part of the application process involves knowing the particular patient's context; identification of the morally relevant features of the case requires an intimate knowledge of the individual's situation. That is at least part of the reason why texts on Jewish bioethics insist that individuals consult their own rabbis rather than rely solely on the information contained in the books: only your rabbi, who knows your situation, can properly advise you on the right course of action. Yet it is not clear how the rabbi is to do this, or why this approach is important.

I argue that this process is precisely what feminists have elucidated in an ethic of care. By attending to particular features of the patient's case, the rabbis are both caring for the individual and helping her to care for others. After an explanation of care theory, I will demonstrate how this works in practice by applying the elements of care theory to Stephanie's case.

An Ethic of Care

Care ethicists recognize that a decision like the one with which Stephanie is faced relies on more than just an understanding of the medical information and applying universal—religious or otherwise—principles to the situation. In fact, because so much of our moral life depends on subjective experience, care ethicists claim that conditions are rarely "sufficiently similar" to justify universal principles.[8] An ethic of care has as its focus the goal of preserving and maintaining relationships, as it recognizes that our moral lives consist in interactions with other people. We are all interrelated in important ways, and the recognition of this is an essential part of the moral life. Hence, care ethicists suggest that in order to make a morally sound decision, our thought process must include the others with whom we are in contact, and it must explicitly address how our actions will affect them.

A proposed course of action that makes even a few people with whom I am in a relationship worse off is a primary concern for me in my deliberations and may on balance prove to be the wrong decision. This at least partly arises because of the primary focus on the connection of individuals, rather than on the separation of one moral agent from another, as seems to be the focus of an ethic of justice. Those operating with the latter perspective "assume separation" from others and only later begin "to explore parameters of connection," while those who operate from within an ethic of care first "assume connection" and only later begin "to explore parameters of separation."[9] This is important because the very orientations lead us to different actions largely because they produce "different images of self and of relationships."[10] According to Carol Gilligan, a scholar often credited with being the founder of the ethic of care, the central insight in the ethic of care is that the self and other are interdependent.[11] It is this realization that leads to moral action. So, unlike the traditional ethic of justice, an ethic of care makes no sharp distinctions between self and other. "My day-to-day interactions with other persons create a web of reciprocal caring,"[12] and it is the recognition of this web that binds us to each other, that constrains and determines our moral action.

So what does it mean to care? Caring involves many things such that it is impossible, and undesirable, to make lists of criteria or categories of caring.[13] At the very least, caring requires some action by the one caring on behalf of the one who is cared for.[14] This points to an important realization: "Caring is not simply a cerebral concern, or a character trait, but the concern of living, active humans engaged in the processes of everyday living. Care is both a practice and a disposition."[15] The disposition evolves from what Nel Noddings, another feminist scholar, calls a "relation of natural caring," which is not simply a matter of projecting myself onto the object of care but rather involves an engrossment with the other.[16] Noddings compares this relationship to the one of mother and child, where the mother shares the infant's feelings, as is exemplified by the impulse to comfort the child even before she ascertains what is causing his distress. When this happens, when the individual becomes truly receptive to the other, Noddings argues that there is a fundamental motivational shift: my motivational energy is now shared with the other.[17] Once this shift has taken place, "the one-caring assumes a dual perspective and can see things from both her own pole and that of the cared-for."[18]

To be more precise, Joan Tronto and Berenice Fisher have developed four "analytically separate, but interconnected" phases of caring that speak to both the disposition and the actions necessary to maintain an ethic of care.[19] In the first phase, moral agents recognize that care is necessary and are attentive to that need. This is more difficult than it appears, as Lawrence Blum comments: "Understanding the needs, interests, and welfare of another person, and understanding the relationship between oneself and that other, requires a stance toward that person informed by care, love, empathy, compassion, and emotional sensitivity. It involves, for example, the ability to see the other as different in her own right, rather than viewing her through a simple projection of what one would feel if one were in her situation."[20] To care for others includes listening to what the other needs and recognizing that this may very well be something different from what I would assess that she needs based on my interpretation of the situation. Caring may require that I act very differently from what I would want in a situation or from what I think the one who is cared for *should* want. Ideally, if the motivational shift takes place, then my desire to meet the needs of the cared-for will arise naturally.

In the second phase of care, the agent takes responsibility for responding to the identified need. Caring is not simply a "cerebral concern" but instead "implicitly suggests that it will lead to some type of action. . . . [T]o care implies more than simply a passing interest or fancy but instead the acceptance of some form of burden."[21]

In the third phase, the moral agent actually meets the need of the cared-for. For this to work, one must be competent to give the kind of care necessary, or else the action is not really caring.[22] If all I do to meet the needs of the other is to offer the care that I would need in that situation because I do not have the ability or skills to offer what the cared-for really requires, then I am not caring; instead, I am doing something like "transferring." Furthermore, if all I do is offer money so that someone else can provide the goods or services, then I am also not truly caring. What I am doing is enabling someone else to care, and this is a commendable action in the instances when I am incompetent to give care myself. I should not, however, assume that this provision means that I am caring, because this is simply not the case.

In the final phase of care, the one who has received care will respond to the care received. It is this element that demonstrates the success of the caring, and it is this element that alerts us to a potential problem in caring if, for example, I have simply transferred my values and needs onto the person of the cared-for rather than cared for her as is appropriate for her in this case. So this element alerts us to insufficient or inadequate care as well.

By acknowledging these elements and completing the four phases of caring, Tronto and Fisher propose that we have completed an act of caring. Often this process is more difficult than it appears, and conflicts in caring do arise. For example, sometimes what the cared-for needs is in direct opposition to what someone else may need, and hence there is a moral dilemma. But when conflicts in caring arise, the approach to solving them is quite different from the procedures in traditional ethic of justice. In cases of a conflict, the agent "does not decide by formula, nor by a process of strict 'rational decision-making.' [Instead] she turns away from the abstract and back to the person for whom she cares. [A decision] is right or wrong according to how faithfully it was rooted in caring—that is, in a genuine response to the perceived needs of others."[23] So to evaluate whether or not a decision that I reached was the right one, I look less to the actual outcome of the decision and more to the process. As Noddings argues, "The test of my caring is not wholly in how things turn out; the primary test lies in an examination of what I considered, how fully I received the other, and whether the free pursuit of his projects is partly a result of the completion of my caring in him."[24]

Application of an Ethic of Care

The right decision for patients like Stephanie and her partner seems to be one that is responsive to the needs and desires of the people involved. They would need to consider what it would mean to care for a child like this, what it would mean to be in relationships with others in this situation, and, perhaps most importantly, what it would mean to care for each other in such a situation.

It is clear that Stephanie will consider how this situation will affect her relationship with her partner, but other relationships may also be important here, including those with her extended family, her colleagues, and her friends. We do not live in isolation, nor are we often successful at compartmentalizing some relationships to the exclusion of others. Recognizing the complex interactions that characterize lived experience will enable Stephanie to consider fully how any action may affect her web of caring.

It is a fact that different people would deal with Stephanie's dilemma in different ways. For some people, caring for a special needs child is simply more than they can bear, while for others, it is an opportunity to express their love to those who desperately need it. Stephanie would do well to consider not just her own feelings about having and raising a child with TSD but also those of her partner and her extended network. A choice that pleases her but makes others much worse off would be the wrong choice. Suppose, for example, that Stephanie's parents had planned to care for the child while Stephanie and her partner were at work. Caring for a child with TSD is completely different from caring for a child without this disorder. It is possible that Stephanie's parents would welcome this challenge, regardless—or maybe because—of the different opportunities that this situation presents. Yet it is also possible that, while they would still agree to care for the child, this new situation would present challenges beyond that which they could bear. They might struggle with becoming attached to a child they know will be taken from them, and this may result in feelings of resentment toward or distance from Stephanie and her partner. Only Stephanie can know the likelihood of either of these outcomes, because only she truly knows the people in her web of caring. Any choice, therefore, must include considerations of how this decision will affect others in her life.

But how does Stephanie do this? That is, even if we grant her commitment to considering the needs of those in her network, it is unclear how Stephanie can both assess these needs and also care for herself. I argue that this is where her rabbi comes in. Only part of the reason why individuals seek clergy counsel when making medical decisions has to do with understanding the religious laws, principles, and precedents related to the case. The other, equally important, function of the rabbi in these situations is in fact to listen to the patient and to help her consider various alternatives in light of her particular context. Recall again the textual caveat for individuals to seek guidance from *their* rabbi, who presumably has knowledge of the patient's relational situation. The rabbi's job is to help Stephanie process the caring needs of those in her network, as well as to discover how those individuals will, in turn, care for Stephanie.

One way the rabbi can do this is by attending to the four phases of caring described by Tronto and Fisher. The first phase has partly been facilitated by Stephanie's appointment with the rabbi; this action demonstrates the existence of a need for caring. Yet to go from here to attend to that need is much more difficult. However, the very skills that are essential for accomplishing this task are those for which the pastoral training of clergy prepares them: really listening to the patient and understanding her struggle as situated within her life and context, recognizing that this is a unique constellation of factors. The rabbi can help to identify resources for Stephanie no matter which decision she makes: resources to assist with the medical and physical aspects of caring, as well as resources for her social, emotional, and spiritual needs and the needs of those in her caring network.

Knowing whether or not the rabbi is successful in caring for Stephanie in this way can only be ascertained by the response from Stephanie herself. Attending carefully to Stephanie's response will alert the rabbi to the extent to which the rabbi accurately heard Stephanie's needs and whether or not more remains to be addressed in other ways.

Objections

Some would argue that other ethical theories, such as utilitarianism, could meet these needs in relation to the family in question. Utilitarians would take account of possibilities and risk, as well as the idea that different people evaluate different circumstances differently. Ascertaining the morally right action in a particular situation relies both on the consequences of such a decision on all those affected and a determination of how the proposed action fits in with our life plan or other goals. Hence, nonmoral considerations are relevant as well.

I will grant that the outcome, that is, the decision eventually chosen, may be the same for an ethic of care and utilitarianism in this case. What is different, however, is the process by which such a decision is reached. Consequences of a proposed action are not evaluated according to utility but rather according to how well an action maintains or continues the relevant relationships. To enter into a caring relationship is to take responsibility for meeting the needs of the other as well as meeting the needs of oneself, and a decision such as this is simply a concrete manifestation of the commitments made in the context of the relationship. Those who care have an obligation to meet the needs of the other in a way very different from that of utilitarians. In fact, the very life plan and values that we hold emerge from this commitment to caring, and hence the assessment of how best to meet these goals is an affirmation of the dedication we have to our moral ideals. To care genuinely involves more than simply choosing the best outcome but instead requires the engagement in a particular kind of process along the way.

It is this relationship of caring that is crucial for moral decision making. Recognizing what it takes to maintain the relationship between spouses will, and should, factor into a decision about whether or not to abort a pregnancy that will result in a child with TSD. Recognizing what it would mean to care for such a child is also important. Recall the four phases of caring and their associated ethical elements proposed by Tronto and Fisher. An individual must be able to work through all of the phases if she is truly to care for another.

The crucial point here is that not everyone will be competent to care for such a child. For those who are not, despite the best of intentions, it is, according to this schema, impossible for them to truly give care to the recipient. "When a 'carer' can no longer respond adequately to the demands in her sphere of activity, she has to turn to those adjacent to her in a chain of caring. The people to whom she addresses her request are then obligated to respond to her, just as she is obligated to respond to those who address her."[25] Noddings here does not specify what the response must entail but simply that those with whom the couple is engaged in caring must respond. It is conceivable that the caring response to the couple is the one that respects their decision to terminate a TSD pregnancy and then to act as a support system for them after making such a choice.

It is also important to remember that a crucial part of the caring relationship is the cared-for's response to the caring. This seems to have significant implications for children with TSD who, after a certain point, will not be able to respond to caring at all (except at the most rudimentary physical level). In such a circumstance, I argue that the truly caring response by parents may be not to subject a child to the pain associated with this degenerative disease by not bringing them into the world in the first place.[26] These considerations must be balanced with other considerations, to be sure, but it is an option that must be considered.

Finally, the duty to heal that can be derived from the requirement to preserve life and health in Judaism can be interpreted in a number of ways. With TSD, no "healing" in the traditional sense can take place; there is no cure for TSD, and physicians can do little to ease the discomforts of these children and their families. I contend that this duty to heal can be interpreted in a broader way to include caring and compassionate responses to individuals as a part of the process. Although we cannot offer a cure to the parents of children with TSD, we can care for them: we can offer our assistance in sitting with the child, in preparing meals for the family, in offering a sympathetic ear or a shoulder to cry on. Because we now know that spiritual and emotional healing are important in physical healing, it stands to reason that the duty to heal can and should be extended in this way.[27]

Because this demonstrates that healing includes nonphysical aspects, applying this expanded notion of healing to TSD is plausible. Caring for the parents of children afflicted with Tay-Sachs is a form of healing, and one that many (although not all) of us are competent to engage in.

So what is the bottom line for Stephanie in this case? Astute readers will notice that there is no clear implication for Stephanie's decision that can be inferred from the discussion. This is because in order to truly understand what the right decision is for her, we would need a much fuller picture of Stephanie's relational situation. Although this may be difficult for pedagogical purposes—it is hard to teach rabbis or others to extend care in its fullest sense—it in fact mirrors real life: the moral life is messy, and it is fraught with contradictions. What an ethic of care will allow us to do in a situation like Stephanie's is not just to identify the traditional religious considerations—the (im)permissibility of abortion, the value of genetic information, etc.—but also to locate those features of her lived experience that contribute to the assessment of a proposed action.

The fact that rabbis are in a good position to perform this act of caring presents another opportunity. Even those individuals who may relegate specific religious considerations to a lower position in a list of priorities for making a medical decision may still be ably assisted by their clergy members in the decision-making process. In today's health-care system, there are fewer and fewer opportunities for the kinds of conversations that are required to process a decision like Stephanie's, and so the role played by clergy becomes even more important. Having individuals well suited to the business of listening to and caring for patients on psychological and spiritual levels adds a dimension to the decision-making process that will ensure patients are making more authentic choices from within their own contextual framework. How Stephanie chooses to proceed, given the information about risk, will depend upon this context and her ability to assess both her needs for caring and the needs of those around her.

NOTES

1. Fred Rosner, "Tay-Sachs Disease: To Screen or Not to Screen?," *Journal of Religion and Health* 15 (1976): 271.
2. Autosomal disorders are to be contrasted with sex-linked disorders, where the affected gene is on one of the sex chromosomes. Autosomes are the twenty-two pairs of chromosomes that do not determine one's sex.
3. Thomas D. Gelehrter, Francis S. Collins, and David Ginsburg, *Principles of Medical Genetics*, 2nd ed. (Baltimore MD: Williams and Wilkins, 1998), 295; also Rosner, "Tay-Sachs Disease," 271.
4. Michael M. Kaback, Ted J. Nathan, and Susan Greenwald, "Tay-Sachs Disease: Heterozygote Screening and Prenatal Diagnosis—U.S. Experience and

World Perspective," in *Progress in Clinical and Biological Research*, ed. George J. Brewer, Vincent P. Eijsvoogel, Robert Grover, Kurt Hirschhorn, Seymour S. Kety, Sidney Udenfriend, and Jonathan W. Uhr, vol. 18, *Tay-Sachs Disease: Screening and Prevention*, ed. Michael M. Kaback (New York: Alan R. Liss, 1977), 13.

5. Ashkenazic Jews trace their ancestry back to central and eastern Europe; the vast majority of American Jews today (approximately six million) are of Ashkenazic descent.

6. For a much more detailed account of traditional Jewish bioethics, see Elliot N. Dorff, "Jewish Theological and Moral Reflections on Genetic Screening: The Case of *BRCA1*," *Health Matrix* 65 (1997); and his subsequent book, *Matters of Life and Death: A Jewish Approach to Modern Medical Ethics* (Philadelphia: Jewish Publication Society, 1998); Immanuel Jakobovits, *Jewish Medical Ethics* (New York: Bloch Publishing Company, 1975); Fred M. Rosner, *Modern Medicine and Jewish Ethics,* 2nd ed., rev. and enl. (New York: Yeshiva University Press, 1991); and Fred Rosner and J. David Bleich, *Jewish Bioethics* (New York: Hebrew Publishing Company, 1979).

7. Dorff, "Jewish Theological and Moral Reflections."

8. Nel Noddings, *Caring: A Feminine Approach to Ethics and Moral Education* (Berkeley: University of California Press, 1984), 5.

9. Carol Gilligan, *In a Different Voice* (Cambridge MA: Harvard University Press, 1982), 38.

10. Gilligan, *In a Different Voice*, 38.

11. Gilligan, *In a Different Voice*, 74.

12. Rita Manning, "Just Caring," in *Explorations in Feminist Ethics: Theory and Practice*, ed. Eve Browning Cole and Susan Coultrap-McQuinn (Bloomington: Indiana University Press, 1992), 46.

13. Noddings, *Caring*, 9-12.

14. Noddings, *Caring*, 10.

15. Joan C. Tronto, *Moral Boundaries: A Political Argument for an Ethic of Care* (New York: Routledge, 1993), 104.

16. Noddings, *Caring*, 30-31.

17. Noddings, *Caring*, 33.

18. Noddings, *Caring*, 63.

19. Tronto, *Moral Boundaries*, 106-8.

20. Lawrence A. Blum, "Gilligan and Kohlberg: Implications for Moral Theory," in *An Ethic of Care*, ed. Mary Jeanne Larrabee (New York: Routledge, 1993), 51.

21. Tronto, *Moral Boundaries*, 102-3.

22. Tronto, *Moral Boundaries*, 133.

23. Noddings, *Caring*, 53. Noddings here is referring to a specific (and realistic) example of a conflict between a woman caring for her child and caring for her husband. The needs of all the participants are important, and an assessment of all of those needs must be a part of the process.

24. Noddings, *Caring*, 81.

25. Nel Noddings, "In Defense of Caring," *Journal of Clinical Ethics* 3 (Spring 1992): 17.

26. It may also be possible to justify this kind of caring response simply on account of the suffering that the child would endure, regardless of considerations of his or her ability to respond to care. Because this action may contradict other important principles in Jewish bioethics that have not been discussed here, I will leave this consideration aside.

27. Of course, the Rabbis have long supported such a notion, and it is operationalized in the obligation to visit the sick. See Dorff, *Matters of Life and Death*, 255–59.

PART 4 *Genetic Intervention*

CHAPTER 13 *Summary of the Science of Genetic Intervention*

ELLIOT N. DORFF AND LAURIE ZOLOTH

Gene therapy was born of the hope that it may someday be possible to change mutated genes that lead to disease in humans. Rather than treating the symptoms of disease, the point would be to go to the cellular source of the disease, the coding mechanisms in the cell that regulate the proteins that cause cells to malfunction, die, or overgrow.

Genes are patterns of chemicals in a long molecule called DNA (deoxyribonucleic acid). These genes, or patterns of chemicals, are copied and inherited across generations. DNA is made of simple units that line up in a particular order within this long molecule. The order of these units carries genetic information, similar to how the order of letters on a page carries information. The language used by DNA is called the genetic code, which allows the genetic machinery to read the information in the genes. This information is the instructions for constructing and operating a living organism. DNA is a nucleic acid. RNA (ribonucleic acid) is another nucleic acid. Cellular organisms use messenger RNA (mRNA) to convey genetic information, and some RNA molecules play an active role within cells by catalyzing biological reactions, controlling gene expression, or sensing and communicating responses to cellular signals. These two acids, DNA and RNA, alongside proteins and carbohydrates, compose the three major macromolecules essential for all known forms of life.

Varying Attempts to Accomplish Genetic Intervention

Genetic intervention begins with the basic idea that it is genetic sequences or groups of molecules, made in different patterns and intensities, that drive the chemical reactions that make the proteins that are the catalysts of all life processes, from cell division, to metabolism, and to the complex electrochemical reactions that are the basis for behaviors. Genes, it was discovered in the 1970s, could be altered not only by deleting or altering them by exposing them to radiation, for example, but by "recombining" them with similar genes found in other species. Genes could be "infected" with a virus that carried a specific genetic sequence, and then this sequence would be inserted into the DNA of an organism, just as viruses are able to do in nature. (Our own DNA carries the mark of centuries of this sort of alteration, and this process of infecting one's DNA with a virus is the molecular description of how vaccinations work.) We can combine a human gene with a gene from another species because genes are highly conserved across species, which, simply put, means that the chemical codes that solve one sort of problem in simple cellular organisms—for example, how to digest food, or how to divide—are retained as one goes up the evolutionary chain to organisms that are more intricate. Genetic solutions that work are retained even if they are now "silenced" or unused by the organism in question.

The genome is the entirety of an organism's hereditary information—one's genetic makeup, as it were. The genome includes both the genes and the noncoding sequences of DNA (or, in many viruses, RNA). Whereas a *genome sequence* lists the order of every DNA base in a genome, a *genome map* identifies the landmarks. A genome map is less detailed than a genome sequence, just as a map of a geographic area that one might use to find one's way to a particular place does not include every house or tree, but precisely because a genome map is less detailed than a genome sequence, it aids in navigating around the genome, just as a regular map does.

Because, at the genome level, so much of the genomic map of all species is virtually identical, parts of the genome will function well if taken from one species to another. Once in the new species, these parts continue to function. A classic example is the Flavor-Saver Tomato, which, to resist freezing damage, uses the gene sequence that allows blood to circulate in fish that inhabit freezing water. Genes that make proteins that spiders use for webs are spliced into goats, using viral vectors to transport the genes, to allow goat milk to have similar properties, creating easily obtainable, extraordinarily strong threads, and important hormones for growth or blood clotting (HGH, or Factor H) can be made in *E. coli*, a bacterium, by human genes spliced into its sequence.

Although animals and plants are now routinely altered and then certain ones of them are selected and bred to produce many more of the species with desired inherited traits, such as the Flavor-Saver Tomato mentioned above, humans cannot morally be altered, selected, and bred. That would be eugenics in its worst form. Thus, the goal of human genetic alteration, as first conceived in contemporary scientific research, was instead to somehow use viral vectors to transport genes into people whose own genes had ceased to function or who were lacking the correct version of the genes to make a protein necessary for normal human functioning. The aim was, in other words, to cure debilitating diseases by changing the genes that were causing them into genes that instead enabled a person to function normally.

In his presentation to the contributors of this volume, Professor Theodore Friedmann of Moores Cancer Center at the University of California at San Diego noted that the early years of gene therapy were marked by a sense that medicine would be swiftly altered once the genetic basis for disease could be found, but "the hopes of patients and their advocates were disappointed." He stated that the slow but eventual success of a genetic therapeutic approach is still unfolding, through slow and careful basic research on signaling, proteomics, silencing RNA, and ontological biology. Despite early ethical objections, the weight of medical urgency and the "obvious correctness" of the approach as a concept led to an outflow of research on genetic modification, the application of gene transfer to clinical uses, the direct correlation of specific mutations to particular diseases, the replacement of a diseased gene with a "wild type," and the means to deliver and "correct the defect." Gene transfer was logical. There had been a revolution in the recombinant technology, a rush of identification of genes responsible for many diseases, and the ability to clone genes to study and produce them, and so the relationship between genes and pathogenesis (the origins of disease) was becoming clearer.

Human gene *therapy*, however, proved elusive, for getting genes into the cells of the body is hard, maintaining the proper placing of the insertion difficult, and knowing how much virus to introduce unknown. At first and for many years, the main route to alter genes was not a manual insertion; it was to use the infectious properties of viruses, which naturally insert genetic material into a cell's DNA. Scientists linked a useful thread of DNA to a virus that successfully invades human cells. But genes can replicate in unexpected ways and overwhelm the immune system, and viruses can insert genes into odd and unexpected places in the human genome, causing other serious diseases in the process of trying to cure one. In one otherwise successful attempt to cure a fatal childhood disease called X-linked SCIDS (X-linked severe combined immunodeficiency disease syndrome), for example, the child developed leukemia after, and presumably because of, genetic intervention therapy used to cure his SCIDS.

Friedmann himself played a role in the history of gene transfer that he described as "Cassandra-like" in that he continually asserted that clinical success was likely to be elusive, but it was not until 1995 that the first publications began to document what had become apparent over a twenty-year narrative: gene transfer was difficult to achieve and to control. Transferred genes' actions were often diminished and transient, and they produced only a scant effect on the underlying pathophysiology. The director of the National Institutes of Health, Harold Varmus, funded two reports that documented this problem, and he then took the unprecedented step of rebuking a scientific experiment in process after a fatal tragedy.[1] In 1999 Jesse Gelsinger, an eighteen-year-old outpatient enrolled in a gene transfer study at the University of Pennsylvania, died through numerous errors, including uncertainty about his understanding of the nature of the risk, clinical issues, and problems in basic science. His widely publicized death closed all such trials for months. As mentioned above, shortly thereafter, a French and British study used gene transfer to treat X-linked SCIDS, and while initially successful for twelve patients, it caused fatal cancer in one and recurrent leukemia in another.[2]

These setbacks, together with decades of basic research, "have slowly raised the project to the level required of all clinical research: a balance of honestly assessed benefits and risks." Friedmann noted that a life-threatening genetic disease has, in the SCIDS case, been effectively treated and perhaps even cured by gene transfer. Yet recently much of the trajectory has been marked by politics, speculation, fantasy, and recrimination about the idea of "designer babies" that has not been warranted. In the last several years, however, as the human genome was more fully understood and then mapped and as new technologies became possible, the field changed again.

Sophisticated molecular biology, using these techniques, created new possibilities. Professor Leor Weinberger of the Institute of Engineering in Medicine at the University of California at San Diego described to the members of our group a project whose aim was to move to another stage in gene transfer by transferring the genes of one strain of virus into another, more easily transmissible strain. This idea extended the traditional account of gene transfer in several directions. First, it expanded the concept to infectious disease, not only inborn errors in metabolism in the human host. Next, it attempted to use the power of a virus to defeat a virus using the tools of molecular biology (gene transfer) and the premise of evolutionary biology. Finally, it attempted to address a core problem of modernity, the new plague of AIDS.

AIDS, as a disease, ravages the poorest in the world. In Africa, eastern Europe, India, and Southeast Asia, AIDS compounds the ongoing public health crisis by attacking young adults, the core of the economy, leaving the elderly and orphans behind. Human immunodeficiency virus (HIV) kills by first infecting a person, then becoming dormant for years until the slowly growing viral load accumulates and grows rapidly. In the early 1990s, HAART (highly active antiretroviral therapy) was developed that suppressed the capacity of the virus to replicate and thus to continually challenge the CD4 T cells in the immune system. (It is the exhaustion of these cells that is deadly in AIDS, as opportunistic infections then overwhelm the body once HIV has destroyed its defensive system.) However, long-term HAART is highly problematic. It is toxic and has side effects, and, worst of all, HIV simply mutates and defeats the regime in about five years.

Weinberger's idea was to convert virus-producing cells into a pseudolatent state rather than trying to purge the reservoirs of virus in infected individuals, similar to how some vaccinations use weakened forms of a virus to immunize a person from more virulent strains of the same virus. The project works by introducing modified viral vector cells, crHIV-1, which contain only regulatory gene motifs but none of the necessary elements for viral packaging; instead, they carry an antiviral gene that inhibits any of numerous wild type HIV-1 functions. A cell that is infected with crHIV-1, then, if it becomes infected with HIV, faces an antiviral vector "payload" and thus will reproduce only crHIV. This only works in cells that already are HIV positive, thereby reducing vector toxicity concerns and blocking mutation because the vectors have the potential to encode for inhibitors of cellular factors needed for such replication. It was his hope that the weakened virus of the model crHIV would evolve to mutate reciprocally to match any mutation that wild-type HIV might make. Thus it could destroy the toxicity of HIV in any of the many forms to which the virus might mutate to escape traditional HIV therapy. Still, noted Weinberger, the idea of gene transfer in infectious disease carries a new sort of risk—an altered form of a pathogen introduced into an infected person. The weakened form of the HIV virus could be transmitted between individuals, and that could be either a potential overall good—spreading a weakened form of the virus to immunize many more people from a lethal form of HIV—or a factor of unknown risk.

However, in the last few years, gene therapy suddenly accelerated and seemed to yield results, and it did so under the very careful supervision of the Recombinant-DNA Advisory Committee (RAC) of the National Institutes of Health. One of the editors of this volume (Zoloth) serves on that national committee, and the unfolding science continues. Human trials are now in their second phase; the first successfully treated babies who have not gotten X-linked SCIDS because they were treated with gene therapy are growing, and, despite the early leukemias, they are alive. Genetic therapeutic interventions in cancer, in blindness, and in other devastating genetic conditions are now beginning to take shape as larger trials are being built on the slow success of early phase one trials. The human leukocyte antigen (HLA) molecule controls part of the immune system. Nonviral vectors that target it directly or that use synthetic molecular "fingers" to alter its genetic expression, or single molecular drugs that target the leukocyte antigen with precision, are now being developed. This idea has rapidly moved into human trials. The success of gene therapy and its normalization have led to a growing acceptance of its clinical use.

In the months just prior to the final editing of this volume, there has been another remarkable advance: CRISPR. Since 2012, the CRISPR/Cas system has been used for gene editing (silencing, enhancing, or changing specific genes) in mice and nonhuman primates. After the insertion of a plasmid containing cas genes and specifically designed CRISPRs, an organism's genome can be cut at any desired location. The first evidence that CRISPR can reverse disease symptoms in living animals was demonstrated in March 2014, when MIT researchers cured mice of a rare liver disorder. Editas Medicine, a forty-three-million-dollar start-up, aims to develop treatments that employ CRISPR/Cas to make edits to single base pairs and larger stretches of DNA. Inherited diseases such as cystic fibrosis, sickle cell anemia, and Huntington's disease are caused by single base pair mutations; CRISPR/Cas technology has the potential to correct these errors. The "corrected" gene remains in its normal location on its chromosome, which preserves the way the cell normally activates or inhibits its expression.

Thus, in the period of time during which this volume was produced, we who watch science have witnessed the transformation of a discipline from speculation to clinical intervention, with new technologies unfolding as we write.

Ethical Questions about Genetic Intervention

The first general ethical question of gene therapy—apart from a sense that it might not work at all—was that if it could be perfected, it could be used to alter all sorts of codes, including ones that are not technically disease mutations. This might allow for the creation of advantages, or enhancements. Should our ability to change genes be used only for therapy or also for enhancements? Although some intended changes are clearly therapies to cure diseases (e.g., cancer), and, on the other end of the spectrum, some are clearly enhancements (e.g., to increase intelligence), there is a broad middle of that spectrum in which it is not clear whether a given intervention is therapy or enhancement. In those cases, how should we define the difference? Does it matter? That is, are enhancements equally as moral as therapies? If not all enhancements are acceptable but some are, which ones are? Why? Finally, who in society should benefit from these interventions, if and when they become available for human beings? Everyone? Only those who can afford the intervention? Should we use some other criterion?

In the early 1970s, molecular biologists themselves as well as ethicists began to be concerned about the implications of such a powerful technology and met at a California state park called Asilomar to discuss the rules that would govern the research. The Asilomar conference established a structure of oversight: a national committee of experts and public members, a set of clinical guidelines that would limit the venues for research, and a set of "bright lines" for the research. Research on human "germ line" genes (i.e., those of the reproductive system) would not be permitted because that would affect not only the present organism but all of its progeny, and research with the goal of mere enhancement would not be supported. All research would be done in level 4 (i.e., the most secure) containment facilities, and there would be a moratorium on certain research until the safety considerations could be arranged and verified. All research in recombinant DNA on humans then and now is vetted by the national board that was established as a result of the recommendations of that conference.

By the 1980s, after a full ten years of debate, a moratorium on the use of *E. coli* in altered form to provide biological platforms for the transfer of genes was considered, as newly efficient viral vectors and other methods were discovered. In animals and cell cultures, the newly introduced "wild type," or normal genes, could restore missing gene function and reverse the deadly metabolic damage done by defective genes.

One significant ethical problem at the heart of gene transfer stems precisely from the fact that we do not and cannot know the risks involved in using genetic mutations to cure genetic diseases. It is not the only challenge. In 1999, the American Association for the Advancement of Science concluded work on the ethics of inheritable genetic alteration. Here, a groups of ethicists, including Laurie Zoloth, the coeditor of this volume, together with scientists, physicians, theologians, and lawyers, reflected on the implications of a human species capable of transforming itself by altering genes not just in a singular human person, and for fatal illnesses, but in the entire lineage of the person so treated. The ethical concern was that this "unprecedented choice, unprecedented power" would tempt individuals to make a choice to enhance traits in an unjust way. Fears included "two classes of persons, enhanced and unenhanced, based on class, race, or privilege"; "the creation of warriors, genetically bred"; or the always-mentioned "designer babies."

We have moved on from considerations that vexed the early years of genetic intervention. New technologies continue to emerge, however, such as the idea of swapping mitochondrial DNA from a donor and adding it to an egg from a mother with mDNA disease, and they rekindle our anxieties about designing our descendants. Given the still very investigatory state of the science and the regulatory barriers to human subject experimentation, such fears seemed then, and still, to be designed for a future that is still extraordinarily distant.

The moral issues described above, though—namely, the distinction, if there is one, between therapy and enhancement; the moral status of trying to induce a genetic change in either one individual, that individual's progeny, or an entire population; and the criteria for determining who will be the beneficiaries of genetic interventions when they become available—are still very much worthy of consideration, even as we continue to provide reflection on the new and remarkable experiments being proposed for diseases once thought unalterable. The chapters in this section of the volume offer some reflections on these issues from various Jewish perspectives, which become imperative questions as the technology increases in power and certainty.

NOTES

1. J. Couzin and J. Kaiser, "As Gelsinger Case Ends, Gene Therapy Suffers Another Blow," *Science*, 18 February 2005, 1028.
2. D. B. Kohn, M. Sadelain, and J. C. Glorioso, "Occurrence of Leukaemia Following Gene Therapy of X-Linked SCID," *National Review of Cancer* 3, no. 7 (July 2003): 477–88.

CHAPTER 14 *Some Jewish Thoughts on Genetic Enhancement*

SHIMON GLICK

It must be stated at the very outset of the discussion that the following presentation is largely theoretical, certainly as far as germ line modification is concerned. At the present time, the techniques for germ line modification are very far from the consideration of their use in humans by ethical scientists. As Francis Collins stated to President George W. Bush's Council on Bioethics regarding the status of necessary safety techniques, "We are nowhere near meeting [the requisite safety requirements] and I think will not be for the foreseeable future."[1] The techniques are primitive, even for somatic cell modification; the potential benefits and certainly the possible side effects are unknown.

In the state of present knowledge, the changes that are meant to be passed on to future generations would essentially be irreversible. These limitations are valid even for germ cell modifications that are meant to eliminate serious disease states. Therefore, there is virtually unanimous opinion among scientists, no less than among ethicists, that any attempt at germ cell modification in normal individuals in order to enhance any of their attributes is absolutely unethical and beyond the pale.

There have been hundreds of articles written about the ethics of enhancement and its implications, in contrast to the treatment of disease. The world cannot permit itself to ignore the infamous era of eugenics in the first half of the twentieth century. This virus, which in its most malignant and extreme form, of course, infected Nazi Germany but did not leave democratic Western countries such as the United States and the Scandinavian countries unscathed, has left its scars—appropriately so. On the other hand, the trauma of this unfortunate era must not yield to a paralyzing neurosis whereby blanket prohibition of genetic modification becomes the norm.

In this chapter I will deal mainly with the theoretical aspects of the issue, since, as I have pointed out, the current state of the science justifies a blanket prohibition of all somatic cell manipulations and, all the more so, all genetic germ cell manipulations. For the sake of the analysis, I will use a thought experiment with certain assumptions that for now may be science fiction and may or may not ever be realistic. I will assume that science has advanced to the stage at which germ cell genetic modification has been perfected and scientists can manipulate the genes safely and predictably. Let us also assume for the sake of the thought experiment that data have accumulated over several generations. Furthermore, techniques have now been developed to enable reversal of the genetic modification if for whatever reason we choose to do so or if the second or third generation so desires. Germ line modification for disease prevention has now become an acceptable and safe procedure. We now face the question of whether and to what extent we should permit genetic manipulation for the sake of enhancement.

Distinguishing Therapy versus Enhancement

The expressed opposition to the use of genetic manipulation for the sake of enhancement is influenced by several factors, only some of which I will mention. First, there is the tragic history of the misuse of genetics based in part on a gross misunderstanding of science and subsequently on evil racism that reached its ethical nadir during the Nazi regime. Second, imaginative writers from Aldous Huxley on have depicted all kinds of potential misuse of genetics in the future as a result of massive application and misuse of genetic modifications. Novels such as *The Boys from Brazil* that fantasize on the creation of clones of Nazis have cast suspicion on all genetic manipulation. Third, misuse of steroids and other drugs in athletic contests, which epitomizes the unethical aspects of unfair competition more than the question of enhancement per se, while unrelated to genetics, has had the side effect of exaggerating the fear of genetic enhancement that might also be used, for example, for creating athletes and other individuals with abilities well beyond the current human norms.

It is important not to fall prey to genetic *exceptionalism*, a term that describes an emotional response to almost all aspects of genetics in a kind of neurotic reaction that treats genetics emotionally rather than dispassionately.[2] An example of this kind of thinking occurred several years ago at a meeting of the Israeli national ethics committee dealing with the use of preimplantation genetic diagnosis (PGD) to eliminate a variety of diseases during in vitro fertilization (IVF). Some ethicists insisted that the Ministry of Health draw up guidelines about the specific diseases and/or malformations for which PGD would be permitted or forbidden. This proposal was considered seriously in a society and at a time when abortions are permitted through most of pregnancy for almost any trivial fetal malformation and by the same individuals who raised no similar demands about abortions. Thus, abortion, which by any standard represents a more serious ethical problem, was unchallenged, while PGD, because of the emotional attitude toward genetics, was questioned.

One of the major issues is whether we can clearly differentiate enhancement from disease treatment. If we can do so, should we adopt a policy under which only the latter should be permitted? I believe that the answer is negative on both issues. Let us take an example in daily use that is widely accepted by almost all. Immunizations clearly are enhancements of human natural abilities, but they do prevent disease, and so they have been accepted by almost all ethicists. Western medical practice encourages behavioral norms and medications that definitely prolong life by preventing atherosclerosis and thus coronary artery disease. One might conceivably develop genetic modifications that would affect the width and pliability of coronary arteries or increase the synthesis of high density cholesterol, the so-called good cholesterol. The effect would be a deferment of atherosclerosis and coronary artery disease, but this would clearly result from an enhancement well beyond the normal state.

In all likelihood, major depressive disorders have a genetic origin, and those who would accept genetic modification for disease treatment might therefore accept gene manipulation to prevent serious depressive disorders. But within the spectrum of normality there are also enormous variations in response to stress and life crises. Aaron Antonovsky, impressed by individuals who had demonstrated an unusual resilience to adversity, such as concentration camp internment, coined the term *salutogenesis* (in contrast to *pathogenesis*) to describe the factors predisposing to health rather than disease. He explained salutogenesis as a "sense of coherence," a feeling of meaning in life in terms of environmental and perhaps cultural factors.[3]

But there is now evidence that there may be genetic factors that predispose individuals to react in a healthy or unhealthy way to stress. Recent work by Avshalom Caspi and colleagues has shown in a study of 847 young people in New Zealand that individuals with two long alleles for a molecule (5-HTT) that affects the transmission of serotonin were half as likely to develop depression in response to life events as were those with two short alleles.[4] Thus, conceivably, by alteration of genes one might create individuals who can handle life stresses better, which is surely an enhancement that may prevent disease (suicide) and suffering. These results confirmed earlier animal experiments.[5]

Caspi and colleagues made similar findings with respect to the interaction of childhood abuse and a gene for an enzyme (*MAOA*) involved in the regulation of several neurotransmitters.[6] The chances of a boy exhibiting subsequent antisocial behavior as an adult was affected jointly by his genotype and by his early childhood environment.

Enhancement of intelligence has been one of those aspects feared by opponents of genetic manipulation. But there are increasing data that enhancement of intelligence within the normal range may result in extending life expectancy. A group of researchers from Scotland recently presented preliminary data suggesting that childhood IQ affects adult mortality.[7] It has also been shown that higher cognitive function in middle age, as measured by three tests, is associated with lowered mortality, a result confirming previously reported data in the elderly by several investigators.[8]

A number of reports presented at a meeting of the International Society for Intelligence Research suggested that higher IQ was associated with greater success in terms of health, longevity, and daily functioning.[9] The reasons for these results are unclear and are probably varied. But again, what at first glance seems merely enhancement can readily be shown to be a form of treatment in that death and disability are prevented or delayed.

At a meeting of the Council on Bioethics, the staff working paper on distinguishing therapy and enhancement pointed out that "the distinction between therapy and enhancement is hard to articulate for three principal reasons: (1) they are not mutually exclusive, (2) the activity involved is often the same, and (3) the standard of health and 'improvement' against which the difference between therapy and enhancement might be measured can be very hard to define."[10] The authors nevertheless maintain that it is important to distinguish the two, but they seem unable to provide the necessary criteria for this distinction. Increasingly, the idea that enhancement and treatment cannot be clearly distinguished is achieving wider acceptance,[11] and it is becoming clear that such a distinction in any but the most egregious examples of enhancement is extremely difficult.

Enhancement as a Goal of Medicine

But even if we were to draw a line between treatment and enhancement, it is not at all clear that medicine's role must be confined to disease treatment. That would, in my opinion, be an inappropriately narrow view of medicine's role, which is really to enhance health. Health promotion is increasingly seen as part of the physician's role. The World Health Organization's definition of health certainly is extraordinarily broad and clearly would encourage actions by physicians that could easily encompass enhancement.

Enhancement has often been described pejoratively in terms of a father who wants his child to be eight feet tall because he aspires to have a son who will be a star basketball player or by a variety of other such examples whose results might well restrict the child's life opportunities in order to direct him or her into the specific pathway chosen by the parents. But enhancement, if used thoughtfully, could much more likely be used, for example, to improve a child's musical talent in a family that has little such ability, or to improve a child's IQ from low normal to high normal, or to enhance physical coordination. All of these various enhancements not only would improve the quality of the child's life but might even prolong life and prevent disease, and they would broaden the child's opportunities rather than narrow them. Making life easier is liberating rather than confining and restricting.

Another report prepared for the Council on Bioethics summarized the issue succinctly: "Relying on the distinction between therapy and enhancement to do the work of moral judgment will not succeed."[12] Just as in current medical practice, in which a great deal of treatment could just as readily be categorized as enhancement, there are enhancements that most would accept as reasonable and other kinds of enhancement that many would label as unreasonable and undesirable. But enhancement in and of itself should not be regarded as a pejorative word.

Certainly, within the American ethos the twentieth century has witnessed what Sheila M. Rothman and David J. Rothman have called "the pursuit of perfection."[13] While it is true that this pursuit is perhaps most strongly apparent in the United States, Françoise Baylis and Jason S. Robert argue convincingly that the use of genetic enhancement technologies is inevitable because of a variety of reasons, but mostly because of our inherent and universal human tendency to strive to be masters of our human evolutionary future.[14] They therefore contend that, rather than engage in futile efforts to stop genetic enhancement techniques, ethicists would do better to try to influence the directions that these techniques will take and ensure proper oversight of these techniques.

Enhancement from a Jewish Perspective

All of the above is written from a general ethical point of view. What might be a specifically Jewish point of view, and would it be more or less positively inclined? What are the goals and duties in life of human beings according to the Torah? Clearly, these goals and duties are to serve God and to act in a way that glorifies his name. Actions that enhance a person's ability to accomplish these goals would be laudable.

Maimonides and a whole series of Jewish scholars and rabbis were physicians, and the medical profession has for centuries had an unusual attractiveness for Jews. Prevention of disease or health promotion was always primary in the traditional definition of the role of medicine.

In the Torah, the Lord promises that "the diseases that I brought on Egypt I will not bring unto you, for I am the Lord your healer,"[15] clearly defining the role of the physician as one who carries out preventive medicine. In his many medical writings, Maimonides places major emphasis on disease prevention and health promotion, and there is no clear demarcation anywhere in the Jewish tradition between treatment and prevention. The latter is to be preferred, and only if it fails does the physician have to treat.

Whereas interference with nature and playing God are some of the objections raised to genetic enhancement, these views find little support in Judaism. Indeed, man is commanded to be a cocreator with his Creator in many areas of endeavor. So, for example, "Fill the earth and conquer it" is interpreted in Jewish sources as a mandate to activism on the part of man.[16] The sentence in Genesis about the creation of the world reads "which the Lord has created, to do."[17] The latter two words are interpreted as a statement that God has left the creation unfinished with the expectation that man is to expend energy and thought to "complete the job," as it were.

In the Talmud and midrashim, there are a number of discussions whereby rabbis use circumcision, bread and wine making, and the prescription of medications as examples of man's accomplishments in the area of human creativity.[18] These activities are seen as a form of *imitatio*

Dei, examples of the command to follow the example of the Creator. In addition, there are references to the fact that in the afterlife, man will be called to account if he has not exploited his abilities for improving the world (*tikkun olam*) through such acts as transacting business honestly, fixing time for Torah study, and establishing a family.[19]

Rabbi Judah Loew (Maharal from Prague) stated: "The creativity of people is greater than nature. When God created in the six days of creation the laws of nature, the simple and the complex, and finished creating the world, there remained additional power to create anew, just as people can create new animal species through inter-species breeding. . . . People bring to fruition things that are not found in nature; nonetheless, since these are activities that occur through nature, it is as if it entered the world to be created."[20] In other words, human creativity is inherent in the creation of the world and is a positive and appropriate activity.

Rabbi Joseph B. Soloveichik, an important twentieth-century Orthodox rabbi, posits that creativity is a human responsibility and not a heavenly monopoly. It is forbidden for man to be merely a passive observer of the universe and its problems. He writes,

> Dignity of man expressing itself in the awareness of being responsible and of being capable of discharging his responsibility cannot be realized as long as he has not gained mastery over his environment. . . . Man of old who could not fight disease and succumbed in multitudes to yellow fever or any other plague with degrading helplessness could not lay claim to dignity. Only the man who builds hospitals, discovers therapeutic techniques, and saves lives is blessed with dignity. . . . Civilized man has gained limited control of nature and has become, in certain respects, her master, and with mastery, he has attained dignity, as well. His mastery has made it possible for him to act in accordance with his responsibility.[21]

In the words of Ecclesiastes, "Man cannot conquer the spirit to confine it."[22] We cannot limit our curiosity and desire for knowledge.

But as the wise writer of Proverbs tells us, "Fortunate is the man who is ever fearful."[23] There is, of course, an ever-present caveat, that in all man's endeavors, he must recognize his limitations; he must make certain that imitation of God does not lead to self-worship and hubris, as occurred in the biblical narrative of the tower of Babel. Man is also commanded to act within the bounds of what is permitted according to the moral code of Judaism and toward goals that are in keeping with the spirit as well as the letter of the law. But it is important to emphasize that in the Jewish tradition that which is not expressly forbidden is generally permitted. Permissiveness is the "default mode."

Jewish sources from time immemorial have recognized the existence of genetic influences in both humans and animals. In the biblical narratives, Jacob tried to influence the qualities of the sheep he tended by genetic manipulation.[24] The Midrash tells us of the striking resemblance of Isaac to Abraham to belie the charges that Abraham was not Isaac's father.[25] There are references that strongly encourage taking precautions to include in one's considerations genetic factors when choosing a mate. A tall person is cautioned not to marry another tall individual for fear of giving birth to a giant.[26] Similar admonitions are given with respect to other physical features. One is to avoid matings that may lead to significant deviations from norms of appearance. One is encouraged to choose a mate positively, with consideration given to positive family features. Particularly, one is advised to check the qualities of the potential wife's brothers, because the offspring of the mating are likely to resemble them.[27] Scholars made great efforts to seek out the very best students in the academy for their daughters. It has been suggested that some of the high intellectual achievements among Jews occurred because the most highly intelligent and educated talmudic scholars had many children to fulfill the biblical mandate, and their children married the children of other highly talented scholars. In contrast, some of the best minds in Christian society in the Middle Ages, the priests, remained celibate by choice.

In an attempt to influence the nature of one's offspring, it is related in the Talmud that Rabbi Yohanan, a handsome talmudic sage, would station himself outside the *mikveh* (ritual bath) so that women emerging from the *mikveh* would see him, and this would (in the beliefs of that time) influence the nature of their offspring conceived on the night following the ritual immersion.[28] Here we see a clear example of an attempt to manipulate the genetic pattern in a woman similar to Jacob's attempt with the sheep in the Bible. Obviously, this episode bears no resemblance whatever to manipulation of genes, but within the knowledge and beliefs of that time we are presented with a positive attitude toward influencing the qualities of a fetus rather than a passive acceptance of the result of natural events.

In Judaism, one has an obligation of *gemilut hesed*, to help others, to make their life more pleasant, to reduce suffering. These efforts fall under the rubric of the Torah's command to "Love your neighbor as yourself."[29] Although it might seem unreasonable to mandate such an obligation to a yet nonexistent entity, to future generations, there is no reason not to consider such an act as meritorious. In the hierarchy of good deeds, the highest is *hesed shel emet* (an act of true loving-kindness), which the Rabbis maintain is done by attending a funeral and preparing the body for burial, for there is no chance for the dead person to repay the favor.[30] I might suggest that in a similar vein, making an effort to help future generations might be called *hesed shel emet*, since, in general, they cannot repay the individual who brought about the improvement in their lives.

Parents have obligations toward their offspring to provide not only basic sustenance, education, and so on but also the tools they need to support themselves and to thrive in the world. Swimming lessons and the teaching of an occupation are specifically mandated.[31] In the Jewish tradition, grandchildren are often regarded as equivalent to children. Thus, while again it is hard to find a specific mandate to enhance the opportunities of generations to come beyond one's children, it would seem that such an action would be praiseworthy.

How does *halakhah* (Jewish law) look upon enhancement in everyday nongenetic medicine? As is well known, *pikku'ah nefesh* (saving of life) is one of the highest values in Judaism, and one is not permitted to risk one's life unnecessarily. Thus, the question has been asked about the permissibility of activities, such as cosmetic surgery, that are not strictly therapeutic and have an element of risk. In many cases, cosmetic surgery has been permitted by leading rabbis, for example, to enhance one's chances of finding a marriage partner or to find employment,[32] because *halakhah* has a broad view of medicine's role and human welfare. This attitude is in keeping with classic Jewish sources such as Maimonides, whose view of medicine encompasses both mental and physical health promotion as a means to a cure.[33]

The data cited earlier suggest, for example, that raising intelligence from low normal to high normal has significant impacts on health and on successful functioning within and on adapting to society to the point of lengthening life. It would also be expected that enhancement of other physical, intellectual, and emotional qualities would probably add health and longevity to humans and would certainly make their life more pleasant. Therefore, these enhancements should be looked upon favorably by the *halakhah*.

Objections have been raised by some that the use of genetic manipulation, like the use of drugs, to achieve stress reduction and more desirable behavior is somehow undesirable, whereas achieving those same goals by sustained human effort is to be commended. One might even suggest that this differentiation can find support in Judaism. For example, the study of the Torah is one of the major duties of every Jew. Not only is study commendable because the knowledge of the Torah in all of its ramifications improves one's ethics and morals, but there is additional great value attached to the *amal*—the struggle and effort. One is specifically commanded to maximize effort in one's studies.[34] It might be argued that by drug or genetic manipulation, thereby making the achievement of the goals easier, one is bypassing the *amal*, and this should be interdicted.

My personal view is that by making the task easier, one merely sets the standards higher, so that the *amal* will be at a more sophisticated level, because there is almost no limit to human aspiration and ability to accomplish in the field of human knowledge and Torah study. Introduction of the printing press by Gutenberg and the computer in our lives has made Torah study much easier. Surely no one would argue that easing the struggle of research by these innovations should be looked upon unfavorably. These innovations merely freed the scholars to devote their efforts to more productive, though no less difficult, activities. So too enhancement of one's intelligence, of one's ability to concentrate or remember, would be a positive accomplishment.

Another argument against genetic enhancement is that it may create two societies: the genetically enhanced, who will naturally come from those segments of society that are already advantaged, the so-called genobility, and the disadvantaged, who will not benefit from this expensive manipulation and will be doomed to be a permanent underclass. This is a serious concern, but such a result is not inherent to the technique but rather results from society's application of the technique. The same charge can be made in almost every area of human creativity and endeavor. The advantaged will in all likelihood be the ones who will be able to take advantage of innovations, as they have almost always been. Thus, it is incumbent on society to see that its resources are distributed equitably, but that is not a reason to deny progress. Theoretically, one could envision a truly just society that would preferentially provide these new services to the disadvantaged in a sort of "headstart" program or affirmative action.

In summary, if indeed the safety and reversibility issues in genetic engineering are solved, there should be no inherent banning of the use of these techniques for enhancement, any more than for treatment. Just as with conventional medicine today, each application of genetics should be evaluated on its merits—its goals, dangers, and consequences.

NOTES

1. Francis S. Collins, "Session 5: Genetic Enhancements: Current and Future Prospects" President's Council on Bioethics, http://www.bioethics.gov/transcripts /dec02/session5.html (25 May 2007).
2. Michael J. Green and Jeffrey R. Botkin, "'Genetic Exceptionalism' in Medicine: Clarifying the Differences between Genetic and Nongenetic Tests," *Annals of Internal Medicine* 138, no. 7 (2003): 571–75.
3. Aaron Antonovsky, "A Call for a New Question—Salutogenesis—and a Proposed Answer—the Sense of Coherence," *Journal of Preventative Psychiatry* 2 (1984): 1–13.

4. Avshalom Caspi et al., "Influence of Life Stress on Depression: Moderation by a Polymorphism in the 5-HTT Gene," *Science*, 18 July 2003, 386–89.

5. Aaron Antonovsky, "The Sense of Coherence as a Determinant of Health," in *Behavioral Health: A Handbook of Health Enhancement and Disease Prevention*, ed. Joseph D. Matarazzo et al. (New York: Wiley, 1984), 114–29.

6. Avshalom Caspi et al., "Role of Genotype in the Cycle of Violence in Maltreated Children," *Science*, 2 August 2002, 851–54.

7. Carole L. Hart et al., "Childhood IQ, Social Class, Deprivation, and Their Relationships with Mortality and Morbidity in Later Life: Prospective Observational Study Linking the Scottish Mental Survey 1932 and the Midspan Studies," *Psychosomatic Medicine* 65 (2003): 877–83.

8. Valory N. Pavlik et al., "Relation between Cognitive Function and Mortality in Middle-Aged Adults: The Atherosclerosis Risk in Communities Study," *American Journal of Epidemiology* 157 (2003): 327–34.

9. Constance Holden, "The Practical Benefits of General Intelligence," *Science*, 10 January 2003, 192–93.

10. Staff Working Paper 7, "Distinguishing Therapy and Enhancement" President's Council on Bioethics, April 2002, http://www.bioethics.gov/background/workpaper7.html (8 June 2007).

11. David B. Resnik, "The Moral Significance of the Therapy-Enhancement Distinction in Human Genetics," *Cambridge Quarterly of Healthcare Ethics* 9 (2000): 365–77; Arthur Caplan, "Is Better Best?," *Scientific American*, September 2003, 104–5.

12. "Human Cloning and Human Dignity: An Ethical Inquiry. Chapter 1: Biotechnology and the Pursuit of Happiness: An Introduction," President's Council on Bioethics, http://www.bioethics.gov/reports/cloningreport/appendix.html (14 June 2007).

13. Sheila M. Rothman and David J. Rothman, *The Pursuit of Perfection* (New York: Pantheon Books, 2003).

14. Francoise Baylis and Jason S. Robert, "The Inevitability of Genetic Enhancement Technologies," *Bioethics* 18, no. 1 (2004): 1–26.

15. Exodus 15:26.

16. Genesis 1:28.

17. Genesis 2:3.

18. Circumcision and bread and wine making: Midrash Tanhuma-Buber (5) Tazria on Leviticus 12:3. The prescription of medicines: A. Y. Karelitz (Hazon Ish), *Letters*, vol. 1, #136.

19. B. *Shabbat* 31a.

20. Judah Loew of Prague (Maharal M'Prague) *Be'er Ha-golah*, 38–39 (Jerusalem 5731).

21. Joseph B. Soloveichik, "The Lonely Man of Faith," *Tradition* 2, no. 7 (1965): 13–14.

22. Ecclesiastes 8:8. The following sentence is a homiletical interpretation of the verse. In it its context, it means, "No man has authority over the lifebreath to hold back the lifebreath [from leaving the body when the time comes to die]; there is no authority over the day of death" (Jewish Publication Society translation in *JPS Hebrew–English TANAKH* [Philadelphia: Jewish Publication Society, 1999], 1777).

23. Proverbs 28:14.

24. Genesis 30:37–43.

25. Rashi on Genesis 25:19.

26. B. *Berakhot* 45b.

27. B. *Bava Batra* 110a.

28. B. *Berakhot* 20a.

29. Leviticus 19:18.

30. *Genesis Rabbah* 96:5.

31. B. *Kiddushin* 29a.

32. S. Z. Auerbach, *Minhat Shelomoh* (Tenina) 6:3.

33. Maimonides, *Treatise on Asthma*.

34. Rashi on Leviticus 26:3.

CHAPTER 15 *Curing Disease and Enhancing Traits*

A Philosophical (and Jewish) Perspective

RONALD M. GREEN

To produce clear and justified guidelines as to whether we may use genetics only to cure diseases or also to enhance abilities or traits, we must first specify what we mean by "disease." In this chapter I will use a definition suggested by one of my colleagues at Dartmouth, explain the importance of its various clauses, and then apply it to the question of whether enhancement through gene therapy is morally justifiable.

The Need to Define "Disease"

A person has a malady if and only if he has a condition, other than his rational beliefs and desires, such that he is suffering, or is at increased risk of suffering, a harm or an evil—namely, death, pain (physical or psychological), disability, loss of freedom or loss of pleasure—and there is no sustaining cause that is distinct from the person.[1]

The complete sequencing of the human genome in 2003 represents the end of one phase of genomic investigation and the start of another. With the full sequence in hand, scientists can now link features of the genome to observable human characteristics. This will provide a powerful tool for studying the genetic factors that contribute to disease and for understanding the biological causes of variation among human beings and between humans and other species.

As science launches into this important effort, it is important that we possess a clear definition of what we mean by "disease." The concept of disease is crucial for medicine, ethics, law, and public policy.[2] The curing or treatment of disease is widely believed to be the principal goal or "end" of medical care.[3] When a person suffers a disease, she is declared to be "ill," "infirm," or "sick" and may come under the care of medical practitioners. The treatment of an illness may be covered by medical insurance, whether private or public, whereas nondisease conditions, such as cosmetic wants, are usually not covered. Illness may require someone to cease work, as when airline pilots are asked to stop flying if they suffer conditions that impair their ability to perform the job. Illness can also entitle someone to exemption from the ordinary responsibilities of his work. Mental illnesses can be an excusing factor in criminal prosecutions. For all these reasons, it is important to have an accurate, objective, and sustainable understanding of what constitutes a disease. A poor definition can lead to misuses of medical care, unjustified discrimination, and exclusion from social opportunities, or, in the opposite direction, unjustified exemptions from social responsibilities.

The philosophical and social science literature is rife with attempts to define disease.[4] Some of these definitions are questionable. For example, students of the issue have noted that societies sometimes regard as healthy or normal conditions states that other societies view as diseased. Thus, some nonliterate traditions venerate as holy men people who experience repeated hallucinatory or seizure behavior, conditions treated in our society as illnesses. On the basis of these differences, some conclude that disease is nothing more than a "socially constructed" concept. Disease *is* whatever a society says it is.[5]

There are several serious problems with this approach. As a description of how societies view disease, it is clearly an exaggeration. From the fact that there is some disagreement about some disease conditions, these observers conclude that there is disagreement about *all* such conditions. But this ignores the fact that most disease conditions are the same across societies and everywhere are regarded as grounds for whatever medical or therapeutic interventions the society offers. Fevers, serious digestive disorders, childbirth problems, broken limbs, skin ailments, and the like almost everywhere are seen as conditions that invite medical care.

This approach is also questionable because it opens the way to the abuse of medical care for purposes of oppression or discrimination. During the Soviet era, it was common for political dissidents to be labeled by authorities as suffering from one or another invented psychiatric disorder and to be incarcerated in mental hospitals as a result. ("Sluggish schizophrenia" was a favorite diagnosis for someone who failed to conform to the rules of Communist society.)[6] In the United States, homosexuals were frequently subjected to cruel "therapeutic" regimens, including castration and electroshock therapy, in order to "cure" their so-called mental illness.[7] If disease is defined as whatever a society says it is, we lose the possibility of criticizing these inappropriate interventions as misutilizations of medical care.

Some years ago, a group of scholars at Dartmouth College led by philosophers Bernard Gert and Dan Clouser and bioethicist Charles M. Culver worked to develop a rigorous definition of disease that drew on cross-culturally shared intuitions and was able to command wide assent. Recognizing that the term "disease" does not adequately embrace all the conditions we think are suitable for medical treatment, such as traumatic injuries (broken bones or burns), the team chose to reintroduce the somewhat archaic term "malady." This also covers the host of "illnesses" for which the term "disease" seems too extreme (e.g., headaches, colds), as well as the variety of mental illnesses.

In what follows I briefly outline the definition of a "malady" developed by this team and try to explain each of the key terms or phrases in the brief definition offered at the start of my discussion. Following this, I will indicate the usefulness of the definition in helping us understand the meaning and moral importance of the distinction between treatments aimed at preventing, alleviating, or curing maladies and those that aim at enhancing nondisease states or conditions. Throughout, I will try to apply these concepts to the area of genetic medicine. Finally, I will explore some of the ways that this definition mirrors Jewish thinking about health and disease.

The Elements of a "Malady"

1. A CONDITION OF THE PERSON. First, according to the definition, a malady is a "condition" of a person. This means that it is an indwelling problem of the person's body. Although it may have been initiated by external causes, it is a condition that now somehow resides within a person's body. A malarial environment is not yet a disease. But if a person becomes infected with the malaria parasite following a mosquito bite and she is likely to suffer from fever and other manifestations of the infection, she has a malady (as we will see, this is true even if she is not yet symptomatic). Not all evil states are maladies. Poverty, war, social injustice, and environmental pollution are all conditions that can lead to people suffering maladies, but only certain adverse conditions that they cause *within* persons are properly termed maladies.

2. OTHER THAN HIS RATIONAL BELIEFS AND DESIRES. I say "certain adverse conditions" because the definition indicates that not all conditions residing within a person are maladies. A person's rational beliefs and desires are part of her mental makeup, and they can be a cause of increased risk of suffering harms or evils. But such rational mental states are not maladies. If I believe that current economic conditions are likely to lead to a sharp decline in the stock market, and I have substantial stock holdings, I will probably suffer great mental distress. If I have the wish to climb K-2, I am at a much higher risk of suffering serious injury. But it is entirely rational to hold such beliefs and have such desires, and I do not have a malady because of them.

On the other hand, if I believe, despite all evidence, that I am about to be abducted and tortured by aliens from Mars and I suffer distress, I have a malady—a mental illness. This belief is irrational, because its falsity is obvious to almost everyone with similar knowledge and intelligence.[8] I am also (mentally) ill if I act on irrational desires, such as the wish to incessantly and perilously lose weight even though I am not fat. People who suffer from anorexia have a malady. They have irrational desires that greatly increase their chances of suffering harms and evils.[9]

3. SUFFERING, OR AN INCREASED RISK OF SUFFERING, A HARM OR AN EVIL. This is a key feature of the definition. Maladies are everywhere regarded as something that people normally want to prevent, alleviate, or cure, because having a malady involves suffering or being at increased risks of suffering harms or evils. Harms and evils are further defined here as involving "death, pain (physical or psychological), disability, loss of freedom or loss of pleasure." The fact that these are universal harms or evils, harms or evils that human beings almost everywhere normally wish to avoid, is shown by the fact that these are routinely used by societies as punishments for, or deterrents to, wrongful conduct. Thus, when a person commits an offense, he may be punished by loss of life (capital punishment), physical pain (lashes or whipping), imposed disability (the cutting off of a hand for thievery), incarceration, or the deprivation of pleasure (as when we send a naughty child to bed without dessert).

It is important to note that someone has a malady not only when, as a result of an indwelling condition, she is suffering any of these harms or evils but also when an indwelling condition puts her at higher risk of doing so. Thus, high blood pressure is a malady—and hence a medically treated condition—even though persons with hypertension may, for long periods, not experience any adverse symptoms of the disease. But it is the nature of this condition that it significantly increases one's risk of suffering a stroke, heart attack, or other deadly or disabling condition.

This is a key point to bear in mind as we consider genetic diseases. The new powers of human genome analysis will probably make possible the early identification of many genetic conditions that cause disease but may not evidence themselves for many years in disease symptomology. This is already true of Huntington's disease (HD), a condition that can be detected as early as the fetal stage but that may not show its grievous symptoms of neurological decline until the fourth or fifth decade of life. New tests for inherited breast cancer are similar. Nevertheless, according to the definition of "malady" being offered here, persons with the HD or breast cancer mutation have a genetic malady, even though they are not yet sick. In practical terms, this means they have a condition suitably subject to medical care, surveillance, and whatever preventive therapies are available. A woman in her twenties who tests positive for the breast cancer *BRCA1* mutation and elects to have both her breasts surgically removed to reduce her risk of cancer is making a decision that fits fully within the scope of medical treatment.

I must say something now about the terms "increased likelihood" and "disability" embedded within this definition. "Increased" raises the question of "over what"? Here, Culver, Clouser, and Gert reply, "Over what is normal for the species." A person has a malady if he has an indwelling condition that greatly increases his likelihood of suffering harms or evils over what is true for the overwhelming majority of human beings. Thus, a person has a malady if she has a severe allergy to cat dander. Most human beings do not suffer from this problem, which is a malady because it increases one's likelihood (over others) of suffering pain, decreases one's freedom (one cannot go into a room where there are cats), and deprives one of pleasure (cat lovers like me would say "great pleasure"). However, a person does not have a malady if she chokes and has breathing problems in a smoke-filled burning room. Although a small number of human beings may be more resistant to the airborne toxins of fire, the vast majority of us are not. A person caught in a burning building has a problem, but it is not a malady.

Similarly, "disability," according to this definition, is the lack of an ability that the overwhelming majority of human beings possess in the prime years of life. If I am blind or deaf, or if I am paraplegic, I have a disabling malady. I do not have a malady, however, if I am unable to run a four-minute mile or regularly win at grandmaster chess. My lack of ability here is not a disability. This point will become relevant shortly when I turn to the issue of genetic enhancement.

As I signal the centrality of suffering harms and evils for this definition of a "malady," it is important to reiterate that suffering (or increased likelihood of suffering) harms or evils is a necessary condition of having a malady. Some efforts to define disease have taken as their cue the idea that disease is a "deviation from the natural functional organization of the species" or a departure from "species-typical normal functioning."[10] This definition is superficially attractive because many physiological or mental abnormalities do lead to increased suffering and because normality plays a role in measuring the increased likelihood of suffering or establishing the presence of disability. But it is important to stress that abnormalities that do not cause suffering or increase the risk of doing so are not maladies. Left-handedness is not a malady. Freckles and red hair are not maladies. Homosexual orientation is not a malady, although for years it has wrongly been treated as such by the medical profession. If we are to believe Kinsey's statistics, homosexual orientation is abnormal, in the sense that only a minority of human beings prefers same-sex sexual partners. But in the absence of external social discrimination and oppression (which, we will soon see, are best thought of as "distinct sustaining causes"), being gay does not seem to increase one's likelihood of suffering harms or evils.

The centrality of suffering harms and evils to this definition is, again, well borne in mind as we enter the era of genomic medicine. Genome research will certainly lead to the discovery of myriad variations in the genome sequence among individuals and among ethnic groups. Some of these variations, such as the one that causes Tay-Sachs disease, are genetic maladies because they greatly increase the likelihood of death. But other variations are not linked to harms or evils, such as the many single nucleotide polymorphisms (SNPs) already used for forensic identification purposes. As genomic research proceeds, we may find that ethnicities characterized by long histories of endogamy, such as the Jews, have common and distinctive genomic sequences that distinguish them in particular ways from the majority of other human beings. But unless these distinctive features are clearly associated with increased likelihood of suffering evils or harms, they are not diseases. Regarding them as diseases because of a misleading definition will only lend medical force to the stigmatization that has too long been a part of Jewish experience.

4. DISTINCT FROM THE PERSON. Finally, this definition excludes from the class of maladies conditions for which there is a distinct, external sustaining cause. This is a cause that, once it is removed, ends the condition of suffering. If I have you in a wrestling hammer lock, you are likely to suffer pain. You certainly suffer reduced freedom of movement. Nevertheless, my wrestling hold is not, for you, a malady. There is a distinct sustaining cause for the suffering and loss of freedom. If I release you, your pain vanishes, and your mobility immediately returns. Of course, it is true that something that starts with a distinct sustaining cause may lead to a malady. If I grasp you hard enough, I may pull your tendons, causing an indwelling condition that increases your suffering of pain. Then you have a malady.

Similarly, homosexuality is not a malady. It is a condition that produces suffering almost entirely as a consequence of social oppression and persecution, all of which are "distinct sustaining causes." Remove these causes, let gay people lead free and open lives without stigmatization, and nothing suggests that they cannot be as happy as heterosexuals. For the same reason, Jewish identity is not a malady, even though it has been historically associated with a greater risk of suffering evils. Nevertheless, it is true that after years of oppression, some gay people and some members of religious or racial minorities come to experience indwelling psychological states (psychological disorders) that cause them to suffer. They then have psychological maladies, although their original condition is not itself the cause.

Treatment versus Enhancement

There are many nuances of this definition that I cannot explore and many questions it raises that I cannot answer here. I recommend that the interested reader explore the original articles and chapters on which I draw. Nevertheless, I think I have given enough of an idea of this definition to suggest its implications both for the emerging issue of genetic enhancement and its bearing on Jewish thought.

One of the great values of this definition is that it affords us a clear distinction between genetic medicine directed at relief or cure of genetic diseases (maladies) and genetic enhancements. The former aim at indwelling conditions of the person that either cause or increase the individual's likelihood of suffering death, pain, disability, or the loss of pleasure. Inherited breast cancer, Tay-Sachs disease, Duchenne muscular dystrophy, and Fanconi anemia are all genetic diseases in this sense and are so even when diagnosed presymptomatically by means of genetic analysis. They are all diseases that are highly prevalent in Ashkenazi Jewish communities, where efforts have been made to reduce their incidence by means of various programs of carrier, prenatal, and adult testing.

Genetic enhancements, in contrast, are medical interventions that do not aim at treating or curing a malady. They do not reduce the risk of suffering evils that an individual faces because of some indwelling physical or mental abnormality. Instead, they aim at *improving* a normal person's physical condition over and above what is normal for human community. This said, however, it is very important to recognize that enhancements come in two forms, the implications of which are very different for medical care.

The first kind of enhancement aims at the prevention of disease. A good example is the polio vaccination. Most human beings are highly susceptible to infection by the polio virus, although a few probably have natural immunity through previous (mild) exposure or for other reasons. Vaccination thus represents an improvement in biological resistance over what is normal for the species. But vaccination has long been recognized as a natural and even laudable aspect of medicine because it aims at the prevention of disease. It draws its moral acceptability from its ability to reduce the incidence of the universally recognized harms and evils that figure in our definition of malady. The same degree of acceptability will likely apply to genetic enhancements that have this same effect. For example, it is recognized that a very small percentage of human beings lack the CD4 receptor that enables the AIDS virus to enter cells. These people are naturally immune to infection. If it were possible to use gene therapy to convey the same immunity to the many men, women, and children now exposed to HIV AIDS, and if the genetic vaccine were reasonably safe, I doubt that anyone would object to doing so.[11]

The second kind of enhancement does not aim at the cure or prevention of disease. It seeks *improvements* in bodily status that confer added benefits on the recipient. An example is the use of blood doping to enable a good marathon runner to achieve the level of an Olympic medalist. So, too, is an intervention designed to raise a child's IQ from 110 to 140. Neither the condition of a good distance runner or a child with a near normal IQ is a malady. Neither individual suffers a disability in the sense of lacking an ability possessed by the overwhelming majority of people. The enhancement in this case does not cure or prevent evils. It provides a benefit.

Some people believe that this second kind of enhancement, an improvement aimed at providing an added benefit, represents an inappropriate or morally questionable use of medical resources. They insist that medicine should confine itself to preventing, alleviating, or curing diseases or maladies.[12] But it does not follow from the definition I have outlined that this is always true. It is important to distinguish between the definition of a malady, which has important implications for medical care, and the moral decisions about the larger uses of medicine. The definition does inform us that medical care draws its importance and nearly universal acceptability from its promise of reducing or avoiding the evils and harms associated with maladies. Since nearly all reasonable persons want to avoid these evils (unless they have a compelling reason based on the avoidance of similar evils),[13] it is quite understandable that effective medical care is widely accepted and highly valued and that medical judgments have great weight in social and even legal contexts.

The centrality of evils and harms to this analysis also tells us that enhancements become morally questionable when they threaten to create an increased likelihood of suffering death, pain, or disability. This is one of the reasons we tend to look askance on the use of performance-enhancing drugs in sports, many of which can have serious health consequences for the athletes involved.[14] It also partly explains the widespread concern about the use of risky silicone breast implants for purely cosmetic purposes.[15] And it underlies the persistent worry that genetic enhancements may themselves become the cause of proliferating iatrogenic (medically induced) genetic diseases.[16] Some enhancements, such as the use of steroids in athletics, only raise the risks for all competitors without, in the end, really providing competitive advantage for those who utilize them.[17] This is a serious problem that may characterize many envisioned human enhancements such as greater height or reduced need for sleep.

Nevertheless, although this definition explains the priority that reasonable people place on avoiding maladies over the pursuit of risky and less universally valued benefits, including those that may someday be achieved by genetic enhancements, the definition alone cannot rule enhancements out. Since enhancements confer benefits, and since it is ordinarily wrong to deny people the opportunity to pursue benefits unless there is a good reason for doing so, we should expect that we would permit medical enhancements and allow medical practitioners to provide them when the perceived risks are minor. Liposuction and Botox injections are examples. Although we usually do not regard such treatments as suitable for insurance reimbursement (a measure of a treatment's moral requiredness), neither do we prohibit people from spending their own money for them (a measure of their moral allowability).

Maladies and Jewish Thought

It is remarkable how many features of Jewish thinking about health and disease conform to the basic insights within this definition. This mutually supportive relationship both upholds the definition and testifies to the rational nature of Jewish ethical teaching in this area.

Foremost among these features is the high priority Jewish thinking has always placed on preventing and curing disease. There is a tendency in all religions that affirm God's sovereignty to adopt a passive or fatalistic attitude to misfortunes. Biblical texts describing God as the "healer" (e.g., Jeremiah 17:14) might have reinforced the tendency to interpret health and sickness as entirely in God's hands. A sectarian movement within Judaism, the Karaites, actually adopted this stance, holding that "God alone should be sought as physician, and no human medicine should be resorted to."[18] Nevertheless, the Karaites were isolated in taking this stance. Normative rabbinical Judaism always affirmed in the strongest terms human responsibility for the promotion of health and the prevention or cure of disease.[19] The firm rabbinical teaching over the centuries was that, while in biblical times God may have directly caused and removed disease, "nowadays one must not rely on miracles."[20] Medical care was always valued, and well-known rabbis like Maimonides often had distinguished medical careers. Although God is the ultimate healer, he works his will through human agency.[21]

The recognition that disease means suffering (or the increased likelihood of suffering) death, pain, and disability and Judaism's inherent opposition to unnecessary suffering contributed to the Jewish insistence that virtually all the commandments of Torah must be set aside to save a human life. This is the solemn obligation of *pikku'ah nefesh*, the duty to save a human life. If necessary, Sabbath restrictions must be set aside to provide lifesaving medical care,[22] and the rabbi of a community who did not make this known is regarded as derelict in his duty.[23] Individuals who caused injury to others are also required to furnish healing medical services.[24] Although Jewish teaching views suffering as potentially ennobling and edifying, as, for example, in the fast days of the Jewish liturgical year, Judaism never made the deliberate pursuit of suffering a religious goal, as ascetic Christianity sometimes did. In conformity with our definition's understanding that death, pain, disability, and the loss of pleasure are evils, Judaism perceived the struggle against disease and injury as a sacred task.

In the modern era, these traditional teachings and the attitudes associated with them have shaped Jewish responses to innovations in genetic medicine. Jews of all denominational backgrounds have been early adopters of prenatal testing as a way of avoiding the birth of children with serious genetic defects. This response is supported by Jewish teachings about the lesser moral status of the early embryo or fetus.[25] Genetic testing also finds acceptance among ultra-Orthodox groups, which, though less willing to resort to genetic abortion, have used preconceptional carrier testing, as in the Dor Yeshorim program,[26] to dramatically reduce the incidence of genetic disease in their communities. The point is that Jews everywhere, by virtue of their historic commitment to the avoidance of the evils associated with disease, have eagerly accepted the information and tools of genetic medicine. This has been true despite the bitter experience of modern Jewry with the abuses of genetic medicine in the Nazi period. And it has been true despite the growing number of disability rights activists, feminists, and others who have criticized prenatal testing as unjust and discriminatory. Ordinarily, we might expect that their historic opposition to injustice and discrimination would lead Jews to give these criticisms a hearing, but the deeply rooted themes in Jewish teaching of resistance to the evils of death, pain, and disability have proved stronger.

We might assume that the resistance to these evils would also render Jewish thought resistant to risky elective medical interventions, including possible medical enhancements. By and large this is true. For example, the Talmud requires that circumcision be deferred or not performed on an infant belonging to a family with members who have been known to bleed profusely or heal poorly.[27] Commentators have noted that this is possibly an early recognition of the presence of inherited hemophilia. If we regard circumcision as a religiously motivated "enhancement," as the midrash suggests,[28] then the message here is that the prevention of death, pain, and disability takes precedence over improvements.

Yet, in all other circumstances, circumcision is not just permitted, it is required. As the Jewish bioethicist Moshe Tendler notes, circumcision is the original Jewish medical enhancement.[29] It attests to the historic Jewish belief that the physical and biological form of humanity was not complete at creation and that human beings, guided by God's law, can actually participate in and improve on God's creation.

It is a leap, of course, from circumcision to genetic enhancements. But we can imagine that in the future, Jewish thinkers, instructed by the normative tradition, will permit genetic enhancements so long as these do not entail significantly increased risks of suffering death, disease, pain, or disability. This will almost certainly be true of what I have called "preventive enhancements," interventions such as genetic vaccines against cancer or infectious diseases that improve a person's resistance to these maladies above and beyond what is normal for the species. Unless the risks of these therapies are very high, I cannot imagine Jewish thinkers opposing them. Since Judaism does not venerate the created order per se and has always perceived human beings as God's cocreators who are appointed to fulfill God's creative intention, it is likely that Jewish thinkers will also support modest and relatively risk-free cosmetic enhancements that confer benefits on children, such as stronger and whiter teeth and a reduced tendency to obesity.[30]

Good definitions are central to the work of philosophy and ethics. They are most helpful when they build on and help clarify the ways that people actually use and understand terms. The task of crafting definitions aims at a kind of "reflective equilibrium,"[31] whereby the definition seeks to conform well to our widely held intuitions about a term while it helps us apply the term reliably to new circumstances and issues.

The definition of "malady" that I have presented here has these qualities. It embodies features that are embedded in our common use of terms such as "disease," "injury," and "illness," and it explains the power and uses of these terms. It explains to us why "illness" is an excusing condition in many circumstances and why "disease" is something worth marshaling social resources to fight. This is because both conditions entail suffering of evils that we all regard as serious. It further allows us to apply this concept in new ways to new issues, helping us to draw the lines between the treatment of genetic diseases, the introduction of preventive genetic enhancements, and the development of genetic enhancements that improve our condition but have nothing to do with disease.

I have suggested that Jewish teaching about disease and health offers support for this definition. Proceeding from sound moral intuitions, Jewish teachers over the ages effectively identified and expressed the same concerns and priorities resident in this modern definition. Their work thus reinforces that of the philosophers. At the same time, Jews can benefit from this definition. They should not hesitate to draw on it as they struggle in the future to adapt Jewish teaching to surprising new developments in genetic medicine.

NOTES

1. Charles M. Culver, "The Concept of Genetic Malady," in *Morality and the New Genetics*, ed. B. Gert et al. (Boston: Jones & Bartlett, 1996), 147. Bernard Gert, Charles M. Culver, and K. Danner Clouser offer another extensive discussion of this definition in their *Bioethics: A Return to Fundamentals* (New York: Oxford University Press, 1997), chap. 5.

2. Glenn McGee, "Ethical Issues in Enhancement: An Introduction," *Cambridge Quarterly of Healthcare Ethics* 9 (2000): 299–303.

3. Daniel Callahan, "The Goals of Medicine: Setting New Priorities" (special supplement), *Hastings Center Report* 25, no. 6 (1996): S1–S26.

4. A full bibliography on this topic would fill pages. In addition to the previously mentioned discussions by Clouser, Culver, and Gert, the following are some leading treatments of the issue: Christopher Boorse, "On the Distinction between Disease and Illness," *Philosophy and Public Affairs* 5, no. 1 (1975): 49–68; H. Tristram Englehardt Jr., "The Concepts of Health and Disease," in *Concepts of Health and Disease: Interdisciplinary Perspectives*, ed. Arthur L. Caplan, H. Tristram Englehardt, and J. J. McCartney (Chicago: Addison-Wesley, 1981), 31–46; Joseph Margolis, "The Concept of Disease," *Journal of Medicine and Philosophy* 1, no. 3 (1976): 238–55; W. Miller Brown, "On Defining Disease," *Journal of Medicine and Philosophy* 10 (1985): 311–28; Eric Juengst, "Can Enhancement Be Distinguished from Prevention in Genetic Medicine," *Journal of Medicine and Philosophy* (1997): 125–42; Norman Daniels, "Normal Functioning and the Treatment-Enhancement Distinction," *Cambridge Quarterly of Healthcare Ethics* 9 (2000): 309–22.

5. Henry Sigerist, *Civilization and Disease* (Chicago: University of Chicago Press, 1943); John Lachs, "Grand Dreams of Perfect People," *Cambridge Quarterly of Healthcare Ethics* 9 (2000): 327; Sara Goering, "Therapies and the Pursuit of a Better Human," *Cambridge Quarterly of Healthcare Ethics* 9 (2000): 334. For a discussion of this "social construction" perspective, see H. Tristram Engelhardt, "Disease of Masturbation: Values and the Concept of Disease," *Bulletin of the History of Medicine* 48, no. 2 (1974): 234–48.

6. Larry Gostin, "Taking Liberties," *New Internationalist* 132 (February 1984), http://www.newint.org/issue132/taking.htm (25 April 2007).

7. Timothy F. Murphy, *Gay Science: The Ethics of Sexual Orientation Research* (New York: Columbia University Press, 1997).

8. Culver, "Genetic Malady," 150. Bernard Gert develops the concept of rational and irrational beliefs and desires in his *Morality: Its Nature and Justification* (New York: Oxford University Press, 1998), chap. 2.

9. Not all desires that increase one's chances of suffering evils are irrational. Only desires that one knows will result in harm and that one lacks an adequate reason for holding qualify as irrational. Adequate reasons include the avoidance of even greater evils. See Culver, "Genetic Malady," 150–51.

10. Boorse, "On the Distinction," 50; and Daniels, "Normal Functioning," 314.

11. Lee M. Silver wrongly draws from this example the conclusion that all enhancements are morally acceptable. See his "How Reprogenetics Will Transform the American Family," http://www.hofstra.edu/Academics/Law/law_lawrev_silver .cfm#Ref31 (30 April 2007).

12. D. A. Hyman, "Aesthetics and Ethics: the Implications of Cosmetic Surgery," *Perspectives in Biology and Medicine* 33 (1990): 190–202.

13. Thus, soldiers in mortally threatening military situations have welcomed wounds as a way of escaping otherwise certain death.

14. For a discussion of the prospects and risks of gene enhancements in athletics, see Jesper L. Andersen, Peter Schjerling, and Bengt Saltin, "Muscle, Genes and Athletic Performance," *Scientific American* 283, no. 3 (September 2000): 48–55. For a graphic depiction of the serious health consequences of using performance-enhancing drugs in athletic competition, see Juliet Macur, "Seeking Her Way out of Infamy," *New York Times*, 10 August 2004, Health. Also Michael Sokolove, "The Lab Animal," *New York Times Magazine*, 18 January 2004.

15. Franklin G. Miller, Howard Brody, and Kevin C. Chung, "Cosmetic Surgery and the Internal Morality of Medicine," *Cambridge Quarterly of Healthcare Ethics* 9 (2000): 353–64.

16. For a fictional rendition of this fear, see the short story by Greg Bears, "Sisters," in *Tangents*, by Greg Bears (London: Orion Publishing Group, 2000), 199–238.

17. Darren Shickle, "Are Genetic Enhancements Really Enhancements?," *Cambridge Quarterly of Healthcare Ethics* 9 (2000): 344.

18. Kaufmann Kohler and Abraham de Harkavy, "Karaites and Karaism," *Jewish Encyclopedia.com*, http://www.jewishencyclopedia.com/view.jsp?art id=108&letter=K (18 June 2007).

19. Fred Rosner, "The Imperative to Heal in Traditional Judaism," in *Jewish and Catholic Bioethics: An Ecumenical Dialogue*, ed. Edmund D. Pellegrino and Alan I. Faden (Washington DC: Georgetown University Press, 1998), 99–104.

20. The eighteenth-century rabbi Chayim Azulai, quoted in Fred Rosner, *Modern Medicine and Jewish Law* (New York: Bloch Publishing Company, 1972), 24.

21. B. *Bava Kamma* 85a.

22. B. *Shabbat* 150a; B. *Yoma* 84b.

23. S.A. *Yoreh De'ah* 336:1, quoted in J. David Bleich, *Judaism and Healing* (New York: Ktav, 1981), 4.

24. Exodus 21:19; M. *Bava Kamma* 8:1; B. *Bava Kamma* 85a–85b.

25. David M. Feldman, *Marital Relations, Birth Control and Abortion in Jewish Law* (New York: Schocken Books, 1974); Ronald M. Green, "Religion and Bioethics," in *Notes from a Narrow Ridge*, ed. Dena S. Davis and Laurie Zoloth-Dorfman (Frederick MD: University Publishing Group, 1999), 165–81.

26. "Dor Yeshorim," http://www.jewishgenetics.org/dor-yeshorim (accessed 6 June 2014).

27. B. *Shabbat* 134a.

28. *Genesis Rabbah* 11:6; *Midrash Bereshit* 11:2; *Pesikta Rabbati* 23:9.

29. Remarks delivered at the conference "Jewish Biomedical Ethics: Comparison with Christian and Secular Perspectives," held at Georgetown University, Washington DC, April 1996. Reprinted in Moshe Tendler, "On the Interface of Jewish Religion and Medical Science: The Judeo-Biblical Perspective," in Pellegrino and Faden, *Jewish and Catholic Bioethics*, 10–32.

30. For a more thorough discussion of the ethics of gene enhancement, see my *Babies by Design: The Ethics of Genetic Choice* (New Haven CT: Yale University Press, 2007).

31. This concept is drawn from John Rawls, *A Theory of Justice* (Cambridge MA: Harvard University Press, 1971).

CHAPTER 16 *Genetic Enhancement and the Image of God*

AARON L. MACKLER

The Talmud relates that a dispute arose during the time of the emergence of Yavneh as a center of Jewish teaching following the destruction of the Second Temple. When the Temple had been in existence, the shofar was sounded in the Temple on Rosh Hashanah, even if it occurred on the Sabbath, when sounding the shofar is generally prohibited. Once in Yavneh, when Rosh Hashanah occurred on the Sabbath, Rabban Yohanan ben Zakkai said to the B'nei Bateira, "Let us sound the shofar"; because Yavneh was now the Jewish center, as the Temple had been, similar practices were appropriate. The B'nei Bateira responded, "Let us discuss this first." Rabban Yohanan ben Zakkai said, "Let us sound the shofar, and we'll discuss it later." After the shofar was sounded, the B'nei Bateira said, "Now let us discuss this." Rabban Yohanan ben Zakkai responded, "The horn already has been heard in Yavneh, and we cannot go back after the deed."[1]

I am reminded of this story in thinking about Jewish ethical perspectives on the genetic enhancement of humans. In some ways, any discussion is premature, and there is good reason to postpone discussion until later. Knowledge about the genome and the ability to manipulate genetic material is impressive and developing rapidly, but genetic enhancement of humans remains in the distant future.[2] In some ways, Jewish bioethics cannot yet say much about the ethics of genetic enhancement. Jewish bioethics, like bioethics generally, requires understanding of the medical and personal facts and circumstances that are relevant in order to formulate an appropriate ethical judgment. Traditional *halakhah* (Jewish law) likewise relies on detailed understanding of facts and circumstances in its case reasoning. It is difficult for Jewish ethics (or other approaches) to formulate a clear judgment on genetic enhancement of human beings, which remains speculative. Still, one must beware of the "horn in Yavneh" problem. Once genetic interventions have been developed, it may be too late to undo them; ethical control would be especially elusive if ethical thought only begins then to react to the new possibilities. It is important to begin thinking about Jewish ethical understandings of the genetic enhancement of humans while remaining aware that the analysis is necessarily provisional and incomplete.[3]

As a way to think proactively about genetic enhancement, this chapter considers the implications for genetic modification of one central concept: that humans are created in the image of God, *b'tzelem Elokim*. First, I will review briefly ways in which this concept has been understood in Jewish thought. I then consider what implications this concept might have for the genetic modification of humans, including genetic enhancement.

The Image of God

Genesis 1:26–27 relates God's plan to create humankind: "And God said, 'Let us make man in our image [*tzelem*], after our likeness [*demut*].' . . . And God created man in His image, in the image of God He created him; male and female He created them." The same idea recurs in Genesis at the beginning of chapter 5: "This is the record of Adam's line. When God created man, He made him in the likeness of God; male and female He created them." The Rabbinic sage Ben Azzai proclaimed that this verse presented the great principle of the Torah. As Rabbi Tanhuma observed, one who disrespects a human is disrespecting one who is created in God's image.[4]

This teaching has been esteemed by Jewish thinkers through the centuries. The principle announced in it supports the value of human life and the duty to preserve life. It supports as well the commitment to *kevod habriyot*, the dignity of human persons and the responsibility of respect for persons. Thinkers have attempted to identify qualities of the human person as representing this image of God within the human.[5]

Some thinkers have understood the human intellect as constituting the divine image or human likeness to God. Moses Maimonides writes that the distinguishing capacity of humans is intellectual apprehension. "In the exercise of this, no sense, no part of the body, none of the extremities are used. . . . It was because of this something, I mean because of the divine intellect conjoined with man, that it is said of the latter that he is in the image of God and in His likeness."[6] In his biblical commentary, Rashi likewise explains "after our likeness" as referring to the ability to understand.[7] Some understand the image of God to refer to the spiritual soul of the human, which God imbues and which, like God, is immaterial and immortal.[8] The image of God also may be taken to refer to choice or free will, especially the ability to make moral judgments.[9]

Abraham Joshua Heschel discusses concern for others, or transitive concern, as an essential characteristic of humans that is shared with God. All living organisms exhibit concern for their own continued existence, or reflexive concern. Human beings additionally exhibit transitive concern. Such concern is not just an accidental attribute found in many individuals but an essential characteristic of existence as a human. God is beyond the need for reflexive concern, so only transitive concern is properly attributed to him. While Heschel does not explicitly use the term "image of God" in this connection, one might see the human capacity to love and to be concerned for other persons as representing the divine likeness within the human.[10]

Although it is difficult to be certain of the way in which "image of God" was understood in biblical times, normative Judaism for many centuries has understood that God does not have a physical form. Nevertheless, the human body is important to the way in which humans reflect God's image. *Leviticus Rabbah* relates that the sage Hillel told his students that he was going to perform a mitzvah (one of God's commandments), namely, to bathe in the bathhouse. They asked, "Is that a mitzvah?" Hillel replied that people are employed to clean the images of the Roman king that decorate theaters and circuses, and those who clean these images are honored to do so. "We who are created in the image and likeness [of God], as it is written 'For in His image did God make man,' all the more so."[11] Likewise, whoever kills a human person is considered as though he or she had diminished the divine image, in the same way that a person who defaces a statue of a king would be diminishing the king's image.[12]

Talmudic rabbis associated this idea with the book of Deuteronomy's requirement (21:23) to bury the body of an executed criminal the same day as the execution, "for a hanged [or impaled] body is an affront to God." The Talmud explains with a parable:

> Two brothers, identical twins, [lived] in the same city.
> One was appointed king, while the other became a highway robber.
> The king ordered him hanged.
> Everyone who saw him exclaimed: "The king is hanged!"
> The king ordered him taken down.[13]

Disrespect for the highway robber would be disrespect for the king, whose image he shares. Similarly, disrespect for a dead body is understood as disrespect for God, whose image humans share. For the Jewish tradition, part of honoring God is honoring humans, created in God's image. Part of honoring humans, and God, is honoring the body, even after death.

Some Jewish authorities vigorously deny that there could be anything physical in the human imaging of God, for God has no physical image.[14] Still, humans are embodied creatures. Our physical bodies are part of the way that humans express concern for one another and perform mitzvot. Even if the image of God is not primarily physical, the human body is part of the way in which humans reflect God's image and God's presence. Accordingly, genetic modifications leading to changes in the human body may be justifiable but could not simply be ignored as trivial and irrelevant to the image of God within us.

Some suggest that no quality of humans can be identified with God's image; rather, the *tzelem* signifies the transcendent value of the human person, as the human is the object of God's concern. The Hebrew word *tzelem* can be understood as "shadow." A shadow does not share essential characteristics with that which casts the shadow, but the shadow does indicate the presence of something greater beyond itself. Grounding human value in God's concern, rather than in attributes that many humans possess, both avoids hubris and keeps us mindful of the transcendent dimension of humanity.[15]

A classic Rabbinic statement on the intrinsic value of humans, and the creation of humans in the image of God, is found in the Mishnah. The context is an admonition to witnesses in capital cases, warning them of the grave consequences of their testimony.

> Therefore was a single man [Adam] created, to teach you that anyone who destroys a single person from the children of man is considered by Scripture as if he destroyed an entire world, and that whoever sustains a single person from the children of man is considered by Scripture as if he sustained an entire world; and for the sake of peace among people, that no one could say to his fellow, my ancestor was greater than your ancestor; . . . and to proclaim the greatness of the Holy One, blessed be He, for man stamps many coins with the same die and they are all alike one with the other, but the King of the kings of kings, the Holy One, blessed be He, stamps every man with the die of the first man and not one of them is alike to his fellow.[16]

God's creation of humanity in God's image supports values of the importance of each individual human life, human equality, and diversity among humans.

Genetic Enhancement

Jewish reflections on the creation of humanity in God's image can be illuminating for thinking about genetic enhancement, though it does not provide precise guidelines. First, it should be noted that an individual's DNA is not the "image of God" within that person. This needs to be kept in mind because of both the biological significance and the popular mystique of DNA and genes. Modifying the DNA of a person (or potential person) would not intrinsically constitute sacrilegious trampling on the divine image, and it would not inherently violate respect for persons. Because of the powerful effect of genes on the human person and that person's characteristics, however, caution is mandated.[17]

Modifications that represent healing would be supported by consideration of humans' creation *b'tzelem*. It is appropriate, even commanded, to intervene in nature in order to promote healing and save life.[18] Interventions that entail excessive risk would not be consistent with that value. More generally, interventions would need to be consistent with respect for persons, who reflect the image of God.[19] While the distinction between therapy and enhancement can be difficult to determine, this distinction is important in ascertaining which interventions are consistent with the value of *tzelem*. Therapeutic interventions are less likely to violate respect for persons and positively promote healing mandated by this value.

It is famously difficult to draw a sharp line between therapy and enhancement.[20] This chapter only seeks to think about genetic enhancement in a general way. Some types of modifications would seem to be clearly therapeutic—for example, correcting a "misspelling" in the DNA of a child with X-linked severe combined immunodeficiency disease syndrome (or adenosine deaminase deficiency, or another genetic disease) to restore functioning. Other modifications would seem to clearly represent enhancements—for example, developing the biological capacity for greater strength than is typical of humans, or developing greater reasoning ability of the sort measured by IQ tests, or giving a person the ability to function well on little sleep.[21] Different genetic modifications might be more or less clearly enhancements, and different enhancements would be more or less problematic. Some modifications might be considered preventive. For example, they might seek to prevent infection from germs, perhaps by slowing the decline of the immune system that typically accompanies aging. Other interventions might seek to prevent the loss of bone density or the decline in short-term memory that typically accompanies aging. Another type of intervention might be considered remediative, helping an individual to be within the species-typical range. An example would be the increase of growth hormone for a child who is likely to be of unusually short stature, though caused by factors other than a deficiency of growth hormone. Both preventive and remedial modifications would pose some of the problems of other enhancements, though possibly in less extreme form.[22]

Enhancement lacks the positive support found for therapeutic interventions, and it poses greater danger of violation. Risks not only are more difficult to justify but are likely greater. A change to the genome of simply correcting a defect is relatively unlikely to have unforeseen consequences, though the therapeutic intervention that attempts to enact this change entails risk.[23] An attempt beyond this to modify human characteristics in a way that would differ from typical human functioning is more likely to have negative unintended side effects. As scientist W. French Anderson has argued:

> To add a normal gene to overcome the detrimental effects of a faulty one is probably not going to produce any major problems, but to insert a gene to make more of an existing product might adversely affect numerous other biochemical pathways. In other words, replacing a faulty part is different from trying to add something new to a normally functioning, technically complex system. Correcting a defect in the genome of a human is one thing. But inserting a gene in the hope of "improving" or selectively altering a characteristic might endanger the overall metabolic balance of the individual cells as well as the entire body.[24]

Beyond medical risks to the individual, there is a risk of leveling differences among people as a popular enhancement becomes widespread and perhaps (de facto) normative. According to the Mishnah, diversity among humans is part of the way in which humanity reflects God's image.

Especially if there are barriers to popular enhancements becoming widespread, enhancement raises more acute concerns of justice than do therapeutic interventions. If an intervention is therapeutic in response to a genetic disease, it is relatively clear who should get intervention—namely, those with the defective gene. There are practical problems in the United States, where those who are poor or lack health insurance often lack access to needed care. These are urgent ethical concerns to address for health care in general. Problems become even more daunting, though, with enhancement. Unless the interventions are provided to all citizens (or all humans) equally, it is less clear who should receive enhancement. It seems probable that, even if the United States better fulfills its responsibility to provide access for all to needed health care and gene therapy, enhancement interventions would tend to go to those who are already more fortunate.[25]

Appreciation of the value of human persons as created in the image of God grounds a general commitment of respect for persons. Care must be taken not to treat a person as an object. Possible interventions that would modify personality or personal characteristics raise special concern, though the line distinguishing these from physical characteristics is not always clear. Special attention should be given to modifications that would interfere with those characteristics most closely associated with the image of God, such as intellectual ability, free will, and the capacity to love and care for others. Even interventions that do not directly impact these particular abilities may represent instances of treating persons as objects. One must be aware of the transcendent significance of the *tzelem* and the danger of objectification and treating persons as a means.[26]

The danger of people being treated like objects is clearest when one person is using another person. *Brave New World* portrays a centralized government manipulating the genetic characteristics of future humans for societal purposes in a way that manifestly violates respect for persons who are created in the image of God.[27] A more likely threat in nations such as the United States would be parents treating their future children as commodities to be designed with traits that meet the parents' desires. Morally, concern should be given even to modifications that an individual might wish to make on himself or herself. As a matter of Jewish ethics, judgments must be made not only regarding the interventions that some individuals impose on others but also regarding choices that an individual might make for himself or herself. Like Kant's value of respect for persons, the value of humans' creation *b'tzelem Elohim* entails responsibilities regarding not only one's treatment of others but also one's treatment of oneself.

NOTES

1. B. *Rosh Hashanah* 29b.
2. Therapeutic genetic intervention to treat the most severe maladies has begun in selected experiments. In one experiment, gene therapy restored immune system functioning for children with X-linked severe combined immunodeficiency disease syndrome who otherwise likely would have died from infections before their first birthday. One child developed leukemia as a result of the intervention, however, illustrating the risks of genetic modification. See Salima Hacein-Bey-Abina et al., "Sustained Correction of X-Linked Severe Combined Immunodeficiency," *NEJM* 346 (2002): 1185–93; Theodore Friedman, "What Kind of Future Does Gene Therapy Have?," presentation to the International Jewish Discourse Project on Religion and Ethics at the Frontier of Genetic Medicine, 8 December 2002. For the present, genetic modification of humans is likely to be limited to treating the most severe maladies, for which the potential life-saving benefits of treatment could justify the risks involved.
3. See, similarly, Maxwell J. Mehlman's counsel that it is important to begin considering the ethics of genetic enhancement and possible regulations now, in advance of the development of genetic enhancement: "Without this type of groundwork, there is no way society will be in a position to act in time" ("Genetic Enhancement: Plan Now to Act Later," *Kennedy Institute of Ethics Journal* 15 [2005]: 77).
4. *Genesis Rabbah* 24:7; see also *Sifra* 89b. Rabbi Tanhuma considered this verse as even more central than the command to love your neighbor as yourself (Leviticus 19:18, espoused by Rabbi Akiba). A person who disrespected himself would not thereby have permission to disrespect his neighbor. Humanity's creation in the image of God is also referred to in Genesis 9:6, supporting the prohibition of homicide.
5. The phrase "image of God" may not refer primarily to characteristics of a human being that resemble God but primarily to the role of humans as God's representatives on earth. In 1979 a statue from the ninth century BCE was found at Tell Fekheriyeh in modern Syria. An inscription identifies the statue as representing Hadad-yis'i, a governor of Guzan in the Assyrian Empire. It was to stand in the temple of Hadad. The inscription in Akkadian refers to the

statue as *tzalmu*, and an inscription in Aramaic refers to it as *tzalma* and *demuta*. These words are cognates of the Hebrew *tzelem* and *demut* (A. R. Millard and P. Bordreuil, "A Statue from Syria with Assyrian and Aramaic Inscriptions," *Biblical Archaeologist* 45, no. 3 [1982]: 135). The image of Hadad-yis'i was not understood to copy him but to represent him and to be his representative in offering praise and petitions to his god Hadad. Similarly, the Bible does not claim that humans are a copy of God but that God created humans to represent him and be his representative in governing the created world. The scope of this chapter does not allow me to explore this idea, though I hope to do so in future work.

6. Moses Maimonides, *Guide for the Perplexed* I:1, in the translation of Shlomo Pines (Chicago: University of Chicago Press, 1963), 23.

7. Rashi on Genesis 1:26.

8. Moses Nahmanides on Genesis 1:26. Nahmanides understands the plural form of "Let us make man in our image" as God's addressing the earth. The earth is responsible for the body of humans, as it is for the body of all animals, but God is directly responsible for the soul, and so the human is in the image of both. See similarly *Sifre Deuteronomy* 306.

9. Obadiah ben Jacob Sforno on Genesis 1:26. On this and other views noted in this paragraph, see also Elliot N. Dorff, *Matters of Life and Death: A Jewish Approach to Modern Medical Ethics* (Philadelphia: Jewish Publication Society, 1998), 19.

10. Abraham Joshua Heschel, *Man Is Not Alone: A Philosophy of Religion* (New York: Farrar, Straus and Young, 1951), 136–44; see also Aaron L. Mackler, "Symbols, Reality, and God: Heschel's Rejection of a Tillichian Understanding of Religious Symbols," *Judaism* 40 (1991): 294.

11. *Leviticus Rabbah* 34:3, quoting Genesis 9:6.

12. *Mekhilta,* Bahodesh, chap. 8, cited in Noam J. Zohar, *Alternatives in Jewish Bioethics* (Albany: State University of New York, 1997), 92.

13. B. *Sanhedrin* 46b. The translation is that of Zohar, *Alternatives*, 124. See also Aaron L. Mackler, "Respecting Bodies and Saving Lives: Jewish Perspectives on Organ Donation and Transplantation," *Cambridge Quarterly of Healthcare Ethics* 10 (2001): 422–23.

14. For example, Maimonides, *Guide* I:1.

15. David Novak, *Natural Law in Judaism* (Cambridge: Cambridge University Press, 1998), 167–73.

16. M. *Sanhedrin* 4:5.

17. Scientist W. French Anderson writes, "We do not really understand what makes up our humanness, nor do we know precisely what role genes may play.

But whatever our humanness is, we fear the possibility of its being tampered with" ("Human Gene Therapy: Why Draw a Line?," *Journal of Medicine and Philosophy* 14 [1989]: 682).

18. See, for example, Aaron L. Mackler, *Introduction to Jewish and Catholic Bioethics: A Comparative Analysis* (Washington DC: Georgetown University Press, 2003), 4–10.

19. We must be especially wary of the dangers associated with eugenics, in the extreme associated with the evils of Nazi Germany, but we should strive more generally to assure that actions are consistent with humans' creation *b'tzelem*.

20. See Erik Parens, "Is Better Always Good?," in *Enhancing Human Traits: Ethical and Social Implications* (Washington DC: Georgetown University Press, 1998); Eric Juengst, "Can Enhancement Be Distinguished from Prevention in Genetic Medicine?," *Journal of Medicine and Philosophy* 22 (1997): 125–42.

21. See, for example, LeRoy Walters and Julie Gage Palmer, *Ethics of Human Gene Therapy* (New York: Oxford University Press, 1997).

22. See Parens, "Is Better Always Good?"; Walters and Palmer, *Ethics*.

23. See above, note 2.

24. Anderson, "Human Gene Therapy," 686.

25. See Parens, "Is Better Always Good?"; Anderson, "Human Gene Therapy," 688.

26. Furthermore, Frances M. Kamm warns of "our lack of imagination as designers. That is, most people's conception of the varieties of goods is very limited, and if they designed people, their improvements would likely conform to limited, predictable types. But we should know that we are constantly surprised at the great range of good traits in people, and even more the incredible range of combinations of traits that turn out to produce 'flavors' in people that are, to our surprise, good" ("Is There a Problem with Enhancements?," *American Journal of Bioethics* 5, no. 3 [2005]: 13).

27. Aldous Huxley, *Brave New World* (New York: Harper and Row, 1932, 1969).

CHAPTER 17 *"Blessed Is the One Who Is Good
and Who Brings Forth Goodness"*

A Jewish Theological Response to the Ethical
Challenges of New Genetic Technologies

LOUIS E. NEWMAN

For Jewish religious thinkers, ethics must always be a reflection of theology. Determining the "good" that we should pursue, the virtues that we should cultivate, the norms that should govern our relationships with one another—all these tasks flow from basic convictions about the sort of creatures we are and the ends for which we were created. Such convictions are, in the context of religious traditions at least, matters of theology, questions of how we conceive of God and our relationship to God, of creation and redemption. In the oft-repeated formulation of Clifford Geertz, within every religious system, ethos corresponds to worldview; our view of how we ought to live and what sort of world we should strive to create reflects our deepest convictions about the nature of the world as it is.[1] Or, in James Gustafson's elegant articulation of the essence of Christian ethics, "we are to relate to ourselves and all things in a manner appropriate to our, and their, relations to God."[2]

If this is true of ethics in general, it is especially so in the area of biomedical ethics. Here we deal with issues that touch most directly on matters of life and death, what it means to be human, how to respond to those situations (ever more common with the advance of medical technology) in which we can affect (or control) human life at the margins—within the womb or in the twilight zone between life and death. All such questions are, at their core, questions about what makes us human. And in the emerging field of genetic technology, this is still more the case. The decoding of the human genome, the newest discoveries in the area of stem cell research, the promise it holds out of "regenerative medicine," and, perhaps most dramatically, the possibility of gene therapy and genetic engineering—all these technologies promise radically to alter the way we diagnose and treat disease, putting at our disposal the power potentially to change our very genetic makeup and perhaps even that of our descendants. It is hard to imagine any medical advance that more directly and powerfully raises the questions of what defines our humanity, of how to balance our responsibility to heal with our responsibility to honor our unique and precious human nature.

These issues are vast and enormously complex, far too complicated to deal with comprehensively in any single essay. My goal here is more modest: to identify two key challenges raised by these technologies, and then to explore some Jewish theological perspectives that might provide, if not clear answers, at least a general perspective out of which moral guidance could emerge. As I will explain further below, this approach, for all its indeterminacy, has at least the virtue of not presuming to represent the tradition as monolithic or as offering more definitive answers than the available sources in fact provide.

At root, the new genetic technologies represent especially dramatic examples of vast new powers that we have over the healing process and, indeed, over our genetic makeup, both individually and collectively. Ethics, as Hans Jonas so persistently argued, deals first and foremost with the responsibilities attendant on our power.[3] Whether that is the power to hurt another through the words we use (or fail to use), to inflict bodily injury through physical force, or to insert new genetic material into the body of a cancer patient in the effort to alter the course of her disease, power always entails responsibility. The questions before us, then, are: What sorts of powers do these new genetic technologies put at our disposal (either at present or in theory)? And what theological resources does Judaism offer us that might help us think about the ethical use of these powers? The next two sections of this chapter will deal with these two questions, respectively.

Possibilities and Challenges of Genetic Technology

The most fundamental challenges posed by these genetic technologies, I think, are (1) the power of vastly expanded knowledge and (2) the power to change our unique genetic code and, potentially, that of our descendants.[4] It is important to notice at the outset that each of these powers, like many other human capacities, is inherently neutral. The scientists and physicians working in these areas—decoding the human genome or experimenting with gene therapy—plainly intend to develop powerful new tools in the treatment of human disease, thereby saving (potentially) millions of people's lives and relieving the suffering of millions more. Yet, these same technologies could be used for plainly unethical purposes.[5] Like the power to split the atom, and perhaps equally momentous, genetic technology could clearly be used vastly to improve the human condition but also to destroy it. It is worthwhile to take a closer look at these possibilities.

THE POWER OF KNOWLEDGE. The decoding of the human genome is but the first step in a long process of determining the precise function and interaction of all the genes in the human body and the proteins they generate. Ultimately, this will give us access to the chemical basis for understanding (and so potentially altering) all the processes and conditions, both normal and abnormal, of the human body at the genetic level. Everything from the way our intestines and nerve cells work, to the way malignant tumors and congenital diseases develop, is in principle traceable to the workings of our genes. Nine of the ten leading causes of mortality in America have at least a genetic component.[6] So, too, in all likelihood, is there a genetic component to intelligence, artistic talent, aggressiveness, and sexual orientation. To the extent that these genetic codes are understood, we will understand "human nature" more fully, and whole new fields of research will become possible. By the same token, our view of what is "normal" and "abnormal" will be altered, as will our way of dealing with various genetically related traits that currently evoke social praise or opprobrium. Understanding our genetic composition holds the potential to change our understanding of "who we are," both as individuals and as a species, at a fundamental level.

But this knowledge will have more immediate, practical applications as well. Because genes contribute to many conditions, such as high blood pressure, that typically only manifest later in life, knowing one's genetic propensity to such conditions long before symptoms appear would enable one to begin treating a disease far sooner than is currently possible. So, too, genes plainly account for some of the differences in the ways people respond to specific drugs. Understanding this more fully will one day enable physicians to prescribe those drugs (and perhaps even to develop new ones) that would work optimally for specific groups of patients, with maximal benefits and minimal side effects. When we consider the possibility that, in the not-too-distant future, all this genetic information and more could be available through a simple test in which a bit of saliva is smeared on a computer chip, we begin to appreciate the scope of the medical revolution at hand.

Of course, what we know may hurt us as much as it benefits us. On the social level, we will soon have to deal with questions about whether insurance companies, potential employers, or even potential partners will have a "right" to know our genetic information insofar as this has financial (and emotional) consequences for them. Colleges and universities, interested in the intellectual and personal traits of their students, may want to require genetic test results in addition to (or in lieu of) standardized, and highly flawed, "aptitude tests." We may also discover that the genes linked to various human gifts are not distributed equally among all social or ethnic groups. How will we deal with the information that one group is genetically more prone to violence, or to alcoholism, while another is more likely to be gifted artistically, intellectually, or athletically? Such information, once known, will surely have an effect on stereotypes and potentially on the development of government policies to deal with early childhood education, drug abuse, and other social problems, as well.

Finally, genetic testing already enables us to trace the lineages of various groups of people. Because the Y-chromosome is only passed down from fathers to sons, and because mitochondrial DNA is passed down only from mother to child, geneticists can determine with some precision how whole groups of people are genetically related. Such information may confirm, but may also challenge, a national or ethnic group's sense of identity, its own narrative about its origins, and its relationship to other groups.

So, the explosion of genetic information is plainly a two-edged sword. Its potential benefits are enormous, but so too are the challenges and risks to which it exposes us. Knowledge is power; genetic knowledge is powerful in a way that, arguably, no previous human knowledge has been, precisely because it is knowledge about who we are, individually and collectively, at the most basic biological level. This is not to say that we are defined by our genes but rather to acknowledge the fact that much about our physical lives, including our susceptibility to various maladies, is genetically determined. Decoding the human genome, then, makes available to us a great deal of information about our most fundamental biological features, especially those features that differentiate us from one another.

THE POWER TO ALTER OUR GENES. If simply understanding our genetic composition is transformative, altering it is still more so. For decades now, scientists have been producing genetically altered food, creating strains of corn and other crops that are more bountiful and more resistant to specific bacteria. Altering the human genome is a far more complex matter, both scientifically and morally. Early clinical trials have focused largely on the possibilities of injecting patients with "corrected" genes to combat a condition such as cancer that is caused by an abnormal gene. And though these trials have as yet met with only limited success, it seems that the time is not far off when we will be able to treat, if not cure, certain genetic conditions by either "turning on" or "turning off" the requisite genes. Genetic therapies hold out the possibility of attacking the very root of certain genetically caused diseases, both those that are currently terminal (e.g., Tay-Sachs, anencephaly) and those that can currently be treated only symptomatically and with limited success (e.g., heart disease, diabetes). And, of course, there are many other less debilitating human conditions, such as myopia and baldness, that might also prove to be susceptible to genetic rectification.

The still more powerful (and so more troubling) implications of genetic engineering emerge when we turn from "somatic" cell therapies, those designed to treat a single patient, to "germ" cell therapies, those that would introduce new genes into the egg or sperm cells of the parents, which would then be transmitted to their offspring. Such therapies have not yet been attempted, and many concerned ethicists have urged caution in this area.[7] Still, the potential draw of germ line genetic engineering can be gauged by looking at ads, readily accessible on the Internet, for donor sperm from fathers with specific traits. Here technology again holds out a two-edged sword: with one edge we might eradicate diseases entirely from the population; with another edge we might begin "designing" children with specific desirable traits, those that will maximize the likelihood that they will become star athletes or famed musicians. The same technology, after all, that could enable us to eliminate deadly diseases could also enable us to "enhance" normal human abilities, extend longevity, and "perfect" human nature. One need hardly spell out the potential implications of this new technology for our notions of parenting, eugenics, the gap between those who are genetically "enhanced" and those who are not, or the fabric of society generally.

Resources for a Jewish Response

How might we construct a Jewish response to the ethical challenges posed by these genetic technologies? A classical, halakhic approach would entail searching for appropriate precedents within the vast corpus of legal literature and applying them to the specific issues at hand. But such an approach here will face serious obstacles. Most obviously, it is extremely difficult to find any classical sources that bear, even remotely, on issues like the decoding of the human genome or genetic engineering. To date, few halakhic authorities have even attempted this, and those articles that exist are sketchy at best and tendentious at worst.[8] This would seem to be an area in which halakhic authorities can proceed, if at all, only by working intuitively from a working knowledge of the tradition as a whole (*da'at Torah*), rather than deductively from earlier cases.[9]

A more promising approach would work midrashically, looking to classical texts not as direct precedents for specific actions but as sources for broad values and theological perspectives. The goal of such an approach would be to offer some perspective on the deep issues—about human power and responsibility, about God, creation, and human nature—rather than to provide a *p'sak din*, a halakhic ruling. I recognize that this aggadic approach is open to the charge of being, like all midrashim, just one random interpretation among a myriad of possibilities, no more or less persuasive than any other. And within a tradition accustomed to providing clear directives in every realm of human behavior, such a theological/ midrashic approach will seem ill-suited to the task of defining one's ethical duties. But I confess that, given the radically unprecedented nature of the technologies in question, I see no alternative. Moreover, although there are no obviously apt legal precedents for these technologies,[10] there are very definitely sources that address the underlying issues—the use of dangerous knowledge and the value of human nature. It is time to turn to some of these sources.

The Power of Knowledge

The Mishnah offers this ruling on the teaching of material regarded as potentially dangerous:

> They do not expound [the passages concerning] forbidden sexual relationships with three [students,] nor [those concerning] matters of creation with two, nor [those concerning] the chariot [the mystical vision of God in Ezekiel 1] with one, unless that individual is wise and could understand it on his own.

And anyone who looks into four things, it would be better for him had he never come into the world: what is above, what is below, what is ahead, and what is behind [variously interpreted as what is beyond the physical confines of the earth, or what is before creation and after the "end of days"].

Anyone who is not duly respectful of his Creator, it would be better for him had he never come into the world.[11]

Certain divine secrets are beyond the scope of human knowledge and should remain so. To look into these things—the process of divine creation, or the experience of God's revelation to the prophet—is risky. One might mistakenly come to think one grasps something fully when in fact one's knowledge is only fragmentary.[12] The wise student who understands these matters with a minimum of instruction is the exception; others are not to be trusted with access to esoteric knowledge that could lead them into erroneous thinking or behavior.

The stark and seemingly extreme phrasing of this ruling is instructive: one who explores such matters would be better off never having been born. The Mishnah seems to be telling us that we forfeit our right to a place in the world if we do not know, and honor, our proper place in the scheme of things. When we investigate divine secrets, we trespass in a realm where we do not belong. We fail to honor the limits that God has placed upon us; we are guilty of hubris. In acting as if the space (both physically and intellectually) in which God has placed us is inadequate, we demonstrate that we do not appreciate why we were put on earth to begin with; better, then, not to exist than to live in such a misguided state. The final line of the Mishnah makes this clear: the ultimate issue is honoring God as creator, which means honoring our place in the created order.

It is worth noting as well that there appears here to be a hierarchy of dangerous knowledge. The opening line of this Mishnah moves from matters of forbidden sexual relationships to creation to the mystical vision of God, as the number of students who may be tutored in such matters becomes increasingly restrictive. The Rabbis of the Talmud understood this as a precaution against the possibility that, while the teacher explained a difficult point to one of the students, the others might proceed to study on their own and misunderstand important matters. It seems, though, that the point is more theological than pedagogical, for the text here leads us through successively more egregious forms of transgression. We move from knowledge about forbidden sexual relationships (which involves spoiling the most fundamental human act of creativity by transgressing the boundaries within families) to knowledge about God's creative power (which determines the nature not only of humankind but of the whole natural order) to knowledge about God's own nature. As we move into ever more sacred realms of knowledge, the Rabbis tell us, the act of transgressing, of trespassing where we do not belong, becomes more serious, and so the restrictions placed upon the teacher of such knowledge become more severe.

Interpreted legally (i.e., as *halakhah*), this source would probably preclude much scientific investigation (at least, astrophysics—"what is above" and "what is before"—and geology—"what is below"). But interpreted metaphorically (i.e., as midrash), it opens up the question of how we might think about the theological implications of investigating the divine mysteries of life—the workings of creation and even the nature of God. One could argue (though I will not press the point here) that contemporary investigations of the human genome and attempts to alter it are indeed the closest contemporary analogs to ancient attempts to unravel these divine mysteries. In any event, the point is that this new science challenges a certain traditional notion of our rightful place in the world; in colloquial terms, it involves "playing God." At the very least, such efforts run the risk of fostering disrespect for God and the naturally created order.

I take this source, then, not as a rule proscribing particular intellectual pursuits but as a caution that any attempt to unravel the mysteries of creation (or to crack the genetic code) should take place in the context of a pervasive reverence for human life and the order of creation. Some forms of knowledge are so powerful that they readily mislead us into thinking we are the masters of creation rather than its subjects. That inversion is a form of idolatry, of worshiping the knowledge we seek and the power that flows from it; it is, in the Mishnah's words, a matter of not duly honoring our Creator. And that may well lead us into a kind of life that would ultimately not be worth living, a life in service of the wrong ends. As the Talmud comments on our passage, dangerous knowledge is rightly transmitted only to one of great stature and learning (*l'av beit din*) and/or to one "whose heart is anxious within him" (*mi sh'lebo do'eg b'kirbo*).[13] This last qualification in particular suggests that those who pursue this knowledge must be fully aware of the accompanying dangers and must be tentative and cautious, not bold and reckless.

The Power to Alter Our Genes

The second set of moral challenges that arises from genetic technology is not about the power of knowledge itself but about the power to use that knowledge to change our genetic nature. The theological issues raised here cluster around the question of how to think about the sacredness of nature, especially human nature. Let us examine some texts that bear on this question:

> Rabbi Simeon ben Yohai, quoting "The Craftsman, His work is perfect" (Deut. 32:4) said: "The Craftsman who wrought the world and man, His work is perfect. In the way of the world, when a king of flesh and blood builds a palace, mortals who enter it say: 'Had the columns been taller, how much more beautiful the place would have been! Had the walls been higher, how much more beautiful it would have been! Had the ceiling been loftier, how much more beautiful it would have been!' But does anyone come and say: 'If I had three eyes, three arms, three legs, how much better off I would be! If I walked on my head, or if my face were turned backward, how much better off I would be!' I wonder. To assure that no one would say such a thing, the King of kings of kings, the Holy One and His court had themselves, in a manner of speaking, polled concerning the placing of every part of your body and set you up in a way that is right for you."[14]

One cannot read such a text today, in light of the promises of genetic technology, without experiencing a touch of irony. No one, perhaps, would set out to design a person with three eyes or a face turned backward. But a great many people do indeed wish to redesign people to be "better" than they are by nature—stronger, less susceptible to disease, more of one thing, less of another. Our text suggests that such efforts are foolhardy, for we are already the products of a "master designer"; all of our essential needs can be met without altering the form in which we have been created.

Moreover, in the contrast drawn to the way we respond to human creations, the text further suggests that we ought not to regard our bodies as akin to a human artifact. Artifacts we can design to our liking, and we can readily redesign them to better suit our needs. But our bodies are not of our own making; hence, we ought not to indulge in idle speculation about how we might improve on them.

It should be noted that this text presupposes that human nature, at least in its physical aspect, is fixed and immutable. The notion that as *Homo sapiens* we have evolved into our current form over a period of many millions of years, of course, is strikingly at odds with the "creationist" view of humankind held by biblical and Rabbinic writers. Moreover, the text does not seem to address the problems of those born with physical deformities but only with the normal physical traits of our species. Presumably, the Rabbis would not have objected to altering or curing physical ailments in those who are abnormal, only to the effort to improve upon the characteristics of humankind as a whole. Any such effort would imply that we are better "craftsmen" than God, which is precisely what this midrash wants to refute.

Viewing God as the master craftsman of our bodies could inform a very skeptical attitude toward genetic engineering, but before pursuing this line of thinking, we do well to consider a text that points in a rather different direction:

It is told of Rabbi Ishmael and Rabbi Akiva that, while they were walking through the streets of Jerusalem accompanied by a certain man, a sick person confronted them and said, "Masters, tell me, how shall I be healed?" They replied, "Take such-and-such, and you will be healed." The man accompanying the sages asked them, "Who smote him with sickness?" They replied, "The Holy One." The man: "And you bring yourselves into a matter that does not concern you? God smote, and you would heal?" The sages: "What is your work?" The man: "I am a tiller of the soil. You see the sickle in my hand." The sages: "Who created the vineyard?" The man: "The Holy One." The sages: "Then why do you bring yourself into a matter that does not concern you? God created it, and you eat the fruit from it!" The man: "Don't you see the sickle in my hand? If I did not go out and plow the vineyard, prune it, compost it, and weed it, it would have yielded nothing." The sages: "You are the biggest fool in the world! Have you not heard the verse, 'As for man, his days are as grass' (Ps. 103:15)? A tree, if it is not composted, weeded, and [the area around it] plowed, will not grow; and even if it does grow, if not given water to drink, it will die—will not live. So, too, the human body is a tree, a healing potion is the compost, and a physician is the tiller of the soil."[15]

Here the operative metaphors—of the human body as a plant, of the physician as a farmer—point toward a more active stance in relation to the body's needs. We must "cultivate" our bodies, intervening in the natural process to ensure that they develop properly. There is here a kind of partnership between what is naturally given (earth/plant, or the human body) and the human activity of developing that (through agriculture or medicine). The Rabbis' interlocutor is made to look simpleminded, though on one level he simply represents a different theological perspective, albeit one that the Rabbis have forcefully rejected. From that man's perspective, what God has created should be left in God's hands.[16] The Rabbis do not directly challenge that view but appear to reject it through a kind of reductio ad absurdum. To preclude people from interfering in God's world entirely would greatly impede normal human development, just as refraining from tilling the soil would impede the growth of a tree. So, too, when we intervene in the natural processes of the human body, we are only "pruning away" the elements that preclude it from developing in the way that God intended.

One could also read this midrash more restrictively, as sanctioning such interventions in the natural order *only* when they are necessary for sustaining human life, as in the case of agriculture. That reading would be supported by the reference toward the end of the passage to the threat that without intervention the plant (or person) will die. Where human life is imperiled, intervention is required; we do not uniformly prohibit tampering with nature on the grounds that it is God's domain.

In light of these two contrasting sources, the moral challenge of genetic engineering might be cast as that of finding some equilibrium between (1) accepting and honoring our bodies just as they are, as the work of a master craftsman, and (2) "cultivating" our bodies, nourishing and pruning them to maximize their potential for healthy development (or, at least, to preserve life). Both sources make powerful claims upon us. The former instructs us that we are, after all, "created in the divine image," and we dishonor our Creator if we presume to improve fundamentally upon God's work. The latter reminds us that human life requires us to intervene in the workings of nature, including in the processes of the human body; we are partners with God, "cultivating" and "tending" the raw material that God has given us. This same tension is reflected in the contrast between the six weekdays, when we are required (not simply permitted!) to transform the world, to master and subdue it, and the Sabbath, when we are required to refrain from interfering in the natural world at all, to honor the created order just as it came from God.[17] In responding theologically to the challenges of genetic technologies, then, we must find a way to mediate between these conflicting demands.

We come still closer to a Jewishly informed view of human health and the natural state of the human body when we consider the following blessing, to be recited upon seeing someone physically deformed: "Blessed are you, Lord our God, who fashions diverse creatures [*m'shaneh ha-b'riyot*]." The notion of reciting a blessing praising God for something that is "abnormal" may seem odd, even perverse. After all, when we see someone with a physical deformity, we tend to feel uncomfortable, to look away; we are repelled, and we feel that this is only a natural reaction. The blessing, though, requires that we look past the superficial features that we find disturbing; instead, we are to see what God sees, another human being bearing the divine image. The theology implicit in the practice of reciting such a blessing is worth examining more closely. All that occurs naturally is from God. The same Creator who is the source of rainbows and cloudless, sunny days is also the source of devastating storms and earthquakes. More to the point, God is behind both the perfectly healthy, wondrously functioning human body and the body that is naturally impaired, less than whole and healthy. The tradition of reciting this blessing, then, challenges us to see beyond "what meets the eye." In doing so, we will have to set aside, at least temporarily, our preconceived notions about what is "normal" versus "abnormal" and, more importantly, to abandon our instinctive inclination to value the former and devalue the latter.[18] If we were to internalize the message of this blessing, we would come to affirm that God is at work in all of creation, including those parts that seem to us "dysfunctional."

And yet, of course, reciting this blessing in no way mitigates our obligation to heal the sick or to remove such impediments to healthy human functioning as can be removed. But the goal of healing is not that healthy, functioning bodies are "better" or "more blessed" than malfunctioning ones, for otherwise there would be no point to the blessing. The goal and meaning of healing is decidedly *not* about making our bodies "better" or "more perfect" or more like some ideal represented by a Greek statue. All bodies, no matter how impaired, are occasions for praising God as creator of all. How, then, are we to understand the responsibility to heal, the power of medicine and its limits, from a Jewish theological perspective?

Healing, Holiness, and Idolatry

I think Maimonides best captured the essence of the classical Jewish view of health and healing when he wrote:

> He who regulates his life in accordance with the laws of medicine with the sole motive of maintaining a sound and vigorous physique and begetting children to do his work and labor for his benefit is not

following the right course. A man should aim to maintain physical health and vigor in order that his soul may be upright, in a condition to know God. . . . Whoever throughout his life follows this course will be continually serving God . . . because his purpose in all that he does will be to satisfy his needs so as to have a sound body with which to serve God. Even when he sleeps and seeks repose to calm his mind and rest his body so as not to fall sick and be incapacitated from serving God, his sleep is service of the Almighty.[19]

Medicine is valuable because it is essential to the human flourishing that alone enables us to fulfill our divine mission. In order to do God's work in the world, we must first and foremost make every effort to preserve life against any and all forces that threaten it. Moreover, we must ameliorate any condition that diminishes our ability to observe God's law and to do deeds of loving-kindness. In short, the human body and its health are valued not as ends in themselves but as the necessary means to the end of serving God and bringing more holiness into the world.[20] Dysfunction and pain—whether physical or psychological—make this work more difficult and, in extreme cases, impossible. Healing the sick, then, is just a way of ensuring that our ability to serve God is restored, as much as that is possible.

But how does this theological perspective on health and healing shape our response to the challenges of genetic technologies discussed above? To answer this question we must note, first of all, that the challenges we face in this area are fundamentally different in degree, not in kind, from those that medical science has always placed before us. The new knowledge we glean from decoding the human genome is powerful, to be sure, but so too was the discovery of germ theory in its day. The power to alter our genetic makeup is extraordinary, but so too is the power of vaccination and of organ transplantation, both of which involve altering the patient's body in dramatic ways. These technologies are radically new, but the moral and theological questions they raise are not. And those questions, again, are: What responsibilities go along with these new powers? How are we to balance our responsibility to heal with our responsibility to honor our unique and precious human nature?

Answers to these questions may emerge from the theological perspective on medicine outlined above. For if medicine as a whole is meant to enable us to serve God, it follows that any particular medical therapy is valid only to the extent that it furthers this purpose. Any medical therapy that made serving God more difficult would be self-defeating. It might serve our bodies, but only at the cost of ignoring the fact that our bodies are meant to serve God. The practical implications of this line of reasoning will become clearer when we consider for a moment what the tradition has to say about imitating God and furthering God's purposes in the world.

The Bible repeatedly proclaims that Israel's purpose is to fulfill God's commandments, thereby bringing the world closer to God and to the culmination of human history, when "God will be one and God's name will be one."[21] The goal of Jewish religious life is the imitation of God, which serves to bring God's presence into the world. When the Torah requires that Jews "walk in all God's ways," the Rabbis comment, "Just as God is gracious and compassionate, you too must be gracious and compassionate."[22] Indeed, the classic articulation of the thirteen attributes of God found in the holiday liturgy (based on Exodus 34:6–7) is: "The Lord, the Lord is a merciful and gracious God, slow to anger and abounding in kindness and truth. He keeps kindness to the thousandth generation, forgiving iniquity and transgression and sin, and acquitting the penitent."[23] To imitate God is to strive to emulate these qualities of compassion and forgiveness. In a more universalist vein, Micah proclaims, "Man, what does the Lord require of you? Only to do justice, love mercy, and walk humbly with God."[24] The Rabbis understood the purpose of all people, Jews and non-Jews alike, as living a life of righteousness, thereby honoring God and God's world. As they put it, "The righteous of all nations have a share in the world to come."[25] Insofar as the purpose of human life is to make of this world a godly place, to imitate God and to sanctify all aspects of life, anything that enables such activity is good, anything that prevents it is bad.

Within this theological framework, we can begin to construct a response to the challenges of genetic technologies. The new powers that these technologies put at our disposal—through the knowledge of the human genome and the potential to engineer our genes—are moral precisely to the extent that they facilitate living a holy life. Such a life will require that we think of ourselves, first, as subservient to God and hence avoid the prideful temptation to think that we are the "masters of our fate." This temptation has become ever more difficult to resist as the power of biotechnology has increased; the tremendous expansion of scientific knowledge itself tends to foster a certain grandiosity in us. In our historical setting, then, it is especially important to heed the message of the Mishnah cited earlier, that dangerous knowledge should always be accompanied by an attitude of reverence and should be vouchsafed to those "whose heart is anxious within them." Living a holy life also means remaining mindful of the fact that a righteous life is lived in community with others. Deeds of righteousness and holiness are, by definition, not deeds done primarily to serve one's self-interest; rather, they further the interest of the community as a whole, perhaps the human race as a whole, bringing everyone closer to God.

Still, we must attempt to translate these general norms into more specific criteria. To the extent that decoding the human genome increases our awe in relation to the complexity of human life or reminds us that all human beings are fundamentally one (since our genes are 99.99 percent the same), it is surely a blessing. It fosters an awareness of the mysterious Source of all life and reminds us of our duty to treat all others with the respect owing to a creature of God. To the extent, on the other hand, that it reinforces prejudices against "outsiders" or is used by any group to claim superiority over others, it surely presents an obstacle to living righteously. Insofar as genetic engineering promotes health and restoration of physical or mental functioning, thus enabling people to serve God more fully, it is good. Insofar as it grows out of (or promotes) a reverence for the power and beauty of the body in its own right or is used for purely self-serving ends, it diverts us from serving God. Indeed, it replaces service of God with service of the body or the self, which can only be regarded as a form of idolatry.

Here, then, we discover the theological foundation that can help us to mediate between the two midrashic texts cited earlier. Our purpose on earth is to live reverently, serving God in all we do. Sometimes this requires that we "cultivate" the body that God has given us; at other times, it requires that we honor the natural order of things just as God has created it. Just as we do God's bidding both when we labor for six days and when we rest on the Sabbath, so too we must distinguish between the times when we serve God through medical interventions and those when we accomplish the same purpose through refraining from such manipulations of our natural state.

This criterion, broad as it is, helps us to see why certain uses of genetic engineering would be illegitimate while others might be permitted, even required. Altering my genes to eliminate a debilitating condition is good, for it enables me to pursue (potentially, if not actually) the work that God put me on earth to do. Altering my genes, or those of my descendants, for the purpose of making them star athletes or gifted musicians is not good, for this puts the technology in the service of a purely human goal; it idealizes, even idolizes, a specific human condition as an end in itself. To put it starkly, having more great athletes or musicians in no way furthers the moral and religious goals for which we were created, which is to do deeds of righteousness and holiness. To cure disease or relieve pain, on the other hand, does promote that goal, for it removes a significant impediment to its pursuit.

This theological perspective points to one further specific issue that deserves attention. If further research into the human genome were to identify a specific gene that encouraged violent behavior and that seemed to have no socially positive effect, I suggest that it would be morally appropriate (and perhaps even required) to alter this gene (again, assuming that this could be done safely, that we knew the long-term consequences of this change, etc.). For this would be genetic engineering in the name of creating people more prone to righteous behavior (or, at least, less prone to destructive behavior), and this would put the technology directly in the service of living a godly life. A genetic change that vastly improved our intelligence, on the other hand, could not be sanctioned on the same grounds, however, because intelligence alone is neither a necessary nor a sufficient condition for either righteousness or reverence.

By these criteria, genetic technologies will certainly be a mixed blessing, both good and bad, depending entirely on the purposes for which they are used and the motivations that give rise to their development. In this respect, too, they are fundamentally no different from any other technologies. To expect otherwise is surely to be naive, for what human artifact is *entirely* or *intrinsically* either good or bad?[26] The power that technology gives us is invariably *neutral*, and our moral evaluation of it depends on its use, actual or intended.[27] What makes genetic technologies special, if anything, is that the power they give us seems dramatically greater, or more far-reaching, than in many other cases. Accordingly, both the potential blessing and the potential curse they pose weigh more heavily upon us. From a Jewish theological perspective, they can be used to facilitate a life devoted to either holiness or idolatry. The challenge we face, therefore, is to maximize the positive potential of these technologies to ensure that in their development and employment, we will "give due honor to our Creator."

It is tempting when confronted with new technologies to declare them either desirable or undesirable, good or bad. But the foregoing analysis has pointed to the need for a more nuanced approach to these moral issues. Genetic technologies are inescapably *both good and bad*, depending on whether the purposes for which they are employed further or hinder a life of holiness and reverence. We should not be swayed in our moral response to these technologies by the fact that their creators intend only that they be used to promote health and healing, uses that Jewish theology would surely endorse. For in the real world it is clear that these technologies can and will be used for purposes that Jewish theologians would just as surely regard as idolatrous. There are not, nor can there be, any guarantees that the noble uses of these new powers will outweigh the ignoble ones, just as no such guarantees exist for the practice of medicine in general.

All of this brings us back to the insight of the Mishnah, that dangerous knowledge should be transmitted (insofar as this is possible) only to those who give due honor to their Creator. A Jewish assessment of the ethics of these dramatic new powers will depend largely, if not entirely, on the religious orientation and the moral sensibilities of those who wield them.

Concluding Midrashic Reflection

In exploring these technologies and the theological resources within Judaism for thinking about them, I have reflected many times on the intriguing blessing that the tradition asks us to recite on the occasion of hearing good news: "Blessed are You, Lord our God, who is good and who brings forth goodness." On the most basic level, this blessing is surely meant to remind us to express our gratitude to God, who is, after all, the source of all fortuitous events in our lives. Accepting good fortune, our own or anyone else's, without acknowledging God would be akin to eating the fruit of God's world without first reciting a blessing acknowledging God as the source of our sustenance.[28]

On a deeper level, though, I think this blessing points us back to the biblical account of creation. Certainly, one of the central themes of that narrative is that, at the conclusion of each day's work of creation, "God saw that it was good." The created order and each component within it are proclaimed to be inherently good. It is not long, though, before the narratives of Genesis turn to the many ways in which human beings are capable of misusing the powers and misappropriating the gifts God has given them. The goodness of the world, apparently, is rather fragile, subject to human manipulation and defilement. Indeed, only a few chapters after the dramatic account of creation in Genesis 1, God "regrets" having created human beings and sets about to destroy them and return the world to watery chaos.

The effect of these stories is to teach us something about the goodness of creation. God's creations, including human beings, it turns out, are good only provisionally, not ontologically. They are "good" in the sense that God created them, if you will, "out of the goodness of God's heart" and with the intention that this world become a place where goodness will flourish. In this sense, they are only potentially good; what becomes of them in actuality depends very much on what we do with them. The following midrash on creation makes this point forcefully:

> When the Holy One created the first man, He took him and led him around all the trees of the Garden of Eden, and said to him: "Behold My works, how beautiful, how splendid they are. All that I have created, I created for your sake. Take care that you do not become corrupt and thus destroy My world. For once you become corrupt, there is no one after you to repair it."[29]

What we make of the world, whether we actualize its potential goodness and holiness or destroy it, is entirely in our hands.

In response to each new development in the field of genetic technology, we might well recite this blessing, for each is surely "good news." But the blessing, I think, contains a hidden warning, as well. It reminds us that while God is the source of what is good, what is good does not always "bring forth goodness"; indeed, sometimes it can be made to bring forth just the opposite. What presents itself to us as good must still be used for good ends. When we are fulfilling the role that God intended for us, we use what is given to us—our knowledge and our power—to "bring forth goodness." Genetic technologies afford us yet another opportunity to decide how we will use our newfound powers. Like everything else that comes from God, the fact that we have these technologies is an expression of God's goodness and is intended for the benefit of humankind. Whether we use them to serve purely human and ultimately self-centered ends, or whether we use them to further our capacity for living a holy life, thereby extending God's goodness in the world, depends on us.

1. Clifford Geertz, "Religion as a Cultural System," in his *The Interpretation of Cultures* (New York: Basic Books, 1973), 89.
2. James M. Gustafson, *Ethics from a Theocentric Perspective*, vol. 1, *Theology and Ethics* (Chicago: University of Chicago Press, 1981), 113.
3. See especially Hans Jonas, *The Imperative of Responsibility: In Search of an Ethics for the Technological Age* (Chicago: University of Chicago Press, 1984).
4. I have not included here the ethical challenges of embryonic stem cell research insofar as the central issue here appears to be only the manner in which these cells are derived, that is, through the destruction of a nascent human embryo, or blastocyst. No one, to my knowledge, considers the use of such cells to generate human tissues for transplantation a morally problematic enterprise in itself. See the articles in Suzanne Holland, Karen Lebacqz, and Laurie Zoloth, eds., *The Human Embryonic Stem Cell Debate* (Cambridge: Massachusetts Institute of Technology Press, 2001).
5. It is important to differentiate between the risks associated with the development of these technologies, especially those related to the safety of patients in treatment protocols, and the risks associated with the use of these technologies once developed and established as safe. The former risks are overseen by federal agencies responsible for medical research and the development of medical products (e.g., NIH, FDA). I am concerned here with the latter risks, those associated with the use of technologies after they have been safely developed and made available to the public.
6. Presentation to Jewish Discourse Project of the American Association for the Advancement of Science Program of Dialogue on Science, Ethics and Religion by Alan Guttmacher, "Human Genome Project: Hope or Hype?," 9 December 2001.
7. See, for example, Mark S. Frankel and Audrey R. Chapman, *Human Inheritable Genetic Modifications* (Washington DC: American Association for the Advancement of Science, 2000).
8. See, for example, the articles by Fred Rosner and Arziel Rosenfeld in Fred Rosner and J. David Bleich, eds., *Jewish Bioethics* (New York: Hebrew Publishing Company, 1979). The authors cite only a few classical sources, and those they do cite (such as Rabbi Yohanan's comment that young women emerging from the ritual bath would look at him and hope that in doing so they might give birth to sons equally handsome) are laughable as precedents for questions of eugenics and genetic engineering. Notably, Elliot Dorff, who discusses this genetic engineering briefly in his *Matters of Life and Death* (Philadelphia:

Jewish Publication Society, 1998), avoids offering an halakhic opinion on the subject but notes the urgency of tackling the "philosophical and moral tasks of defining the line between therapeutic and non-therapeutic uses of this technology" (164).

9. The notion of *da'at Torah* is that rabbinical authorities are sometimes in a position where they issue a halakhic ruling not on the basis of precedent but based on their own extensive knowledge of the tradition. This presumably happens in a case, much like the one under consideration here, where no precedents exist. Walter Wurzburger, in his *Ethics of Responsibility: Pluralistic Approaches to Covenantal Ethics* (Philadelphia: Jewish Publication Society, 1994), characterizes this as an instance where "the residual influence of exposure to *halakhic* categories of thought makes itself felt in areas where the law itself cannot be applied" (33).

10. Another, oft-cited source in discussions of Judaism and genetic engineering is the prohibition against *kilayim*, the interbreeding of diverse species (see Leviticus 19:19). But this is plainly no precedent for the matters at hand. The biblical prohibition concerns animals (not people), and its rationale would seem to be the desire to honor the distinctions among species, not (like the rationale of genetic engineering) to heal life-threatening illnesses. Moreover, any attempt to prohibit these new medical technologies on the grounds that they are "unnatural," or that they introduce "foreign" elements into the human body, must explain how these techniques differ from the introduction of vaccines, artificial limbs, or pacemakers—all of which are deemed permissible by halakhic authorities.

11. M. *Hagigah* 2:1.

12. According to the Talmud's discussion, this is the reason for the restriction concerning the teaching of laws regarding forbidden sexual relationships. For while the teacher is engaged in responding to one student's question, the remaining two might continue studying on their own and mistakenly come to believe that some sexual act is permitted when in fact it is not. See J. *Hagigah* 2:1.

13. B. *Hagigah* 13a.

14. *Sifre Deuteronomy* 307; *Genesis Rabbah* 12:1. The midrash plays on the word *tzur* (rock) in Deuteronomy 32:4, reading it instead as *tzayyar* (craftsman).

15. *Midrash Samuel* 4, as cited in Hayim Nahman Bialik and Yehoshua Hana Ravnitzky, eds., *The Book of Legends*, trans. William G. Braude (New York: Schocken, 1992), 594–95.

16. The Rabbis skillfully point out the man's inconsistent application of this theological principle. But we need only consider the views of contemporary Christian

Scientists or Jehovah's Witnesses with respect to medical interventions to appreciate that such a view can be held consistently.

17. For a penetrating, and poetic, exploration of these themes, see Abraham Joshua Heschel, *The Sabbath* (New York: Harper Torchbooks, 1966). Irving Greenberg has argued that this dialectic between work and rest, repairing the world and honoring creation, is integral to the concept of covenant and can be useful in formulating positions in biomedical ethics. See his "Toward a Covenantal Ethic of Medicine," in *Jewish Values in Bioethics*, ed. L. Meier (New York: Human Sciences Press, 1986), 124–49.

18. The same theological point, I think, is expressed through the notion that "one must bless God over the evil [that occurs to us] just as one blesses God over the good." See M. *Berakhot* 9:5.

19. M.T. *Hilkhot De'ot* 3:3, quoted in Elliot Dorff, *Matters of Life and Death* (Philadelphia: Jewish Publication Society, 1998), 26.

20. It is standard in many presentations of Jewish health-care ethics to suggest that the preservation of life is an absolute value in Judaism, not relative to any other purpose or goal. But classical sources affirming the value (in extreme cases, even the requirement) of martyrdom belie such simple formulations. In certain cases, one is required to sacrifice one's life for a higher good, that of serving God and fulfilling the most important of God's commandments. The same is true, I suggest, in the realm of health care.

21. Zechariah 14:9.

22. Deuteronomy 11:22; *Sifre Deuteronomy*, Ekev, Piska 49.

23. Translation from Philip Birnbaum, *High Holyday Prayer Book* (New York: Hebrew Publishing Company, 1951), 278.

24. Micah 6:8.

25. T. *Sanhedrin* 13:2; B. *Bava Batra* 10b.

26. There are, perhaps, certain exceptions to this generalization, as, for example, instruments of torture.

27. I am mindful, as Alan Weisbard pointed out to me, that technology often feels seductive, that it seems to exploit our vulnerability in that it holds out the promise of making our lives better in ways that we can resist, if at all, only with great effort. In that sense, it may be misleading to talk of technology as morally "neutral." Yet, I want to suggest that this seductiveness is, at least partly, a product of our own misguided values. If we could adopt the sort of religiously grounded perspective on life, health, and medicine that the classical Jewish sources seem to promote, we would readily forgo the pursuit of that knowledge and those technologies that threaten the religious values of paramount importance to us. We see examples of this in the willingness of

Christian Scientists to forgo much of modern medicine, and also (to take a less extreme example) in the willingness of many religious people to forgo life-extending treatments in the face of imminent death.

28. Some prayer books indicate that this is the blessing recited if the good news affects others besides the one who hears it; otherwise, one recites *sheheheyanu*. See *The Hirsch Siddur* (New York: Feldheim Publishers, 1972), 718. The classic expression of acknowledging God as the source of our sustenance is: "A man should taste nothing before he utters a blessing. Since 'The earth is the Lord's and all that it holds' (Ps. 24:1), a man embezzles from God when he makes use of this world without uttering a blessing" (T. *Berakhot* 4:1).

29. *Genesis Rabbah* 8:1; B. *Sanhedrin* 38a.

CHAPTER 18 *Jewish Reflections on Genetic Enhancement*

JEFFREY H. BURACK

As twenty-first-century Westerners in love with technology, many of us are nonetheless wary and fearful of genetic interventions that might make people taller, faster, stronger, or more aggressive, that might make eyes bluer or hair blonder. As Jews, some of us may be especially hostile to such engineering because of its echoes of the recent past's murderous eugenics, and because of the fear that it portends ever greater veneration of a model of humanness that looks ever less like our image of ourselves. But what about enhancements that are less superficial but rather aim at enhancing what we, as Jews, genuinely value in ourselves? What if genetic engineering could make us smarter, more thoughtful, better learners and scholars? More honest, hardworking, and cooperative? What if we could be made more compassionate and fair-minded, more beneficent and loving? More dutiful, humble, and devoted to seeking holiness? What, if anything, would be wrong with seeking to reshape ourselves in these ways?

There has been a good deal written in secular bioethics on these questions. My aim is to reflect on how specifically *Jewish* perspectives might inform our approach to them. What are the sources of the suspicion and revulsion that genetic enhancement arouses in us? And what is added by reflecting on them from a Jewish perspective? Many questions arise: What are appropriate or laudable ways to think about our relationship with our embodiment? How ought we to think about what it means to make ourselves better? Is it meaningful to talk at all about human enhancement? If so, how does enhancement differ from, say, medical therapy? How are we to think about whether and when to undertake such engineering? What are permissible or even obligatory ways to do this? How ought we to think about potential consequences, including not only risks and side effects, but also repercussions to the ways future humans will live?

Disclaimers

I will devote some attention later to the principle of *anavah*, or humility, so let me begin with some disclaimers about myself and my aims. I am a scholar neither of *halakhah* nor of Jewish history and tradition. I am a physician, trained in philosophy and biomedical ethics, but also a Jew, steeped in and responsive to the Jewish and secular North American cultures in which I participate. I have been privileged to listen to and learn from participants in the project on which this volume is based, the International Jewish Discourse Project on Religion and Ethics at the Frontier of Genetic Medicine of the American Association for the Advancement of Science. Rabbis, scholars of Jewish thought, bioethicists, and social scientists, as well as leading scientists in genetic and regenerative technologies were involved in the project. This essay reflects my efforts to recast questions frequently raised in secular bioethics through a lens constructed of the Jewish and scientific thought I have absorbed.

I use the terms "Jewish" and "Judaism" here in Mordecai Kaplan's sense, to refer to the diverse and evolving religious civilization in which North American Jews currently participate, including not only our various theological beliefs and ritual practices but also our shared traditions, literature, culture, and values.[1] My intent is to avoid ideological and denominational divides but instead to apply what I take to be widely shared Jewish attitudes, beliefs, and sources. I make no claim to speak for any Jewish group or movement, and I thus accept the risk that I am speaking for no one but myself.

Goals

Of what use can it be to offer Jewish reflections on such social policy issues as our stance toward genetic engineering? This type of analysis can typically serve one or more of four purposes:

1. advancing halakhic or intracommunity deliberation as to which positions to adopt on matters of policy and personal behavior;
2. contributing to public policy making by offering a "Jewish position" to be considered alongside other religious and ethical perspectives;
3. contributing to public understanding and discourse by reflecting on contemporary issues from the perspective of one, or one *version* of one, wisdom tradition; and
4. helping Jews, whether they view themselves as halakhically adherent or not, make choices for themselves regarding personal behavior and support for policy directions.

My purposes include neither of the first two: I make no claim to be generating either *halakhah* or a "Jewish position." Rather, I believe that refracting this contemporary issue through a Jewish lens may add clarity and substance to the wider, secular discussion and will also be of interest to individual Jews.

Example: Modafinil

To return to our questions about human enhancement, let me borrow a pharmacologic example from Paul Root Wolpe's important paper on neuroenhancement.[2] Modafinil, a drug marketed since 2002 under the name Provigil, is a psychostimulant that improves wakefulness and alertness. Unlike earlier stimulants, modafinil appears to have few side effects and low addiction potential, and it does not interfere with sleep. In fact, it is prescribed to reverse the excessive sleepiness associated with conditions like narcolepsy and sleep apnea. But modafinil's potential uses go far beyond treating these disorders. Healthy users report increased alertness and attention to tasks. The U.S. military tested the drug in sleep-deprived helicopter pilots and found that it improved not only performance but also measures of well-being. Clearly, there might be a wide market for nontherapeutic uses of such a drug—surely by those who need to stay awake, but also by those who simply could benefit from greater alertness. And who among us could not?

Wolpe asks a number of questions regarding the appropriate uses of such a drug: "Should modafinil be prescribed solely as a medical drug, properly used only for those suffering from sleep pathology? . . . If so, third party payers, including government programs, should cover it. Or, should it be classified as an over-the-counter drug, available to anyone who wants it? Or, should we create a class of drugs available only to those who can show legitimate social need, those whose fatigue might put others at risk, such as airline pilots, or truck drivers?"[3] Wolpe goes on to ask whether we should restrict modafinil's use in other defined social settings, in which it might confer unfair advantages to some, prevent acquisition of important learning skills in others, or lend itself to abuse. The yeshivah and the rabbinical seminary might be two such settings.

Why choose a pharmacologic example to frame an essay about *genetic* enhancement? For several reasons: first, to make the point that concerns about enhancement are not unique to genetic interventions. Second, because the sort of enhancement at issue, of mental clarity and acuity, is especially challenging from a Jewish point of view. Efforts to make our children taller or our athletes stronger are easy targets for critique; not so for serious efforts to improve qualities we value. And third, because such chemical enhancements are already with us, while for the moment their genetic analogues remain in the realm of science fiction. Our deliberations about *available* therapies will prepare the way for how we think about now-hypothetical genetic interventions.

And they may not remain hypothetical for long: genetic intellectual enhancement has already happened in mice. In 1999 scientists in Joe Tsien's Princeton lab inserted additional copies of the gene *Nr2b*, which codes for a receptor for the neurotransmitter NMDA. The engineered mice remembered objects and spatial relations better, and performed better on learning tasks.[4] In all the ways that we measure mouse intelligence, the *Nr2b*-enriched mice were smarter.

Consider the laudable goal of learning more Torah. I am acutely aware of my modest, and ever deteriorating, powers of attention, concentration, and memory and the limits they place on my ability to learn. In the service of learning Torah, what could be wrong with taking modafinil, or with availing myself of the analogous gene therapy, such as getting myself an extra *Nr2b* gene once it becomes available to humans? Like other religious and moral traditions, Judaism offers the background beliefs and principles against which such questions are posed, including:

1. an ontology of the meaning and purpose of human life;
2. recommended aspirations and constraints on human desires;
3. specific constraints on action in the pursuit of personal or collective goals in the form of positive and negative commandments; and
4. mechanisms for deliberation about action, including rules about argument, appeals to authority, and the weights to be assigned to different consequences—and in particular, about duty and responsibility.

Secular Critique

It is against these background Jewish beliefs that I want to highlight the chief concerns of the secular bioethics critique of genetic enhancement. I will draw a few examples from the Hastings Center's Enhancement Project, which produced a 1998 report incorporated in a subsequent volume of essays.[5] Just as an aside, I note that the cover of the report juxtaposes two drawings of a woman—presumably "before," with a big, perhaps Semitic nose, and "after," with bump and hook deleted, looking rather like Bette Davis. This should serve to remind us, lest we are tempted to feel morally superior, that as Jews we have been quick to avail ourselves of supposed enhancement technologies and have done so at least sometimes in part in order to seem or be less "Jewish." It is also therefore a reminder, superficial perhaps but clear, of the potential homogenizing power of these technologies to eliminate ethnic and other ancestral differences, whether voluntarily or coercively. But my purpose in turning to the Hastings project is not to criticize their graphics but to summarize what I take to be the major sources of their disquiet with enhancement. These include:

1. that genetic technologies tamper inappropriately with what makes us essentially human and individual;
2. that enhancement is an abuse of biomedical technology properly reserved for treatment, which aims to restore health and normal functioning;
3. that the very idea of enhancement valorizes specific conceptions of which human characteristics are normal and desirable, thereby devaluing persons who do not have them;
4. that the availability of such technologies within our present market-based economy would offend standards of distributive justice by exacerbating existing inequities in capacity and opportunity;
5. that using technology to enhance ourselves may undermine the virtue of living and struggling with our imperfections;
6. that the availability of enhancement technologies promotes human aspirations that are ignoble and destructive and that conflict with or undermine important social goals; and finally
7. that genetic interventions risk incalculable harm, both foreseeable and not, and different in kind from the risks of other biomedical interventions.

For readers who might benefit from modafinil right about now, I will cut to my conclusions so that attention might be allowed harmlessly to wander. While all these concerns find resonance in Jewish thought, I want to assert and try to explain some contrasts and suggest that *Jewish* worry focuses in somewhat different places. A Jewish view, for example, may be more permissive about social uses of technology but more restrictive about personal aspirations and behavior. I will suggest that the concept at the core of Jewish apprehensions about enhancement is *anavah*, or "humility." We must be humble in our appreciation of human diversity and struggle, about our ability to know in detail what is genuinely expected of us, and about our ability to foresee the consequences and implications of our actions. Humility is threatened by the appeal of enhancement, but in Judaism it may be held in place by our sense of duty to God, self, and others. There: feel free to nod off now!

Naturalism versus Cocreation

Despite avowedly secular roots, bioethics harbors an anxiety that it is not our business to tamper with our genetic essence. From the Christian and Greco-Roman traditions that preceded it, philosophical Anglo-American bioethics has inherited a strong naturalist bent that has only been enhanced by its self-assigned role as critic of scientific biomedicine. Naturalism holds that we fundamentally are as given by God, Nature, or evolution and that it is wrong to interfere with or try to change our quintessence. This attitude is reflected in the Edenic and Promethean myths, in which ultimate punishment is reserved for the crime of arrogating the creative spark or knowledge that should be reserved for the Divine.

Oddly, Judaism, though it acknowledges a Creator and our own createdness, does not seem to share this naturalist inclination. In Jewish tradition, what is fundamental to us, our creation in the divine image, is something we cannot change no matter how we might try—so the essentialist concern is taken off the table. And regarding everything else, we are not only permitted but *commanded* to participate with God as cocreators, improving the world—which, after all, is also God's creation. We are reminded at every ritual occasion of our obligation to perform *tikun olam*, the healing and improvement of the given world. The first blessing to mark sacred time, on Shabbat or festivals, is on the light, which God has created but which we re-create in smaller, human ways by kindling candles. The second is on wine, a self-conscious symbol of cocreation. God creates the vine and its fruit, but it is human intervention that turns it into wine, symbol of celebration and conviviality. As an interesting side note, this particular cocreation is one we then use to alter our own consciousness. Far from either the cocreation or its use for enhancement being forbidden, both are commanded. And the natural, God-given raw material alone will just not do it.

Medicine is, of course, also fundamentally antinaturalist, in the sense that medicine's most basic function is to *not* allow disease processes to take their "natural" course. By no accident, then, have Jews and Western medicine shared a close historical relationship, as Elliot Dorff and others have pointed out.[6] It is a basic attitude of Judaism that God has given us the power to heal each other, just as much as God has given us anything else, and we are commanded directly to use that healing power. We do not worry so much about "playing God" with the natural world, because by working to improve God's world, what we are actually playing is *human beings*, reverentially and according to the script given us. We are, nevertheless, acutely responsible for the consequences of those manipulations—all the more so because the objects of our actions never belong to us alone.

Stewardship

A central tenet of the Jewish attitude toward healing is that we do not own our bodies, or even our lives, but rather hold these gifts from God as stewards. The fiduciary obligation of stewardship implies duties both to heal others when we can and to accept healing when in need. The obligation of *pikku'ah nefesh*—saving a life—is far-reaching and so fundamental that for its sake all but the most sacrosanct commandments must be violated.[7] As a result, when the goal is the promotion or preservation of human life, Jewish authorities have tended to be relatively tolerant of the means. A striking recent example is Rabbi Moshe Tendler's 2000 testimony to President Clinton's National Bioethics Advisory Commission, in which he surprised many by arguing that, from a halakhic perspective, pursuing stem cell research is not just permissible but obligatory, even though it requires the destruction of human embryos. Tendler concludes unequivocally that "the moral obligation to save human life, the paramount principle in biblical law, supersedes any concern for lowering the barrier to abortion by making the sin less heinous."[8] So Jewish authorities tend to be fundamentally more permissive with respect to biomedical policy in that interventions generally are not ruled out merely because of the *types* of interventions they are. Yet at the same time, and equally fundamentally, among the range of socially tolerable options, Jewish tradition may be *less* permissive with respect to the options open to the individual, whose decisions after all are not private but fiduciary.

If something I have borrowed from you breaks while under my care, I am responsible for seeing it fixed before returning it. As Benjamin Freedman points out, whether or not to undergo medical treatment is likewise viewed in Jewish tradition not as a matter of personal choice but rather as an obligation.[9] From a perspective of stewardship, we ask whether we are permitted to put what is not ours at risk, or, conversely, whether we are permitted to refrain from accepting that risk. Just as we are forbidden to mutilate the body, which is on loan from God, we may not frivolously take risks with our health. The question of what makes such a risk frivolous suggests one vantage point from which to view the permissibility of enhancement.

As we move along one continuum from clearly life-saving to clearly optional interventions—optional, that is, with respect to biological survival—or along another continuum from less risky interventions to riskier ones, our obligations to preserve life weigh increasingly heavily against intervention. Even a genuine enhancement may be something for which one may not put one's life at even small risk. In this vein, for example, Freedman asks the following questions about cosmetic surgery: "First: Is it consistent with the task of a reasonable caretaker to undergo self-wounding and its accompanying pain . . . on behalf of the promised improvement in appearance? Second: Is he allowed to undergo risk toward this end?"[10] On Freedman's view, most authorities agree that, in answer to the first question, Jews are permitted to undergo self-wounding and pain for what they reasonably judge to be greater goods. But writers divide on the second question, of permissible risk—especially since surgery may entail the risk of death. A recent spate of highly publicized deaths during liposuction tragically illustrates these hazards. But many modern cosmetic techniques involve little risk. And what of enhancement interventions that do not involve surgery at all? Freedman asks: "Is there a level of risk that is so low that it may be disregarded by the reasonably conscientious caretaker?"[11]

Substantive Moral Vision

The most basic difference between Judaism's moral framework and the dominant secular one is that while both are aspirational, Judaism tells us in some detail *to what* we ought to aspire. That is, it presumes and promotes a substantive vision of the purpose of human life and of what gives life value. While there are probably at least as many versions of this vision as there are Jews, I will work from a general formulation that I hope is relatively uncontroversial, following the prophet Micah: "He has told you, O man, what is good, and what *Adonai* requires of you: only to do justice and to love goodness and to walk humbly with your God."[12] We are commanded to be just, kind, and humble and to strive toward holiness through the imitation of God and through lives lived in consciousness of the Torah. We are called to act, to participate with the Divine and with our communities in *tikun olam*. It is fundamental to religious thought, and certainly to Jewish thought, that we strive toward betterment of ourselves and of the created world and that the characteristics of bettered selves and a better world can at least in rough outline be agreed upon and described. A substantive moral framework, then, gives the concept of enhancement a meaning that is independent of our personal desires. This may seem an obvious point, but its impact on how we understand enhancement is enormous. The result is an explicit vision of what an enhanced human might be that is less ambiguous, and therefore more contestable, than what emerges from secular deliberation.

Enhancement implies a standard, norm, or goal against which the value of an object is judged. Sometimes the implied predicate seems obvious and straightforward: a better tool is one that performs its specified task more easily or effectively. But what does it mean to enhance a *person*? What *can* it mean, without an agreed-upon set of standards for what makes a better person? Virtue or character ethics attempts to establish, or at least to query, such standards. But in contemporary biomedicine, virtue ethics is typically invoked only in reference to physicians and other professionals, of whom we ask what might be desirable or appropriate dispositions. Such language is generally absent from discussion of the sort of choices individuals or communities should make about, for example, health care. This is, arguably, as it must be. Secular bioethics aims primarily at settling disputes and adjudicating competing claims among stakeholders in a deeply pluralistic society. Of necessity, then, it has difficulty establishing anything but *procedural* principles for collective deliberation and *prudential* principles for personal deliberation. The usual assumption is that individuals choose so as to maximize their ability to pursue their personal ends. Contrast, for example, the standard secular argument as to why one should not smoke—forgoing short-term pleasure in the service of one's longer term, but still personal, goals of a longer and healthier life—with the Jewish argument that the body belongs to God and that the individual, as steward of the body, may not casually pollute it or cause it harm.

It seems obvious that even if every desired modification were permissible, *not* every desire should have its fulfillment socially guaranteed. But with no point of view from which to answer the ontological question of what genuinely enhances us, secular bioethics has been preoccupied instead with the distinction between enhancement and treatment.[13] Treatment is understood as the use of biomedical technology to remedy what we agree to consider a disease or injury and is contrasted to *mere* enhancement, by which is meant the use of that technology toward any other desired end. This heuristic is then put to work to decide what we should research, guarantee, pay for, and so on. We *think* we are deliberating about what are enhancements and what are not, but we are in fact worrying about which demands impose legitimate duty claims on the rest of us. To put it differently, the bioethical treatment-enhancement distinction tries to draw important normative implications from a conceptual and at least superficially value-neutral distinction—a project that Hume's is-ought problem would suggest may be impossible.

From a Jewish perspective, though, to be enhanced means to be better suited to fulfill our human purpose. We have at least a general sense of what this is—and we have instruction manuals. From a secular perspective, we ask: Is this treatment, or is it mere enhancement? From a

Jewish perspective, we ask instead: What is the goal of this intervention, and does it seek holiness? Does it serve us in our roles as caretakers and stewards of our bodies and our world?

Diversity and God's Image

Let me turn to the "expressivist" critique of enhancement. This position, eloquently argued by Anita Silvers and others, holds that the very pursuit of biomedical enhancements, and even the willingness to view nontherapeutic interventions *as* enhancements, expresses negative attitudes toward some existing persons. By valorizing specific human characteristics, seeking enhancement devalues and discriminates against the disabled—or even just those who have less of those characteristics.[14] As human traits begin to appear optional, subject to being acquired or improved rather than just allocated by the genetic lottery, there might be even greater stigma attached to not having them.

Judaism establishes a set of aspirations for human life and would therefore seem inevitably to favor some forms of life over others. Does it therefore imply negative attitudes toward certain human characteristics and toward the persons who display them? Yes, and no. A substantive ethical system certainly must attach positive moral valence to some characteristics and negative worth to others. But it seems to me that the sort of threat Silvers and others see is minimized by three features of the Jewish view of human purpose and value. These are, first, our universal creation *b'tzelem Elohim*, that is, in the image of God; second, the nature of the actual qualities considered desirable, holy, and constitutive of that image; and third, the spirit of *anavah*, or humility, in which we must approach this analysis and our own self-improvement.

As Bernard Steinberg has put it: "In the cosmic context, differences of biological strength, social stature, economic power, and cultural achievement are reduced to secondary importance. These differences are marginal to the authentic center of human identity. Moreover, aware of the essential limitation of his finitude, the *hasid* knows, objectively speaking, that he is no more worthy than his neighbor. The standard of objectivity is not social convention but relation to God."[15] The source of human worth is the reflection of God's image, not any constellation of observable characteristics.

What precisely *is* God's image is, of course, ineffable, but its manifestations are manifold. A lovely passage from Mishnah *Sanhedrin* (4:5) tells us that when we mortals cast coins, usually with the face of a human ruler, we see precisely the same face on every coin. But the divine ruler has cast countless images from a single original mold, and, like snowflakes, every one of us is different. Furthermore, that original mold was of course itself fashioned in God's *own* image. Judaism's pantheon does not reflect human perfection, physical or otherwise: Jacob limps, Moses stutters, and families by and large are pretty dysfunctional. It may sound trite in our multicultural world, but, as Elliot Dorff has pointed out, Jews are commanded to *celebrate* difference and to respond to disability with gratitude rather than with revulsion or disdain, praising God *meshaneh ha-briyyot*, who creates us different. (I am indebted here to Avram Israel Reisner, who has cautioned that this celebration of difference is by no means uniform across canonical Jewish texts, many of which contain elements of nationalism, chauvinism, and outright scorn for different "others." The risk of selective emphasis notwithstanding, I would maintain that Judaism aspires in principle to attitudes of awe toward, and gratitude for, difference, even if frequently falling short of them in practice.)

Holding a shared conception of the Divine is not an invitation to discriminate when that very conception embraces the belief that all people share in it equally. In fact, such a belief should motivate social action to reduce discrimination. Take, for example, obesity and very short stature. Both can constitute forms of social disability, resulting in unpleasant and discriminatory treatment by other people and often a sense of diminished personal worth. Our obligation is not to change body shapes but to change social attitudes *about* body shapes, since these very attitudes belie disrespect for some manifestations of the divine image.

Enhancement and Social Justice

Furthermore, there is a powerful Jewish norm to be especially concerned for the weak and the disenfranchised. This appropriately leads to caution in approaching any technology that will preferentially benefit the already privileged and potentially enhance the gap between them and the disadvantaged. This *particular* concern about genetic enhancement technologies is not one of kind but of degree when viewed alongside other biomedical technologies or, for that matter, any social goods. Indeed, differential access to enhancement technology is surely no more unjust than inequitable access to basic health care. The possibility, even the likelihood, of maldistribution does not seem an adequate reason to reject an entire category of intervention.

Like the expressivist critique, though, this concern about abuse and injustice seems to turn on the expectation that the sought-after "enhancements" will be traits that confer competitive advantages over other individuals. The expressivist argument is that we will enshrine aspects of physical and intellectual abilities, appearance, and other traits that our social structures have made useful in competition among humans and will come to value them even more than we do now. The social justice argument is that giving an extra portion of such traits to the few would further disadvantage the many, while giving some to the many would even further marginalize the worst-off few.

But from a Jewish perspective, can these traits properly be considered enhancements at all, and is it virtuous or even permissible to desire them? Competition-enhancing attributes are not the qualities emphasized as desirable or godly in Jewish tradition. We are *not* talking, in other words, about enabling yourself to hit home runs or your children to score higher than your neighbor's kids on their SATs. At the beginning of this chapter, I asked what we would think of interventions that could make us, among other things, more thoughtful, honest, compassionate, just, and humble. These are not self-regarding qualities, conferring an edge in the rough-and-tumble of the capitalist marketplace, and they simply do not work the same way with regard to the attitudes they express toward others. Indeed, it would seem on its face that enhancing the list of traits I have suggested among any subset of humans could only *improve* the lot of everyone else. And the very desiring of them seems rooted in an appreciation and respect for the equal worth of others.

Another diversity-related concern about genetic engineering is the risk of homogenizing away some of what we value among ourselves. The composer Felix Mendelssohn had bipolar affective disorder. He produced hardly any music during long periods of profound depression but was extraordinarily prolific during manic episodes. With genetic intervention, even with Prozac, Mendelssohn might have had a happier life, but the world would have less beautiful music. This is *not* an argument for withholding treatment from suffering individuals based on others' interests in their creativity. Rather, it serves as a caution that we do not fully understand how our more complex human traits—like emotional lability and creativity—interrelate, and we can never be certain of what will be lost and gained in trying to engineer any one of them.

A similar caution comes to us courtesy of the Human Genome Project. We have long recognized that complex traits are governed by many-to-one genetics—multiple genes influence a given trait. But we now recognize a *one-to-many* relationship as well. Our bodies are made up of hundreds of thousands of functioning proteins. Yet to everyone's surprise, our genome turns out to include only perhaps thirty thousand genes. This means that genes are pluripotent in their functions. Many, perhaps all, genes participate in coding for multiple proteins, which in turn may serve entirely unrelated physiological roles. This design is miraculous in its elegance and efficiency, an object of awe regardless of whether one believes it to have occurred by immediate design or through eons of evolution. But the lesson for enhancement is that we cannot even begin to appreciate all the possible effects of changing a given gene.

Humility

This speaks to what is ultimately the most important reason that Judaism's conception of the good cannot be oppressive of diversity: humility sets limits to how we are willing to imagine improving ourselves. Humility, according to Steinberg, "is at once both a *halakhic* prescription . . . and an anthropological description of the religious personality."[16] We are taught that no one was ever as humble as Moses, the central figure in the development of Jewish nationhood and monotheism. According to

Sol Roth, "In rabbinic literature, humility is not counted as merely one of a number of virtues—each with equal status. Rather it is judged to be a central virtue."[17] Indeed, Roth cites Bahya as claiming that humility is a necessary component of all the other virtues; through it one comes to be appropriately reverential and to avoid sin. Humility's antithesis, arrogance, is compared in the Talmud to the cardinal sins of idolatry and immoral sexual relations.

Recall that Micah tells us not simply to walk with God but to walk *humbly* with God. What can this mean? Perhaps it means not only recognizing that we can never perfect ourselves but also acknowledging that we can never even be sure what doing so would mean. We interpret how the Torah tells us to live; *humble* interpretation admits that we might be wrong. How, then, can we know what would enhance us? Doesn't thinking that we know, or even that we might know, constitute arrogance? Surely the illusion that God's image looks more like me, or even more like what I aspire to, than like anyone else is inconsistent both with the belief in the universal reflection of godliness and with the demands of humility. We have seen that the principle that all humans are created equally in the image of God confers the ontological problem of how moving from one human state to another can possibly represent an improvement. Humility demands that we add the further epistemological and moral problems, that we are not sure where exactly we should go and forbidden to assume that we know how to get there.

This presents a conundrum: I have suggested that, in contrast to secular bioethics, Judaism seeks to distinguish, among all the modifications that we might seek, between true enhancements that make us better people and those that do not. I have listed some traits that the tradition suggests are desirable. But I have also just argued that the very belief that we know which traits make us better people may be arrogant, even idolatrous. How to reconcile these ideas? I want to suggest that it is the *effort to make* the distinction between what improves us and what does not, not the claim to have *correctly made* it, that characterizes a religious perspective. Jewish tradition demands that we seek self-perfection even while humbly acknowledging that we can never arrive at it nor even know just exactly what it is we seek. Certainly much of the virtue, then, is in the seeking.

Striving and Industry

In this way, humility about our ends may increase the relative weight we attach to the means of trying to get there. This speaks to a more subtle version of the naturalist bias: the concern that genetic enhancement is simply too easy. That is, there might be nothing wrong with the *goal* of enhancement, or with the particular qualities we have chosen to enhance, but rather with technological intervention as the way of enhancing. Is there something important about reaching toward goals of self-improvement that gets short-circuited by simply "flipping a switch" rather than by working at it?

Here we uncover the cultural belief that it is morally preferable for people to adapt to the unpleasant consequences of the genetic hand they are dealt than to rely on technological fixes. Everyone is entitled to our help with prevention and treatment of disease, but mere enhancement should be a function of concerted personal effort—there should be no gain without pain. How else to explain the fact that we are a society obsessed with dieting and exercising but look askance at surgical treatment of obesity or hormonal enhancement of muscle mass? This attitude may be gradually undermined by the growing popularity of cosmetic interventions like liposuction and Botox injections. But for now, the dominant ethos remains that it is okay to want to be better, but you had better be prepared to work for it.

If this sounds like the so-called Protestant work ethic, recall that Jewish tradition values hard work no less and may likewise display suspicion of quick fixes. I am indebted to Elliot Dorff for some examples. In Exodus 20:9 and 23:12, we are positively commanded to work six days a week, right alongside the commandment to refrain from work on the seventh. In *Pirke Avot* 2:2, we read: "The study of Torah is commendable when combined with a gainful occupation, for when a person toils in both, sin is driven out of mind. Study alone without an occupation leads to idleness, and ultimately to sin." In M. *Ketubbot* 5:5, we are told that even if a woman brings one hundred maidservants into a marriage, she must do *some* work in the household, again because idleness leads to sin. And finally, to my chagrin, my own rabbi, Rabbi Stuart Kelman, informs me that even were I fortunate enough to have servants, I would not be permitted to have them put up my entire sukkah; I am positively required to put some effort in myself. Likewise, in the life's work that is the creation of a mature self, we are not permitted to simply let someone else, or even some *One* else, do all the work. To some extent, it is the striving itself that is praiseworthy and that constitutes our goal.

Likewise with scholarship: we are commanded not to *know* Torah but to *study* Torah. There are at least two ways to think about this. One is that what we seek is wisdom, which is more than knowledge and perhaps resides in or arises from the process of acquiring knowledge. Another is that the very model of human fulfillment from a Jewish perspective is one of habit and practice rather than one of achievement or attributes. Either way, our attraction to modafinil, or for that matter to that extra copy of *Nr2b*, may be illusory. Being better able to attend and remember, even to understand, is not the point of study—the point is study itself. At least *some* of our most characteristically Jewish aspirations, then, are inherently tied to the process gone through in seeking them, to the experience of having worked toward them. We can readily see that engineering away this sort of quest deprives us of precisely what is of greatest value.

But perhaps this is not true of *all* that we aspire to. Micah does tell us to *do* justice and to *walk* humbly—skills that *may* and perhaps must be acquired by practice. But what about *loving* goodness or walking *humbly*? What of dispositions or attitudes—like humility or loving-kindness—that some of us might wish to cultivate but that others may be simply better endowed with to start? We might well wish that we, and everyone else, were kinder. Do we really care how that happens? Is kindness bred of conscious effort and practice any better than, or indeed in any way different from, innate kindness? Immanuel Kant might have said yes: that benevolence that is forced and reasoned is morally superior to plain natural niceness, precisely because the first results from pursuing a moral imperative while the second just *is,* effortlessly. But for most of us, this clashes with our commonsense intuitions. Personally, I would happily take a world in which people were naturally kind and just. What I believe we are often most concerned with is not how we get *to* what we want but with what it is that people want and what effects might flow from their getting to have it.

Self-Fulfillment

We fear that in trying to refashion themselves, people will choose bad things: characteristics that in the end will make themselves and others less kind, compassionate, and humble and the social world less just and holy. There are genuine grounds for fearing such an inclination in contemporary American life. In his thoughtful book *Better Than Well*, Carl Elliott explores the American fascination with self-enhancement and self-creation. He links this orientation to our founding hostility toward authority and tradition and the subsequent emphasis on self-sufficiency. Elliott concludes: "We need to understand ourselves as inheritors of a cultural tradition in which the significance of life has become deeply bound up with self-fulfillment. We need to understand . . . the paradoxical way in which a person can see an enhancement technology as a way to achieve a more authentic self, even as the technology dramatically alters his or her identity."[18] Furthermore, with self-fulfillment as the only end to which we all can universally agree, enhancement becomes more than just a desirable option. Again, Carl Elliott: "Once self-fulfillment is hitched to the success of human life, it comes perilously close to an *obligation*—not an obligation to God, country, or family, but an obligation to the self. We are compelled to pursue fulfillment through enhancement technologies not in order to get ahead of others, but to make sure that we have lived our lives to the fullest."[19]

There is a fascinating irony, though, in this worship of self-fulfillment. As opportunities for enhancement proliferate, and as the changes we can potentially make become more profound, we are likely to become increasingly *dissatisfied* both with the lot we are given in life and with our own efforts to improve it. In *The Paradox of Choice*, Barry Schwartz depicts the ways in which we are made miserable by the explosion of options in today's life—from which pair of jeans or television to buy, to which career to pursue and whom to marry. As social affluence has progressively guaranteed that our basic needs are met, we attach greater and greater value to scarce goods and circumstances, regardless of their inherent or practical value.[20] We can adopt one of two attitudes toward the multitude of choices we face. Schwartz cites Herbert Simon's distinction between "maximizers"—those who must make the very *best* choice of all those available—and "satisficers"—those who approach a choice with a list of the features they most want or need, go through the immediately available options until they find one that satisfies their requirements, and then stop there. Schwartz points out that the commitments and social ties that typically make us happiest are, in fact, ones that *constrain* our choices, including marriage or other long-term love relationships and child rearing. He then demonstrates how people inclined to be maximizers are made increasingly unhappy by a growing range of choices. After all, they must review and compare every feature of each possible option to be sure the one they end up choosing is unsurpassed. Maximizers also take the results of their choices, and particularly whether or not they ultimately find reason to regret them, as evidence of their personal worth. Finally, Schwartz points out, "the proliferation of options not only makes people who are maximizers miserable, but it may also make people who are satisficers into maximizers."[21]

So the substitution of self-fulfillment for an externally based conception of human worth, coupled with our increasingly affluent material culture, has led to an expanding and ultimately self-defeating sense that we not only *can* have and be all that we want but that we *must* do so. The failure to strive for maximal self-fulfillment has come to be seen as a moral failing, a failure of imagination, will, effort, or talent. How much more this must be the case when we are considering opportunities not to buy blue jeans or TVs but to enhance ourselves or, worse, our children.

Giftedness

It is against this background that we must read Michael Sandel's penetrating essay "The Case against Perfection."[22] Sandel describes the allure of genetic enhancement as representing "a kind of hyperagency—a Promethean aspiration to remake nature, including human nature, to serve our purposes and satisfy our desires. . . . And what the drive to mastery misses and may even destroy is an appreciation of the gifted character of human powers and achievements."[23] Sandel wants to find "nonreligious" grounds for trying to preserve this sense of giftedness, and he does so by pointing to the potential moral consequences of coming to view the shaping of ourselves as simply one more domain of consumer choice. "If bioengineering made the myth of the 'self-made man' come true, it would be difficult to view our talents as gifts for which we are indebted, rather than as achievements for which we are responsible. This would transform three key features of our moral landscape: humility, responsibility, and solidarity."[24] Sandel worries that, besides making us arrogant, the conviction that we can control our own and our children's destinies will unduly burden us with an excessive sense of our own responsibility for what we are, what we do, and what happens in the world around us. And ultimately, as a political philosopher, Sandel is most concerned with the impact of such hubris on social solidarity. Our secular social contract hangs on the belief that our circumstances are not entirely under our control. Why otherwise would the prosperous, the powerful, and the beautiful owe anything at all to the less advantaged? Sandel concludes: "A lively sense of the contingency of our gifts—a consciousness that none of us is wholly responsible for his or her success—saves a meritocratic society from sliding into the smug assumption that the rich are rich because they are more deserving than the poor."[25]

I share Sandel's concern but want to raise two questions in response. First, the features he enumerates—humility, a limited sense of personal responsibility, and social solidarity—are all important, attractive, and resonant with Jewish and other religious traditions. But why these three? Why are *these* the moral attitudes particularly threatened by the rise of a maximizing attitude toward self-enhancement and the concomitant loss of a sense of giftedness? And second, if the reality is that we are in fact rapidly gaining at least the technological means to this sort of control, what if anything else could preserve these moral features?

Is the ground on which they stand really so shaky, or do they share some common foundation that might survive our growing ability to engineer ourselves?

Most simply, it may just be that Sandel has put his finger on the three most important facets of moral thinking generally. Indeed, these correspond to what Richard Shweder describes, based on anthropological and cultural psychology studies, as the "big three" domains of moral reasoning across human civilizations: divinity, autonomy, and community.[26] I make no claim that these are exclusively or even especially Jewish concepts, but the fact that they *are* central to Jewish thought suggests potential Jewish answers to the concerns I raised above. I want to suggest that once again we turn to the idea of humility. What may stop us from sliding clear off our moral landscape is the belief that there *is no* state of exaltation, perfection, or blessedness to which we can aspire and upon reaching which our work will be complete. In this sense, Judaism's strength may lie in the fact that its messianism is a *process*, an obligation of continuing effort and improvement, rather than a *goal*, a state to someday be reached. (I am indebted to Rabbi Stuart Kelman for this formulation.) We should not fear improving ourselves; on the contrary, we are required to try to do so. But no matter how we may change, there will be no less left to do. As Rabbi Tarfon was known to say: "You are not obliged to finish the task, neither are you free to neglect it."[27]

Duty

Participating in this messianic project implies humbly accepting guidance as to what we do and why and how we do it. Judaism is a tradition of duty. Jews live by mitzvot, or commandments, that detail specific ways of living and of being. These are given to us, depending on one's theology, either by God directly or through the evolution of our traditions and values. But their ground in either case is outside us. Sandel's constellation of moral grounds, which might otherwise seem arbitrary, is supported and interwoven by a Jewish emphasis on *duty* that, following Freedman, is manifest specifically as concern for our relationship with and responsibilities toward God, the self, and one another.[28] In

Sandel's terms, these are, respectively, duties of humility, responsibility, and solidarity; in Shweder's, of divinity, autonomy, and community. It is precisely these three domains that arrogant self-engineering threatens, because they are the categories of duty that, from a humble posture, we accept as commanded.

Freedman's key insight about Jewish moral reasoning is that problems of action look different from a perspective that regards them in terms of adjudication of duties, as opposed to in terms of rights and the assignment of decision-making authority.[29] When we take seriously the notion of being bound by duty, we may find ourselves no more certain as to what course of action to take, but we ask different questions, look to different sources for answers, and view differently our obligation to live by what they tell us. We are not asking merely where the bounds might be to what we might choose to do. In Freedman's words: "One oriented toward duty seeks more, rather than less, moral guidance."[30] As matters of duty, Sandel's three domains constitute the very fabric of Jewish ritual, tradition, and personal guidance, arising out of our covenantal relationships. As Freedman points out, "the nature of obligations to the other is determined by one's life, not simply by one's own choices, by a rich persona existing within and defined by that person's life and much of its surroundings."[31]

Accidental Consequences

Our duties to others demand that we take responsibility for the consequences *to them* of our actions. What is unique about genetic enhancement efforts is the nature of some of those potential consequences. At the same time that we are commanded to heal, we bear enormous responsibility for the impact of biomedical interventions. In M. *Sanhedrin* 4:5, we are told that the very narrative of human creation is intended to reflect the infinite worth of human life and the consequent gravity of taking and saving lives: "Therefore but a single man was created in the world, to teach that if anyone has caused a single soul to perish, Torah imputes it to him as if he caused a whole world to perish; and if anyone saves alive a single soul, Torah imputes it to him as if he saved alive a whole world."

Concern about genetic manipulation, and especially about enhancement, often invokes worry about drastic consequences. Some of this worry is doubtless once again the naturalist anxiety, that in playing God we end up playing Frankenstein. But not all of it.

In 2000 a Louisiana boy named Damian Rodriguez traveled to France to undergo gene therapy for severe combined immunodeficiency (SCID), an invariably fatal single gene disease.[32] The treatment seems to have worked, and Damian and the other children in the French study regained normal immunity. But in December 2002 Damian developed leukemia, directly attributable to the insertion of the gene therapy molecule next to a tumor promoter sequence in his DNA. Damian underwent chemotherapy and appears to be in remission. But when a second child who developed leukemia in the same trial died, the FDA imposed a temporary moratorium on American gene therapy research. Four of the nine successfully treated infants in the French trial ultimately developed leukemia. These results can be (and have been) used to argue two very different points. On the one hand, genetic manipulation may have catastrophic, unforeseen, and perhaps unforeseeable consequences. On the other, many are more than willing to accept the risk of these consequences—including Damian's parents, who, given the choice, say they would do so again. Childhood leukemia is disastrous but potentially curable. There is real hope that Damian will survive his ordeal and live a normal lifespan, whereas there is no such hope for a child with SCID.

This argument would of course not hold water were the genetic intervention directed at making Damian taller, stronger, smarter, or even nicer. Any gene intended to effect such an enhancement would have to be delivered in just the same way as the adenosine deaminase gene that cured his SCID. It would therefore carry the same risk of going as badly awry. How can a Jewish perspective help us think about the relationship between an intervention and the potential severity of its aftermath? The key notion again is stewardship, with its concept of permissible risk, for our responsibilities to others are of a kind with those to ourselves. Freedman's analysis of risk and self-wounding would be apt when considering an enhancement. But while we might speak with some confidence about what might constitute enhancement to ourselves or even to our own children, we become more circumspect as that change echoes into the future. *Germ line,* as opposed to *somatic,* genetic interventions take on an added layer of complexity because we must consider risks to future generations.

Permanent Consequences

Somatic gene therapy is aimed at changing the DNA of an existing person and not their gametes, the eggs and sperm that go on to create the next generation. Any changes wrought by somatic gene therapy are expected to die with the individual. The risk-benefit calculus at issue in treating an individual, like Damian Rodriguez, may be relatively straightforward, as is the question of who ought to weigh the consequences, as they accrue overwhelmingly to the affected individual and his or her loved ones. These decisions are properly viewed in the same way as other medical choices.

Germ line therapy, in contrast, aims to permanently alter the DNA passed down through reproduction, so that a defect can be prevented from arising not only in treated individuals but also in all their subsequent descendants. Judging the safety of germ line therapy is much more difficult: Who can know what the long-term consequences will be of a permanent change in one's DNA? Who ought to be empowered to represent the interests of untold future generations, and how should they do so?

For these reasons, and appropriately, there has been much greater reluctance to experiment with germ line than with somatic gene therapy. Unfortunately, the lines between these types of interventions may not always be so simply and sharply drawn, for biology and chance may overwhelm our efforts to keep them apart. Already, there is evidence from some trials that the viral vector used to carry somatic gene therapy DNA into cells may be detectable in semen, suggesting the possibility of "leakage" of the therapeutic DNA into the reproductive cells themselves. In other words, with present technology, there is no absolute guarantee that a somatic gene therapy intervention will not inadvertently result in germ line changes.

Responsibility for Future Generations

Judaism is fundamentally collectivist not only across individuals and communities at a given time but across generations, and it is deeply concerned with the endowments, the responsibilities, and the world we leave to our children's children. Judaism is not a faith based on the concept of an afterlife in which the deeds of this life are rewarded or punished. Rather, we are duty-bound both personally and hereditarily. One performs mitzvot not only because one is commanded to fulfill them "in every generation" but also because the covenant into which our ancestors entered continues to bind us. The central covenantal promise made to Abraham and later to Jacob is that their offspring will become a great nation. Neither expects to reap substantial reward in his own lifetime. We are promised that the consequences of our actions, while perhaps imperceptible in our own lifetimes, will be visited upon our descendants, for better or worse.

This responsibility for offspring is of the utmost seriousness, and we are reminded directly in one of the Torah's most powerful passages of the repercussions that our actions may have in future generations. In Exodus 34:7–8, in God's one appearance to Moses, we hear *Adonai*'s self-description: "A compassionate and gracious God, slow to anger, and abounding in kindness and good faith, keeping kindness for the thousandth generation, bearing crime, trespass, and offense, yet God does not wholly acquit, reckoning the crimes of fathers with sons and sons of sons, to the third generation and the fourth." Surely the effect is to remind us of the power of our own actions today to drastically affect the lives of our descendants. The signal human action marking personal acceptance of the covenant is circumcision—an act of at least symbolic violence (and, it should be noted, of altering God's creation) that we are commanded to carry out not on ourselves but on our infant sons. The entire covenant is forward-looking. We impose identity and commitment upon the next generation, preempting to some extent the development of their free will by making change upon them. At the same time, we accept responsibility not only for irreversibly marking and inflicting pain upon them but also for protecting, nurturing, and raising them, for orienting our efforts and our world around the furtherance of their well-being.

What guidance does this offer in weighing the potential benefits and harms of genetic enhancement to future generations? Accepting covenantal responsibility does not argue against enhancement per se. Rather, it suggests that the value of undertaking an intervention must always be weighed seriously against profoundly unknowable potential ramifications in future generations. Again, risk must be calibrated to benefit. As the relief of suffering and disability recedes in favor of what looks more and more like "mere" enhancement, future-oriented prudence demands that we exert greater restraint. And it becomes all the more important to distinguish true enhancements from the pursuit of competitive advantage. With appropriate humility about making this distinction, we will be very cautious indeed. A germ line intervention that eliminates a type of cancer *must* at least be contemplated, even if it carries the possibility of creating disease in future generations. One that might make people more generous is one that *should* be taken seriously for its potential to enhance *tikun olam*. One that alters eye color or height is not to be attempted or done, simply because it fails this basic proportionality test.

Unanticipated Effects

We can imagine health consequences, like leukemia, resulting straight-forwardly from interventions we undertake. But there may be other, unforeseen effects. Tsien's "smart mouse" experiments had another, surprising outcome. The engineered mice were more fearful than normal mice when placed in an environment associated with painful shocks, showed more pain avoidance behavior, and appeared to experience pain more intensely.[33] Meanwhile, parallel studies showed that the *Nr2b*-enhanced mice showed stronger responses in two pain-related areas of the forebrain.[34] Tsien's team has argued that the mice do not necessarily *feel* more pain but rather learn more readily about potentially painful experiences and therefore react to them more strongly.[35] Whether these phenomena—feeling more pain versus learning and reacting more strongly to it—are subjectively distinguishable is an interesting philo-sophical question. But even if the actual subjective experience of pain were no different for the engineered mice, one is led to wonder whether, for example, memory-enhanced animals or humans would be more prone to posttraumatic stress disorder, in which distressing elements of past trauma are vividly reexperienced. Being better able to remember, in other words, may be accompanied by being less able to forget—and forgetting is sometimes beneficial.

So besides technological accidents, like Damian Rodriguez's leukemia, we see that enhancement may carry other sorts of risks. About some we can speculate, in that we can see their relationship to the intended enhancement. The way that enhanced memory may lead to an unwanted inability to forget is an example. Others, though, may be altogether unfore-seeable, because we cannot adequately imagine what a community of humans engineered in this way would be like. Now, Jewish tradition often expects us to take action without perfect information about the consequences of our acts. But Judaism is also fundamentally cautious about new situations to which precedential authority cannot be unam-biguously applied. And given the way genetics has already stretched our imaginations, we have to ask: How much do we really know about what we are doing when we intervene to change specific genetic traits? Are the sorts of complex traits that I have suggested might be true enhance-ments of the *right type* to be conceivably improved upon in this way? Is this merely a technical question to be answered as we gain genetic sophistication and computing power, or are the traits we care about so complexly determined, so interwoven with other human fundamentals, that the very project is senseless?

To summarize, then: Jewish traditions emphasize cocreation and positive responsibility for the natural world, including for ourselves, in contrast to the naturalism that characterizes many other approaches. Judaism is not especially concerned that the *means* of enhancement may abuse technologies designed to serve medical ends but rather asks whether the right *goals* are being aimed at. Different Jewish scholars will come up with different lists of desirable qualities; but the fact that there *are* such lists, and better and worse ways to live a human life, opens the possibility of making sense of enhancement. Because we are all created in God's image, we are of infinite worth in our very diversity, and the details of human perfection are inaccessible to us. False "enhancements" that actually represent the idolatrous pursuit of self-fulfillment must be resisted, while the desirable qualities that will help us perfect the world are not likely to be the kind that promote success in the competitive marketplace. Absent knowing our ends more precisely, we find value in the very striving for self-improvement, an orientation not aided, and sometimes hindered, by technological fixes.

We have seen that the appeal of enhancement technology threatens the moral universals of humility, personal responsibility, and social solidarity. We worry that as our sense of giftedness slips its moorings, it will carry these moral foundations away. Judaism, though, continuously reinforces giftedness and sets specific prescriptions for duties to self, to others, and to God. In accepting these duties, we practice a stabilizing humility in a number of ways. We acknowledge being unable to know fully our ultimate ends, let alone how our ways of life contribute to them. We admit uncertainty as to how to characterize cherished but complex traits and dispositions, and we practice prudence with respect to new ideas for improving those traits. And when in doubt, we turn for practical guidance to the accumulated wisdom of those who have come before us.

So what about popping modafinil or inserting that extra *Nr2b* gene? I cannot say whether taking a Jewish view should make us more or less enthusiastic about such enhancements. Rather, I have suggested that Jewish worries lie along somewhat different axes. We can agree that better mental capacities, for example, may facilitate understanding and are therefore desirable ends from a Jewish perspective. But the questions we must ask include:

- What are our purposes in availing ourselves of nongenetic enhancements: Are we doing so for self-promotion, material success, or honor, or are we trying to improve the world?

- Does what is enhanced fully or adequately substitute for the hard work of study and the process of struggling toward understanding?
- Is pursuit of such enhancement consistent with our duty of stewardship and with duties to others and to God?
- Are we confident that we know the impact of these interventions on other elements of ourselves that might be critical, in ways seen or hidden, to fulfilling our duties?
- Have we adequately considered and studied the potential impact on future generations?
- Finally, does this pursuit reflect appropriate humility about our goals and our judgment?

Coda

After I finished this chapter, I watched my young children sleeping and wondered, as many others have, why anyone would think we could engineer a better version of them. Despite their quirks and flaws—or perhaps because of them—how could we aim to improve on this work? I have come to believe that this sense of loving awe reflects one of the central struggles of Judaism: How do we reconcile our positive duty to work toward repairing the world with appropriate humility about its unknowable ends? This tension is captured in a very familiar closing prayer. We dream of an indescribable world in which false gods are swept away—a Messianic world, which, as I have noted above, may be understood more as a journey than as a destination. In any event, we *begin* the prayer by accepting our own responsibility for bringing that world about: *aleinu*, it is to us. I close with part of poet and rabbi Rami Shapiro's version of the *Aleinu*:

> Therefore we bend the knee
> And shake off the stiffness that keeps us
> From the subtle
> Graces of life
> And the supple
> Gestures of love.
> With reverence and thanksgiving

We accept our destiny
And set for ourselves
The task of redemption.[36]

NOTES

1. Mordecai Menahem Kaplan, *Judaism as a Civilization: Toward a Reconstruction of American Jewish Life* (1934; Philadelphia: Jewish Publication Society, 1994).
2. Paul R. Wolpe, "Treatment, Enhancement, and the Ethics of Neurotherapeutics," *Brain and Cognition* 50 (2002): 387–95.
3. Wolpe, "Treatment."
4. Yaping Tang et al., "Genetic Enhancement of Learning and Memory in Mice," *Nature* 401, no. 6,748 (1999): 63–69.
5. The report is Erik Parens, "Is Better Always Good? The Enhancement Project," *Hastings Center Report* 28, no. 1 (1998): S1–S17; the subsequent volume is Erik Parens, ed., *Enhancing Human Traits: Ethical and Social Implications* (Washington DC: Georgetown University Press, 1998).
6. Elliot Dorff, *Matters of Life and Death: A Jewish Approach to Modern Medical Ethics* (Philadelphia: Jewish Publication Society, 1998).
7. Immanuel Jakobovits, *Jewish Medical Ethics* (New York: Bloch Publishing Company, 1975).
8. Moshe David Tendler, "Stem Cell Research and Therapy: A Judeo-Biblical Perspective," in *Ethical Issues in Human Stem Cell Research*, vol. 3, *Religious Perspectives* (Rockville MD: National Bioethics Advisory Commission, 2004).
9. Benjamin Freedman, *Duty and Healing: Foundations of a Jewish Bioethic* (New York: Routledge, 1999).
10. Freedman, *Duty and Healing*.
11. Freedman, *Duty and Healing*.
12. Micah 6:8.
13. Parens, "Is Better Always Good?"
14. Anita Silvers, "A Fatal Attraction to Normalizing: Treating Disabilities as Deviations from 'Species-Typical' Functioning," in Parens, *Enhancing Human Traits*.
15. Bernard Steinberg, "Humility," in *Contemporary Jewish Religious Thought*, ed. Arthur A. Cohen and Paul Mendes-Flohr (New York: Free Press, 1987).
16. Steinberg, "Humility."
17. Sol Roth, "Toward a Definition of Humility," in *Contemporary Jewish Ethics and Morality: A Reader*, ed. Elliot N. Dorff and Louis E. Newman (New York: Oxford University Press, 1995).

18. Carl Elliott, *Better Than Well: American Medicine Meets the American Dream* (New York: W. W. Norton, 2003).

19. Elliott, *Better Than Well.*

20. Barry Schwartz, *The Paradox of Choice: Why More Is Less* (New York: Harper-Collins Publishers, 2004).

21. Schwartz, *The Paradox of Choice.*

22. Michael J. Sandel, "The Case against Perfection," *Atlantic* 293, no. 3 (2004): 50.

23. Sandel, "The Case against Perfection."

24. Sandel, "The Case against Perfection."

25. Sandel, "The Case against Perfection."

26. Richard A. Shweder et al., "The 'Big Three' of Morality (Autonomy, Community, Divinity) and the 'Big Three' Explanations of Suffering," in *Morality and Health,* ed. A. M. Brandt and P. Rozin (New York: Routledge, 1997).

27. *Pirke Avot* 2:21.

28. Freedman, *Duty and Healing.*

29. Freedman, *Duty and Healing.*

30. Freedman, *Duty and Healing.*

31. Freedman, *Duty and Healing.*

32. National Public Radio, "Gene Therapy Giving Hope to Those with Genetic Defects," *Morning Edition*, Renee Montagne, host, broadcast 27 December 2004.

33. Tang et al., "Genetic Enhancement."

34. Feng Wei et al., "Genetic Enhancement of Inflammatory Pain by Forebrain *Nr2b* Overexpression," *Nature Neuroscience* 4, no. 2 (2001): 164–69.

35. Yaping Tang, Eiji Shimuzu, and Joe Z. Tsien, "Do 'Smart' Mice Feel More Pain, or Are They Just Better Learners?," *Nature Neuroscience* 4, no. 5 (2001): 453–54.

36. Rami M. Shapiro, *Aleynu*: alternative version, in *Kol Haneshamah: Shabbat Vehagim*, 3rd ed. (Elkins Park PA: Reconstructionist Press, 2004).

CHAPTER 19 *Mending the Code*

ROBERT GIBBS

Perhaps the greatest challenge the new genetics represents is the need to reconceive not only scientific practices but also biological ones. While we are used to thinking of medicine as a therapy that cures or manages disease, that attempts to restore our bodies to their normal or healthful condition, gene therapy intervenes at a deeper level in our bodies: it alters the genetic code. The idea that our bodies are governed by a code, by strings of DNA, is a fundamental change in paradigm. The code itself must be copied and recopied, and its governance depends on a translating mechanism, whereby our cells take instruction and become able to do their healthy functions. Therapy then becomes a way of fixing a code or even fixing the translating mechanisms. But biological evolution now gains a different possibility, because the selection of one trait or one capacity, even one chemical or one propensity, now seems to appear on the horizon. Evolution will be augmented by interventions at the level of the code. We require serious thinking about how writing, copying, translating, and reading work in order to grasp the new logic of genetics.

I hope in these pages to bring old light to our new problems by a somewhat awkward analogy. I wish to offer the basic analogy that gene therapy may be compared with repairing a Torah scroll. The Torah scroll is also a code that governs life (albeit not at a molecular and cellular level), and it also is copied meticulously. I am *not* saying that the human genome is predicted or contained or even within the bounds of imagination of those who have copied the Torah scroll for centuries. But what I hope to show is a parallel kind of reasoning about the limits of mending. And I will emphasize the reasoning about limits as the key for this chapter. Besides the pleasant shock of thinking about the scroll and the genetic code in one frame, reflection on repair here has a distinctive feature: the "rules" about repair are themselves a matter of dispute. Moreover, the dispute about these rulings shows us a great deal about how a code works and indeed some ways we will want to think about repairing codes.

I will turn to three pieces of talmudic reasoning, all from one chapter in the tractate *Menahot*. Each has its own topic (which bears some analogy with issues in gene therapy), but more importantly, each has its own logical move. Mending a Torah scroll is repairing the text that is most important for Jewish life and thus has a dignity above mending a fence or a shoe. It offers us some sense of the value and the danger in mending genetic code, but because of its importance, the reflection by Rabbinic sages on how to conceive and how to limit mending offers us some needed insight into thinking about our own novel science with its serious and dangerous possibilities.

In order to undertake this analogy, first, I will indicate briefly the nature of the parallel between these two codes and indeed the fascinating matter of copying and mutation. There is a long tradition of understanding the scrolls themselves to be like people (and they are buried when they become too riddled with errors or simply worn out). Second, I will read the three texts and try to make sense of the issues in each, with a glance over our shoulder at what this might mean in the parallel with gene therapy. The task is complex because, while the three cases seem to be simple rulings on fixing things, they actually contest and illuminate many of our assumptions about what would make a scroll holy. Finally, I will conduct a second reading of the texts in section 3, offering a much more challenging task: an analysis of the talmudic text's own way of reasoning. At the heart is the process of correction and exploration of different positions, including ones that are clearly held to be incorrect. In this second reading, a doubling occurs: on the one hand, we have the more widespread quality of legal reasoning with multiple opinions that have to be entertained and respected. But more striking is the question of the genealogy of ideas or reasons. Our final text concludes with the remarkable claim that old is not really old nor the new really new—that the lines of filiation may flow backward or forward. Thus, for the task of reconceiving gene therapy, we may have to rethink our past thinking about it, and we may even find that the most stable opinions are the newest, not the most traditional. The three texts themselves seem more preoccupied with authority and filiation than with the more evident question of repairing scrolls. But that points, in its own way, at the task of repairing reasoning—for mending the scroll is also comparable to talmudic reasoning as a mending of Rabbinic code.

We may extend the parallel, because since the genetic code is a matter of a code for the production of proteins that regulate the functions of our bodies, and if its repair happens both at the level of DNA code and in the proteomics that function in the cells, then we model repair not only of a Torah scroll but also of the Rabbis' interface with it. The application/translation of the code itself may need mending. Moreover, the analogue of how to think about a code that is copied and repaired, that requires lines of filiation and mutation, may inform our larger picture about how we wish to think and to intervene in the new genetics. Talmudic reasoning then might model both the social relations we require in forming policies and research goals and something of the interventions that are still on the horizon of the new genetics.

Codes of Life

At the risk of simplifying basic genetic science, I want to call attention to the usefulness and specificity of the vocabulary we use to describe the processes of genetic biology. There are three processes of interpretation of the genetic code: replication (copying), transcription, and translation.

The secret of the double helix was that it made both replication and transcription readily intelligible; that is, by peeling apart the double helix, DNA is able to replicate itself over and over again. Each line of proteins is originally matched to its complementary pair, and as the helix unwinds, two new complete helixes can be produced by matching up other pairs. Copying reproduces DNA, and it reproduces it with highly successful accuracy.

Transcription is the production of RNA from DNA. It also is produced by matching base pairs, but RNA comes in different lengths and kinds and is produced in abundance from DNA: not two copies from one (or one from one in meiosis) but many copies from each.

The third process is the most complex, because from some of the RNA, the cells regulate proteins that then perform various functions in the cell. Here we are not copying the code itself, either to make a copy of the code (replication) or to make a messenger or working copy (transcription); instead, the code is being put to work to regulate the proteins and the functions of the cell.

We have a great interest in how the code works in these various processes and in the basic task of reading the code (of finding out what it says). This basic sense of reading is best represented by the Human Genome Project, of making a map of the DNA code. But the processes of copying, transcribing, and translating are themselves chemically regulated and are the subject of intense research. Proteomics is the study of translation, of how RNA regulates the production of complex proteins in the cell. It is more complex than the copying process and is, of course, the way that genetics gain biological application—it is the process of controlling life.

With all of this copying and transcribing, errors do happen. The rate of mutation (or copying errors) is very low, but the sheer numbers of base-pair copies in each replication of DNA is staggering. Once there is an error in the DNA code, then every cell reproduced from that cell will also bear the error. And every transcription and every translation will bear the error as well. Many of these copying errors do not seem to have any consequence for the functioning of our bodies. But some are catastrophic. Some errors seem to interconnect with others at the level of translation in a way that is still unclear to us, while others make no difference, and others still, even an error in one base pair, are expressed and are enough to cause disease. As we increase our knowledge of the DNA map and have greater success understanding the proteomic translation of errors in the code, we begin to discern the causes of certain diseases, propensities to bodily characteristics, and so on.

The bold next step, however, is intervening. What if we can correct the copying errors? Or even the translation errors? In the latter case, we might fix one or another cell where something has gone wrong. But if we repair the code itself at the level of DNA, then every cell that is reproduced from the cell we have fixed will now have the correct code. (If we intervene on germ line cells, on the cells that carry DNA to the next generation, we could eradicate an error not only in one cell or one body but in a whole line of descendants.) There is an immediate practical problem, because most genetic diseases and illnesses are not linked to a single site problem. The complexity of translation and expression makes repairing the code a daunting task. But what about the errors that are identifiable and local and require fixing?

In the introduction to this section of the book, there is some account of the progress being made in developing therapies to mend the DNA code. It involves replacing faulty bits of code with correct bits. The ethical questions that arise for us with regard to such interventions include these: What are the implications and what should be the limitations of these therapies? The obtrusive question is whether we can find a limit to therapy or whether we should be looking for upgrading and improving the DNA code, not just returning it to "normal." I will not resolve that question here, but I do wish to bring an analogy into view.

A Torah scroll is a text written in the Hebrew language. While Orthodox Jews believe that the whole text was revealed to Moses at Mount Sinai around 1300 BCE, transcribed by Moses, and faithfully copied since then, biblical scholars and many other believing Jews regard some of the materials in the Torah to be almost of that antiquity but regard the editing of the Torah to have been completed in the fifth century BCE. The text is still written as a scroll today (one of the few books from the ancient world not to lose its status as a scroll after being printed also as a codex, or book). As a scroll, it is read in a way now familiar to all of us from computers, by scrolling and not by turning pages. In any case, the text is written in ink on parchment (cleaned and dried skins), and the pages of the parchment are sewn together and rolled up. The scroll is written in columns in an alphabet borrowed from the Assyrians, with the letters written from right to left without punctuation. (Hebrew distinguishes its consonants, which are letters, from its vowels, which are marked by diacritical points above and below the line of consonants.) The text is cantillated, or chanted while read aloud, but neither the vowels nor the cantillation marks appear in the Torah scroll. Readers must therefore prepare in advance the section they will be chanting so that they know which vowels and notes to use.

Of course, the Torah scroll is more than a physical object: it is a text that tells the story of the birth of the people of Israel, including the prehistoric stories of the Creation and the Flood, of Abraham and of wandering. It follows the story of the people into the wilderness of Sinai, where God gives Moses a teaching (or in Hebrew, *Torah*), and ends with people at the edge of the Promised Land. Much of the Torah is taken up with legal and moral matters and with ritual requirements for sacrifices and offerings for God. Jewish reading of the Torah focuses on commandments, on the teaching of how to live as a people with God. Readers can now access the Torah as a bound book, with the rest of the Hebrew Scripture, including books of prophecies and books of writings such as psalms, proverbs, histories, lamentations, and other diverse genres and matters.

For Jews, the Torah "is a tree of life for them that grasp it" (Proverbs 3:18). It is the center of the cult (now that there is no Temple with its sacrifices). All of Jewish religious thought and normative practice are bound up with the Torah. Through its interpretation it guides Jews in how to live. Some Jews are more tightly bound to the Torah text, others more loosely. The Torah also requires a series of processes in order to function as the code for Jewish life.

First, the scroll itself is copied by scribes. Parchment and ink must be prepared, and the scribe, too, is trained and prepared. The text is copied with great care. There are many regulations for the scribes. These regulations are not included within the scroll but arise from later strata of interpretation and custom.

Second, the scroll is read aloud. It is proclaimed in synagogues in a lectionary cycle (in some synagogues the scroll is read in a one-year cycle, in others a three-year cycle). This practice is linked to the book of Ezra, but there are some presages of the practice in the Torah itself.

Third, the scroll does appear in book form and is studied as such by many Jews. It is not clear whether the transcription of the contents of the scroll into another form is parallel to RNA transcription, but it is true that the second function (the proclamation of the text) does depend on a scroll proper.

But fourth, and most importantly, there is a body of interpretation of the Torah, starting at the time of its redaction and continuing through the Rabbinic sages, including those of the Mishnah (approximately 220 CE) and the Talmud (sixth century CE), and leading through medieval interpreters and on to the current day. Much of Jewish traditional study is devoted to these later interpretations, and indeed the Torah itself is largely read through the lens of later interpretive traditions.

But at this fourth level the term "Torah" also begins to mean all the teachings of the tradition. And here we see the "application," because much of the teaching is moral, ritual, and even legal, as in the matter of contracts, damages, and so on. What is translation and occupies pro-teomics in molecular biology is conducted as *halakhah* and *aggadah* (legal rulings and philosophical/theological reflection) in the Jewish tradition. How to translate the text we copy carefully is often a major source of dispute, and study in itself turns into a practice of life. (This ultimate comparison lies beyond the range of this chapter.)

What is relevant, however, is the need to maintain the accuracy of the text and to repair a faulty text. When a text becomes worn out or turns out to be disfigured or filled with mistakes, it no longer can function as the governing code for the community. A mistake in the Torah scroll would be copied into other scrolls and would also lead to faulty interpretations/translations into the practices of life. A faulty scroll is then hidden away in a safe room or buried in the ground like a corpse. The analogy of the scroll with a human body is complex and recurs in part to recognize that the scroll has the dignity and authority of a human teacher, that it has life within it. Most holy of all are the words that are names of God, and these must be hidden away or buried even when they appear in texts other than the Torah that are defaced or defective.

For a community, the cost of a Torah scroll is large, and so there is a strong incentive to retain a scroll, not to have to hide it away. The scroll needs to be reparable, so that mistakes can be corrected. This will, again, have the benefit of not replicating the errors in copies made from this scroll and, in addition, preventing flawed interpretations based on the erroneous text. Hence, mending the code as found in the Torah scroll is particularly desirable—but again, there must be limits. And it is in the context of discussing the repair of faulty Torah scrolls that we find a Rabbinic discussion of what the limits should be; and, more importantly, within the Rabbis' own texts we find reflection on the method and the principle for setting such limits. This legal debate of what disqualifies a text helps us understand just what the code is and how it works.

Three Kinds of Mending

Near the end of the third chapter of the tractate *Menahot* in the Babylonian Talmud there is a lengthy discussion of the limitations of repairing a Torah scroll. But the context is a commentary on a mishnaic discussion of other cultic objects and what would make such an object disqualified, in particular what part or portion would disqualify it. The Mishnah mentions both a mezuzah and tefillin (two enclosed spaces that contain texts from the Torah, the former for doorways and gates, the latter for binding on the hand and on the forehead during prayers). The talmudic commentary then moves to the question of a Torah scroll itself.

I have pulled out three texts, depending largely on the Soncino translation, with some adjustments.[1] I will record them twice in this chapter and will identify the lines of text by two sets of citations (letters and numbers). The first number is the text (1, 2, or 3); the second is a ruling (1.3). A letter after the number refers to the sages' role in the construction of the argument (1B). The first reading will focus only on the rulings and be only numbered. My second reading (in the section below titled "Rereading Texts") will focus on the lettered interventions. In the appendix the three texts are copied in their context, without interruptions to explain and comment on them.

Text 1: Corrections

The first text concerns the number of mistakes that can be corrected:

R. Joseph said:

    1.1) Rab gave two rulings in connection with scrolls, but to each there is a refutation.

    The first is this:

Rab said:

    1.2) If a Torah scroll has two mistakes in every column, it may be corrected, but if three, it must be hidden away.

And the refutation—

    1.3) It was taught: if three, it may be corrected, but if four, it must be hidden away.

A Tanna taught:

    1.4) If there was one column free from mistakes, it saves the whole scroll.

R. Isaac B. Samuel b. Martha said in the name of Rab:

    1.5) Provided only the scroll was for the most part written correctly.

Abaye asked R. Joseph:

    1.6) How is it if in that column there were three mistakes?

He replied:

    1.7) Since it is permitted to correct them, then correct them.

The topic of this text is the number of mistakes in the column of a Torah scroll that is reparable. Rab (a sage from the mid-third century CE) ruled (1.2) that one may correct only two errors per column. If there are more errors, the scroll must be hidden away and, as such, not be used for reading or copying again. It turns out, however, that this is an example of a bad ruling. The standard in a prior ruling (1.3) was that three mistakes are reparable, and only when we get to four must we hide the scroll. The text will continue shortly to discuss Rab's second mistaken ruling, but in the context of this first one it adds a further discussion of mistakes in a scroll. There is a further old ruling (1.4) that if one column in the whole scroll is flawless, it justifies fixing any number of errors elsewhere. But another sage cites a further ruling of Rab (1.5): that this would require also a scroll where on average the text was correct. But one more question is asked, this time in relation to the second older ruling (1.4): If the one pure column has three mistakes, is that adequate to save the whole scroll? The answer (in 1.7) is that it is reparable, and then it is pure, and so it is fixed, and then it saves the whole scroll.

I will devote more attention to the rhetoric and construction of this text in the third section of this chapter, but here I wish to focus on the matter itself. The conflict in the first part (1.1-1.7) focuses on how many mistakes disqualify a scroll. A column holds about forty lines, or about one thousand characters. We can mend the scroll, replacing a defective or incorrect letter with its proper letter, but only to the limit of a few mistakes per column. What is noteworthy about the argument is that the quarrel is about two or three versus three or four. The two choices are very near each other, and the contrasts that are missing are, on the one hand, the choice of ten or twenty-five mistakes that can be repaired and, on the other hand, the choice of no mistakes at all—a possibility that becomes clear in 1.4.

This dispute presumes that any given text will have some errors. Indeed, if the text were immaculate, if there were a total accuracy to be expected, we would find an argument between zero or one error versus three or four. This particular argument is thus bad news for the fundamentalist, literalist vision of the Torah scroll. In copying the scroll, we aim for accuracy, but we tolerate a small margin of error (two or three out of a thousand). Because both sides tolerate a small number of errors, there is no requirement of perfection, and, indeed, the implication is that there is a small amount of imperfection to be expected in replication. Moreover, this means that some measure of variation is not only inevitable but, religiously speaking, tolerable. The community can function with a Torah scroll with a small margin of error.

What, then, is the argument? Rab takes a stricter line. Not that the earlier ruling is so lenient, but still, he positions himself as more strict. He seems more worried about errors being tolerated, but his position still cannot insist on inerrancy. The thrust of the text, of course, is that he was wrong. The older standard position seems to be a quite strict requirement, no more than three in a thousand. This differs from the "fundamentalist" position of inerrancy (zero), and it also is a choice that is more tolerant than a respectable position (Rab's). We learn, thus, that even a strict requirement can be situated against a stricter one and so seems to be more generous.

Of course, the second discussion, beginning with 1.4, then comes to offer a further reflection on the pure text. While there are many columns in the Torah scroll, the presence of even one without error saves the whole scroll. The position here is less obvious. On the one hand, even a solitary perfect column is so worthwhile that it makes us want to save the whole. Alternatively, one perfect column is a very good sign that most of the scroll will be accurate. Notice, however, that although we can imagine a perfect column, we are not required to have a complete scroll at that level of perfection. Indeed, it seems that 1.4 actually makes it possible to tolerate a lot of errors (more than three per column) and that having one "gold standard" column will redeem the whole maculated text.

But then a further ruling of Rab is cited to indicate the limits on this reading—and indeed, to incline to the pragmatic thought that if one column is perfect, the rest will likely be good. Again, Rab is in the role of a stricter authority—he insists that most of the scroll be right (fewer than four [three?] mistakes). And then Abaye asks R. Joseph (the one who initially cited Rab) about a column that has three mistakes. At first this looks like returning to the initial argument ("What are you going to do about the case we disagree about, namely, where there are three mistakes?"), but really, he is taking that point as settled. The question is rather this: What counts as a perfect column? Is it one with no mistakes, or one with correctable mistakes?

The final answer (1.7) ties together the ruling in 1.3 with the problem in the second discussion of part 1. The solution is, quite simply, that a text with up to and including three errors can be corrected, and once corrected it counts as a perfect column. A repaired text is pure and can function to redeem the rest of the scroll (even one with a large number of errors). Here the power of repair is striking: the pure text is not one that has to begin or even remain pure at all times but rather is one that is reparable. We are not, then, in the first part (1.1-1.7) arguing only about how many errors to tolerate but also about the authority of an almost correct text to save a whole scroll.

Rather than explore this first argument further, let me now merely nod in the direction of gene therapy. Mutation is inevitable and, to our surprise, is even positive, *sometimes*. Most errors in replication of genetic code are irrelevant, and our bodies can tolerate these few errors. Some, of course, are much worse than others, but in general, our bodies do not require absolute accuracy, just very high accuracy (of the order of 2.5 mutations in 10 million). Given that there are errors, we might be able to repair them to restore the text to a condition of perfection. We also may be able to tolerate some lingering errors. But there seems to be a limit to how much we may fix, or how much we should be willing to fix.

The irony is that we will depend on long stretches of perfect code to justify working on the bit that has errors. Any person with major stretches of botched code would die or perhaps not even be born. Think about the people who were exposed to high levels of radiation from nuclear blasts. We are, after all, only considering therapy on a very small part of the whole code, and we can only intervene in small zones at any one time. Indeed, the interventions depend to some extent on our not holding ourselves to a measure of perfection in replication but only to a small but discernible level of error. But we might need to intervene, if we could, to offer therapy for some of these errors, to mend the code, possibly to save a life.

Text 2: A Missing Word

The second text deals with a different kind of error, but the stakes are much the same. The question is this: Would this mistake disqualify a scroll? But the specific problem seems to be this: In the midst of copying, what happens if a person leaves out one of the names of God? For of all the words that make a text holy, a specific set of names of God are those that impart the character of holiness. The most important name, the Tetragrammaton (YHWH), hereinafter referred to as "the Name," is not even pronounced, and often it is only written by an abbreviation, but it also is the proper name of God and as such cannot be erased either. If it were lacking, there would be a fundamental flaw in the text. I now cite the second text, which is a bit longer than the first:

> 2.1) If the scribe omitted the Name of God, he should scrape off what was written and insert it above the line, and he should write the Name in the scraped-off place.

This is the opinion of R. Judah.

R. Jose says:

> 2.2) He may even insert the Name above the line.

R. Isaac says:

> 2.3) He may even wipe away [the written word] and write [the Name in its place].

R. Shimon of Shezur says:

> 2.4) He may write the whole Name above the line but not a part of it.

R. Shimon b. Elazar says in the name of R. Meir:

> 2.5) He may write the Name neither on the scraped-off place nor on the place where [a word] has been wiped away, nor may he insert it above the line.
>
> 2.6) What must he do then?
>
> 2.7) He must remove the whole sheet and hide it away.

It was stated:

R. Hananel said in the name of Rab:

> 2.8) The halakhah is that he may insert the Name above the line. [2.2]

Rabbah bar bar Hanah said in the name of Isaac b. Samuel:

> 2.9) The halakhah is that he wipe away [the written word] and write [the Name in its place]. [2.3]

Why does not R. Hananel say that the halakhah follows this Master? And
   Rabbah bar bar Hanah say that it follows the other Master?
Because there is another reading that reverses the names.

The problem is how to repair the text: how to insert the Name of God that
is missing. So we start with an opinion ascribed to R. Judah, according
to whom the scribe should (1) scrape off the word he wrote in the place
of the Name, (2) give that place to the Name, and (3) insert the word that
occupied that place above the line. We then hear five other opinions
on correcting the error: (1) the Name itself may be inserted above the
line, (2) the wrong word can be wiped off and the Name put in its place,
(3) the Name written above the line must not be abbreviated, (4) the
Name can only be inserted above the line, and (5) none of these options.
Ultimately, almost in frustration with this wealth of choices, the answer
is 2.7: the scribe must hide away the section of the scroll on which the
error occurred. I will turn to the sequel (2.8–2.9) shortly, but here let us
consider this range of options.

   We have a copier, a scribe, who passes by an instance of a Name of
God and then discovers the problem. The Name of God is the holiest
combination of letters, and it seems to require its own proper spot. The
text, like all writing, has lines and spaces between the lines. R. Judah's
position is that the space above the line is usable to repair the code but
that the specific spot where the Name was supposed to have been must
be used by the Name itself. Above the line is not going to be good enough.
Although other words can be moved around and can even function in the
above-the-line space, the Name of God requires its proper place. The other
word is now scraped off—removed from the chain of letters—in order
to yield a bare space into which the Name is inscribed. The scraped-off
place is now as good as new, suitable to house the holy Name (it seems
to bear no trace of having been already used). The Name is in its proper
place; the other word hovers above the line, functioning up there.

   R. Jose offers a very different view of the code: place is irrelevant not
just for most words but even for the Name of God. If the Name is missing,

then write it in wherever it will go. Do not scrape, do not rewrite the other word (which, of course, then might fall prey to a further error); instead, take only one step, namely, write the Name above the line. The sequence is important, but not in absolute space; we just need every word in the vicinity to follow a clear sequence. And we reduce the intervention here: instead of removing and reinscribing the one word and then inserting the Name, we just insert the Name.

R. Isaac offers a different moment of replication: the ink is still wet. We do not need to scrape off a finished word. Just wipe away the wrong word and inscribe the Name. Copying is now more of a process than a settled result. It may be that R. Isaac's position is the same as R. Judah's, but if there is a difference, then we can imagine it not requiring the interlinear addition of the word that was wiped off; instead, once the wrong word is wiped off, the copying just continues from the name of God forward. Here place is important again, but the correction does not require a word above the line, which is out of place for any letter, word, or phrase.

Fourth, R. Shimon of Shezur reverts to R. Jose's position but insists that if we write the Name of God above the line, the name must not be abbreviated, and it must not be written so that some of the letters are on the line and some above it. The inscription must all be in one place and thus be clear and complete. Here completion of the Name is more important than placement (on the designated spot or above the line). The Name (like most words) requires a full inscription to function. Its dignity or holiness or mere capacity depends on the complete sequence of letters in it. Because above the line is a makeshift solution, one might be tempted to write an abbreviation or dangle the name above and down onto the line. The integrity of the Name depends not on the place it appears but rather on the clarity and cohesion of the letters.

But the fifth and most blunt position is R. Shimon B. Elazar's, who cites it in R. Meir's name: no writing above the line nor on a scraped-off place or wiped-away place. Indeed, the conclusion is brutal: the scribe must hide away the whole sheet. Here the sheet is a piece of parchment that holds several columns (between three and eight). R. Meir does not disqualify the whole scroll, in part because the scribe seems to still be in the process of copying it out. If the Name of God is missing, then the scribe cannot fix the sheet. This is not like the errors that could be fixed above; this is an omission that is irreparable. Why? The sheet is not hidden away because it contains the Name of God (it is missing it, although one might think that the Name would already be written somewhere else). It seems that neither an above-the-line addition nor rewriting the Name in a place where something else had appeared is appropriate. The Name has its own place and cannot dwell in a "used" spot. It certainly does not belong out of order.

Briefly, I will summarize the issues and the sense of the Name and its place in the code. There is the sense that it can only appear in its proper place—that part of its signification is holding its own place, on the line, in its sequence. On the contrary, there is another sense that it can appear above the line and that its physical location is not important to how it signifies. Further, there is a sense that the word must be whole and cohesive. And last, there is a sense that it cannot occupy a previously occupied spot. Depending on the way we interpret these different characteristics of the Name, we have different remedies, different kinds of corrections. For some, there is no repair. For others, the scribe can wipe away or scrape off the word that has taken the Name's place and then put it back in its own place. For still others, the scribe can simply insert the Name above the line. The very act of copying is here made thematic precisely by an important error. Here the scribe is also an editor, correcting his own mistake—either with a scraper or a cloth or just by inserting the Name above the line. The scribe may be forced, as editor, to disqualify a whole sheet of parchment.

The text continues into a second part, where two of the previous positions are repeated and announced as the *halakhah* (the valid ruling). The two positions are (1) to simply insert the Name above the line and (2) to wipe away whatever was put in the Name's place and write the Name in its proper place. Thus the irreparable option is abandoned, and in the doubled options for above the line and replacement, one is chosen. Still, we have a clear disagreement in the ruling. The scroll is reparable, and the positions of R. Isaac and R. Jose are retained. I will return to the form of this last part and its punchline in the following section of this chapter. But here I notice that there is a desire to retain views that allow for repair and that the focus of the remaining debate is on the importance of the proper place.

The parallel to gene therapy here depends in part on how we understand the role of the Name of God in relation to genetic code. Other chapters in this book will also consider this analogy, and indeed it seems that in some important way genetic code has a semblance to what we might label the image of God in us. The holiness of the Name as an image of God is matched by the value of the code as the key to our life, as what gives us our possibilities. Clearly, we must avoid a reduction of our image of God to the material strings of the human genome, but once we realize that the human genome is a code, a text that requires replication, transcription, and translation, as well as editing and other processes, we begin to see the material not merely as some stuff but rather as signifying, like a written text. The Torah scroll is not the image of God, but it bears within it many mentions of the Name of God, all of which have attracted a possibility for holiness.

Clearly, R. Jose's position is that the Name of God is just one more word that is necessary but that has no special privilege over other words. The Name might refer to something very important, but its inscription is not different. (In passing, I should add that the actual copying of the Name of God requires special intentionality on the part of the scribe, but R. Jose's point is that in terms of its physical location, it is not unique.)

Now when we can discern a fault in the code, particularly an omission, perhaps of something that is vital in the processes of our living, if we could repair that bit of code, we would need to consider the kinds of repairs that are desirable (and also possible). Some might take R. Meir's position, that an error in the code should not be repaired. Others might hold that the therapy requires a strict replacement of correct code in the same location as the faulty code. And others might be willing to flood a nucleus with loose correct code and take the risk that it might trigger other effects but that it also might serve to generate the proper transcribed RNA. Even if the goal is strictly to replace a defective piece of code, there might be a sense that it can be done during replication (while we still can wipe away the error) or that it should be done with a completed replication. The risks of above the line and of scraping and wiping away are different, as are different modes of replacing faulty code with correct code. Mending here revolves around the different interventions that we choose to make to replace the missing name.

Text 3: Torn Texts

The third and final text I will interpret is specifically about sewing and about tears. Compared to paper, parchment is not easy to tear, but it does tear through normal use (particularly after repeated rolling and rerolling of the scroll). Hence, there is again a question of limit:

R. Zeiri said in the name of
R. Hananel who said it in the name of
Rab:

> 3.1) If a tear extended into two lines, it may be sewn together, but if into three lines, it may not be sewn together.

Rabbah the Younger said to R. Ashi:
Thus said R. Jeremiah of Difti in the name of
Raba:

> 3.2) The rule that we have laid down, namely, that if it extends into three lines, it may not be sewn together, applies only to old scrolls; but in the case of new scrolls it would not matter.

> 3.3) Moreover, "old" does not mean actually old, nor "new" actually new, but the one means prepared with gall-nut juice, and the other means not so prepared.

We are still working with some of our now-familiar sages, except that the frame of this piece is very late, into the fifth century CE. Rab is still the focus of our concern, and it is his ruling (3.1) that goes back to two or three, this time not about errors in orthography but about a tear in the text. If the parchment is torn through two lines of text, then we sew it up and keep using it; if through three lines, it is irreparable. Again, we have the logic of small repairs. This is only limited mending, but it is not the case that any tear disqualifies the scroll. Nor can we tolerate a large tear that is mended.

But then we read another ruling (3.2), a later one that cites and qualifies the earlier one: new scrolls are unlimitedly mendable. The earlier ruling only dealt with old scrolls, and it holds, but with new ones we can mend as much as we want. This makes sense if we think of the newer parchment as able to bear repair more easily.

All is going well until the kicker that probably belongs to Rabbah also: old is not a question of age but a question of whether the scroll was prepared with gall-nut juice. That preparation darkens the parchment. Just what the stakes are is not clear: possibly that preparation "antiquates" the scroll, making it look old; or perhaps the opposite—it preserves the parchment and makes it suitable for mending. In either case, "old" and "new" are now clearly terms that relate to the preparation of the scroll and not to its actual age. The principle seems to be that a scroll that can bear being mended, whether old or new, can be mended, but a fragile scroll can only be mended in a limited way.

I will return to this text when I explore the reasoning in the next part, but here we do see that Rab seems to have won and then, in winning, lost. The central principle is not that only a small tear should be mended but rather that any manuscript that functions as a "new" one can be mended. We have, in addition, a nice parallel to the previous texts, because now we are talking about the material upon which the words are inscribed, the architecture or substrate. Even as we need to fix the letters in order to mend the code, we also need to fix the parchment, too, for the code is embodied in ink *on* parchment.

It is characteristic of Jewish thought that the materiality of communication is explored. A code is not a disembodied word but is materially created, with stuff, and it functions to instruct us only insofar as the code is written correctly and embodied in its entirety. There is no limit to the concreteness of the topic of inquiry—the most important aspects of Jewish life, the Name of God, the holy Scripture, the Torah, all are also things subject to the vicissitudes of things and in need of care as things. But I should be more precise, for their needs as things are transformed because they are signs, encoded realities. To function as signs, to be copied, read aloud, and interpreted into rules for life, they must appear with the right spelling, with the right order of words, and on a page that is not torn or tattered.

At the deepest level the parallel with the genetic code lies here: that the sheer chemical materiality of the base pairs and chains of pairs is not foreign to the function of the code. Fully embodied, the chemicals are not mere things, but they are signs, a code that can be replicated, transcribed, and translated into the orders for the functioning of the proteins in the cell. A tear in the Torah scroll is like a mistake or a damaged gene in the DNA code. We might think we can only sew up little tears (very little tears), but it is also possible to mend the "new" ones as best we can, not just the sequence itself but also the structure upon which the relevant code rides. Perhaps this is about the large part of the DNA that is copied reliably but is largely irrelevant in the functioning of the cells. Still, those large zones carry the part that does signify, and they require copying, and possibly mending, too.

The challenge for a chapter like this, of course, is that we are trading only with analogies that are in principle not suited for collapsing. The rules for mending a Torah scroll are not in themselves binding in our reflections on what the new genetics can do for us in gene therapy. And they should not be binding. But a careful reading of these three texts has shown a series of reflections that are parallel to those we share in viewing the promise of gene therapy. If Jewish bioethics is not simply a task in setting limits, in giving rules for government or science to follow, then it might find a further task in exploring this more fundamental question, about how codes work and what sorts of repairs make sense, of how to limit the repairs and how to gauge the risks.

Rereading Texts

One strand of DNA is transcribed many, many times in strands of RNA. Just as replication is not a singular event, transcription has its function in its repetitions. Reading the Torah scroll is also repeated, as is copying it into other texts.

I hope I can bring you along as I reread these three talmudic texts with a slightly different intention. I wish to focus on the construction of these texts, although not at the expense of the topic and the rulings we have already considered. On the contrary, there is a specific repetition of the thematics, but this time at a different level: at the level of Rabbinic discourse. Hence, what were problems in treating the Torah scroll now will become problems in the sages' own practices. This might be compared to the questions for proteomics, but it also points to a different task for us in reading them, for it shows how the code that is mended is not only the DNA code but also the bioethics discourse about gene therapy. That is, I hope to display how the self-conscious, even playful, reflection on the mistakes and omissions and tears in the Rabbinic discourse can point out for us some more general issues—not on what the limits for gene therapy should be but on how we think and argue about the setting of those limits.

Let us begin again with the first text, the text about how many errors the scribe can correct. And this time, let us look at the frame of the text: the comment that Rab gave two refuted rulings (made two, not three, errors) in connection with scrolls. This time I label the other lines of the text with letters in order to focus on the construction itself of the Rabbinic argument.

1A) R. Joseph said:
  1.1) Rab gave two rulings in connection with scrolls, but to each there is a refutation.
1B) The first is this:
Rab said:
  1.2) If a Torah scroll has two mistakes in every column, it may be corrected, but if three, it must be hidden away.
1C) And the refutation—
  1.3) It was taught: if three, it may be corrected, but if four, it must be hidden away.

1D) A Tanna taught:

    1.4) If there was one column free from mistakes, it saves the whole scroll.

1E) R. Isaac B. Samuel b. Martha said in the name of Rab:

    1.5) Provided only the scroll was for the most part written correctly.

1F) Abaye asked R. Joseph:

    1.6) How is it if in that column there were three mistakes?

1G) He replied:

    1.7) Since it is permitted to correct them, then correct them.

Our focus now is on a text (and there are many like this in the Talmud) that identifies and refutes a ruling by an important Rabbi. R. Joseph is the source of a teaching about two mistakes of Rab (1A). We did not explore the second one, but the point of citing Rab's ruling (1.2) is to refute it (1C). Of course, the content of the refuted ruling is important, but at its own level, the focus is on citing and then refuting the ruling. Correcting a ruling, however, turns out to be a bit more complex: Rab makes a comeback under the citation of R. Isaac (1E), and that is because the cited position (three versus four) is complicated by another piece of code (1D) about the column that is perfect. Typical of talmudic dialectic, there is even a directly hostile question of Abaye to R. Joseph (1F). You would think that if R. Joseph set out to show how Rab was wrong, the correction would follow in a direct manner, but the insertion of another early ruling (1D) complicates matters.

To determine the rule for the upper limit of errors that can be corrected, it is not a simple matter of substituting one ruling (three versus four) for another (two versus three). Instead, there is a further dimension to the issue, the question of a perfect column. The debate that seems so close actually hides the deeper question: Can we require immaculate copying? And the promise of an immaculate column then leads to the questioning by Abaye (1F). The ultimate conclusion points in an interesting way to the importance of repair: the faulty text can be considered as a corrected one as long as it remains within the small limit of fault. The faulty position of Rab can be corrected, but it requires not just a counterstatement but a set of reflections on the way that a repaired text can function like one that had no faults at all.

When ethicists try to set limits to gene therapy, then, they also will need to explore not just the "right" rulings but also ones that are refutable. We need to examine not only the limit we like but also ones we reject. And the exploration of rejected views may indeed require us to consider secondary issues that we otherwise might ignore. Setting the limits is not, therefore, a matter of deductive philosophic reasoning, nor is it a matter of trumping arguments by having a tradition. In order to make the ruling of three versus four, the Talmud both engaged Rab's mistaken opinion and struggled with the question of a flawless column. Even citing a false opinion to reject it requires a further reflection on the possibilities that are adjunct to that opinion.

Thus, the reasoning was not monological. It was not simply a question of setting up a principle and deriving the consequences. It depended on opinions and rulings that have names affixed to them (see below) and a plurality of these named opinions. But bringing in the older ruling (1C) as a refutation is not quite adequate. Here tradition is qualified by requiring further reasoning. If there is one older ruling, then there may be another ruling (1D). Tradition is not simply the end of argument but rather a web of positions that force us to consider not only different positions but different axes of concern.

Hence, to mend the text of Rabbinic argument, to correct the mistake of Rab, we need other opinions, a secondary concern, and an interrogation of R. Joseph, who wishes to make the correction. Setting the limits means, then, a social reasoning process, where tradition has not a unitary voice but plural voices, and where the discussion involves a free engagement with related but diverting matters.

The second text addressed the missing Name of God and explored five alternatives, but in its construction there is a key question about the names of Rabbinic authorities. To present the opinion of a sage, it is desirable to include an ascription of his name. The problem of replication of the code with the scribe now becomes a problem of ascription, even genealogy, among the Rabbis. I will try to move quickly to the decisive moment for this text.

Text 2

2.1) If the scribe omitted the Name of God, he should scrape off what was written and insert it above the line, and he should write the Name in the scraped-off place.

2A) This is the opinion of R. Judah.

2B) R. Jose says:

2.2) He may even insert the Name above the line.

2C) R. Isaac says:

2.3) He may even wipe away [the written word] and write [the Name in its place].

2D) R. Shimon of Shezur says:

2.4) He may write the whole Name above the line but not a part of it.

2E) R. Shimon b. Elazar says in the name of R. Meir:

2.5) He may write the Name neither on the scraped-off place nor on the place where [a word] has been wiped away, nor may he insert it above the line.

2.6) What must he do then?

2.7) He must remove the whole sheet and hide it away.

It was stated:

2F) R. Hananel said in the name of Rab:

2.8) The halakhah is that he may insert the Name above the line. [2.2]

2G) Rabbah bar bar Hanah said in the name of Isaac b. Samuel:

2.9) The halakhah is that he wipe away [the written word] and write [the Name in its place]. [2.3]

2H) Why does not R. Hananel say that the halakhah follows this Master? And Rabbah bar bar Hanah say that it follows the other Master?

2I) Because there is another reading that reverses the names.

Again, not uncharacteristically, we begin with a ruling (2.1), and only afterward does the Talmud ascribe this ruling to R. Judah (2A). We then collect the set of rulings, each ascribed to a different sage (R. Jose, R.

Isaac, R. Shimon of Shezur, R. Shimon b. Elazar: 2B, 2C, 2D, 2E). Only the last ruling is actually ascribed by one sage to another sage, for R. Meir is responsible for the irreparable ruling on the missing Name. Things get more complicated in the second part, when R. Jose's and R. Isaac's rulings are not ascribed to them but only one step back to Rab and to Isaac b. Samuel (2F, 2G). That is, R. Hananel and Rabbah bar bar Hanah both report sayings in the names of their teachers but do not cite them in the names of the yet earlier sages that the Talmud cited above. We followed the argument about inscribing the Name of God, but now the text takes a decisive turn to inquire about the names of sages (2H).

There is some halakhic issue here, because given the preference for two of the five rulings (2.2 and 2.3), we might need to know how to justify these positions to determine the ruling. The focus of the talmudic text, though, is precisely the question of why these rulings are not ascribed to R. Jose and R. Isaac. We are, it seems, looking at how to find the name of the sage that is missing. In the places of the names of the sages are now other later sages. I am not claiming that the sages thought that their predecessors stood in for God, that the Name of God was signified by the name of a sage, but it is hard not to trip up here on the question of how we read the absence of a name that should be there in the midst of a discussion of a missing Name of God in a Torah scroll. So what is the question (2H)?

I suggest that because the ascription to an earlier sage carries special authority, and both R. Hananel and Rabbah bar bar Hanah have failed to include that ascription, in their own way they have cited only a limited authority. The text had no quibble with ascribing the rulings to R. Jose and R. Isaac (2B, 2C), so a reader is confronted by this gap. But the solution (2I) is that the later sages did not have a reliable tradition; in fact, there were contrary traditions. One reading is that Rab had cited 2.8 in the name of R. Isaac, not as in our text in the name of R. Jose. If there is a reversal of the names of the early sages in the teachers' mouths, then the students are stuck. They can cite their own teachers, but not the earlier sages.

Rabbinic tradition is now revealed to be underdetermined. Some of the links of descent are contested. The rulings retain their validity, but the genealogy is uncertain. That is, there is a claim of descent back one generation, but the original sources are obscure. The missing name of the sage is a sign that we cannot recover the source of the ruling.

This has a fascinating parallel with the eclipse of the Name of God and the need to get back to the origin. In the Torah scroll, the missing Name of God requires mending. The only question is whether we go above the line or in the very place it was to have been written. The Name of God has a specific unsubstitutable authority without which the Torah does not signify, does not bind its readers. But a missing link in the chain of authority disturbs the Talmud. The rulings can be entertained and even validated without a complete genealogy, without replacing the missing name. The talmudic text is edited to set up the punchline (21), because citing authorities, particularly of the earlier sages (*tannaim*), is vital. Indeed, the lineage or genealogy of an opinion is a whole rich topic of its own, because a ruling that is established in the second generation of earlier sages, for instance, becomes part of the code of Rabbinic reading and rulings and is replicated and translated forward. Once again, we see a question of replication (copying) and of transcription now become part of the historical sweep of Rabbinic argument. Because the Rabbinic tradition is organic and developing, it depends on its own faithful reproduction through teaching and through citing previous authority.

The talmudic text here, then, argues for a gap in transmission, a possibility that reasoning about the Torah admits of a lack that is *not* made good. Once again, we are looking at a maculation, but this time not of the Torah scroll (which tolerates a certain number of errors) but rather of the articulation of a line of tradition. Once again, tradition turns out not to be emphatically determining, even though it is largely normative, this time because we are lacking some of the links. Although the whole passage gives us the proper ascriptions, it also points to the toleration for a lack of such ascriptions. At the end of the day, we may have conflicting accounts of how we got our rulings from the past lines of tradition, and then we are left to negotiate the rulings on our own. Instead, we are left to interrogate our recent past and to respect its authority as we retell and re-create the more distant past.

Thus, for reflection on how to mend the code, how to repair the Torah scroll, the role of a tradition is not merely not univocal (five opinions, or even reduced to two), but, more importantly, the tradition itself is not understood to have unambiguous lines of descent. The various opinions we might need to entertain will each have a line of tradition, but those lines might not be whole. We need to consider the sources of the positions in the debate, but we also must face the possibility that we cannot validate the positions exclusively through an account of their origin and transmission. We just might not know where this or that position originally comes from.

For comparison, consider a position that is articulated as natural law but that might come from the Talmud, or from Stoic philosophy, from a more recent engagement with neo-Aristotelian philosophy, or even from neo-Thomist Catholic ideology. If we could locate its origin, that would reflect a great deal on its "traditional" authority, but, lacking that genealogy, we can still entertain and respect the position. Perhaps the greatest difference here between the lacking Name of God and the lacking name of a sage is that the human name, lacking or present, is not irreplaceable for the function of the reasoning and ruling (the code).

The third text addressed the question of sewing a torn text, and it is not hard to see that the claim I wish to explore in this second reading is the question of what is new and what is old.

Text 3

3A) R. Zeiri said in the name of
R. Hananel who said it in the name of
Rab:

> 3.1) If a tear extended into two lines, it may be sewn together, but if into three lines, it may not be sewn together.

3B) Rabbah the Younger said to R. Ashi:
Thus said R. Jeremiah of Difti in the name of

Raba:

3.2) The rule that we have laid down, namely, that if it extends into three lines, it may not be sewn together, applies only to old scrolls; but in the case of new scrolls it would not matter.

3.3) Moreover, "old" does not mean actually old, nor "new" actually new, but the one means prepared with gall-nut juice, and the other means not so prepared.

In my earlier reading, I explored the materiality of the code and the general halakhic upshot that mending is permitted. Here, we need to focus on the interaction of the very late sages in 3B as a transformation of the earlier tradition of 3A. We stretch about one hundred years from R. Zeiri (310 CE) back to Rab (240 CE) in order to set a limit. That limit has nothing to do with time. But we then find ourselves another hundred years later with Rabbah the Younger (410 CE) stretching back to Raba (340 CE). I am not a great scholar of the histories of the talmudic sages, but we can see that both lines of attribution are attempts to go back three generations. And Raba clearly is presenting a reinterpretation of the old ruling of Rab (and the two intervening generations).

In itself this is a regular model of change and innovation: we cite an earlier ruling and alter the scope of the ruling with a qualification. But look again at the structure: strings of attribution claiming that a given ruling is "old," but in the competition of the strings, the "newer" one now comes to replace the "older" one. At that point, our attention is drawn to the new alteration of the principle "new versus old," and for new scrolls "it would not matter." The principle repeats the structure of the reasoning: for new ones what was a limitation is irrelevant—for the "newer" sages, too. Here citation is reversal in order to make space for novelty.

But of course, we need to look at the punchline here: old and new are not temporal categories at all (3.3). It is unclear whether this is advanced as Raba's opinion (to my mind less likely), as R. Jeremiah's addition to his citation of Raba (still unlikely), as Rabbah the Younger's comment on Raba's ruling (a reasonable inference), or even as the anonymous editor's comment to the juxtaposition of Rabbah the Younger to R. Zeiri (similarly likely). In any case, that comment subverts the premise not only of Rab but also of Raba and, indeed, makes the notion of tradition quite troubled. If a ruling is restricted in scope, and then the scope itself becomes reconstrued, the original ruling becomes harder and harder to recognize. The mending of the ruling begins to undo the ruling—or, from the perspective of the newest interpreters, the ruling of Rab now looks like a tear that requires two stitches to fix. Without the first stitch (3.2), we would not be able to make the text whole (3.3). Moreover, to uproot or sew over the "tear" that Rab made, we require the complex lineage of 3B to match that of 3A. But we are then left with the question of what will be reparable in Rabbinic discourse. Anything that we can fix with two stitches? Or is it that the antiquity of the rulings makes them irreparable (following the principle of Raba)? In its own way, the text seems to argue that whatever can be repaired, no matter how rigorous the line of ascriptions, no matter how early the ruling, should be repaired. Tears that can be sewn should be.

For bioethics this is not a carte blanche for gene therapy. Rather, it is a metaprinciple about the ability and the procedure for altering our own rulings. There is for one part a traditional ruling, and then for another a qualification of that, and then for a third the wild card that disrupts the traditional ruling by dissolving the qualification. Again, this is not an utterly unusual kind of Rabbinic reasoning, but it points to how we need to think about both our past reasoning in bioethics and how the future will treat what we do. Within limits, older traditions can be repaired, and if they are treated properly, the limit is merely the limit of what we can repair. New rulings can all be repaired.

I do not want to stake out a position that simply undermines tradition; rather, what I want to emphasize here is that tradition is more complex and has within it confusions and incompleteness. Drawing upon "*the* Jewish Tradition" when approaching the new genetics requires a more subtle interpretation of how that tradition understands itself (and by extension other traditions). New and old are not literally new and old, but "new" (and I would say alive) applies precisely to those aspects of the tradition that are capable of extensive repair. Traditions of reasoning need mending, even as the Torah scroll itself does (and as DNA does).

(I would add, but only in parentheses, that the distinction between new and old at the level of gene therapy also raises the question of the gap between therapy and enhancement. We might have thought that therapy was restoring the old and enhancement developing the new. Although I do think that the Rabbinic tradition understands itself as developing and moving forward [and not only restoring the old], it seems more likely that the new/old distinction points to the need to revise and review that Rabbinic tradition than to change the code of the Torah scroll. We will need to alter our interpretations and rulings on the new genetics, but it is less clear that doing so gives us a similar license to enhance the DNA code—even by my strict analogy. Mending the code is repairing, not innovating.)

The process of this chapter has been to gather resources for our own task of thinking about how we can set limits to mending the DNA code. As Jewish thinkers, we draw upon Jewish traditions, particularly Jewish textual traditions, but we find there a remarkable analogy on how to mend a Torah scroll. The very processes of copying and applying the Torah have an interesting analogy with the processes of replication, transcription, and translation with genetic code. And the questions of how to and if we may repair the genetic code receive light from this much earlier problematic.

One real gain in this chapter is a better sense of the complexities of how the Jewish tradition understands its own reasoning. Here is a direct application to the question of how to understand, as well as how to limit, the new genetics, for our reasoning will require diverse opinions, careful consideration of the sources of those opinions, and a willingness to disrupt the received opinions through new reasoning. We will have to explore the reasonings we reject, as well as those we accept. The resources of a tradition, in this sense, are more than a body of rulings but much more a way of reasoning and mending not only the original code but also the code as transcribed and translated. And this chapter has offered yet a further translation from a talmudic text into a context of reflection on the challenges awaiting us in the new genetics.

APPENDIX:

Menahot 29b, 30b, 31b

Text 1 (29b)

1A) R. Joseph said:

    1.1) Rab gave two rulings in connection with scrolls, but to each there is a refutation.

1B) The first is this:

Rab said:

    1.2) If a Torah scroll has two mistakes in every column, it may be corrected, but if three, it must be hidden away.

1C) And the refutation—

    1.3) It was taught: if three, it may be corrected, but if four, it must be hidden away.

1D) A Tanna taught:

    1.4) If there was one column free from mistakes, it saves the whole scroll.

1E) R. Isaac B. Samuel b. Martha said in the name of Rab:

    1.5) Provided only the scroll was for the most part written correctly.

1F) Abaye asked R. Joseph:

    1.6) How is it if in that column there were three mistakes?

1G) He replied:

    1.7) Since it is permitted to correct them, then correct them.

Text 2 (30b)

    2.1) If the scribe omitted the Name of God, he should scrape off what was written and insert it above the line, and he should write the Name in the scraped-off place.

2A) This is the opinion of R. Judah.

2B) R. Jose says:

    2.2) He may even insert the Name above the line.

2C) R. Isaac says:

    2.3) He may even wipe away [the written word] and write [the Name in its place].

2D) R. Shimon of Shezur says:

    2.4) He may write the whole Name above the line but not a part of it.

2E) R. Shimon b. Elazar says in the name of R. Meir:

    2.5) He may write the Name neither on the scraped-off place nor on the place where [a word] has been wiped away, nor may he insert it above the line.

    2.6) What must he do then?

    2.7) He must remove the whole sheet and hide it away.

It was stated:

2F) R. Hananel said in the name of Rab:

    2.8) The halakhah is that he may insert the Name above the line. [2.2]

2G) Rabbah bar bar Hanah said in the name of Isaac b. Samuel:

    2.9) The halakhah is that he wipe away [the written word] and write [the Name in its place]. [2.3]

2H) Why does not R. Hananel say that the halakhah follows this Master? And Rabbah bar bar Hanah say that it follows the other Master?

2I) Because there is another reading that reverses the names.

Text 3 (31b)

3A) R. Zeiri said in the name of

R. Hananel who said it in the name of

Rab:

    3.1) If a tear extended into two lines, it may be sewn together, but if into three lines, it may not be sewn together.

3B) Rabbah the Younger said to R. Ashi:

Thus said R. Jeremiah of Difti in the name of

Raba:

3.2) The rule that we have laid down, namely, that if it extends into three lines, it may not be sewn together, applies only to old scrolls; but in the case of new scrolls it would not matter.

3.3) Moreover, "old" does not mean actually old nor "new" actually new, but the one means prepared with gall-nut juice, and the other means not so prepared.

NOTES

I owe a debt to the members and the group as a whole of the International Jewish Discourse Project. In addition, the many genetics specialists who joined with us taught me my ABCs of the new genetics. In addition, my colleague here at the University of Toronto, David Novak, patiently shared his talmudic expertise in my reading of the texts. And finally, Laurie Zoloth not only recruited me for this project but intellectually recruited me to this task. The errors that remain in this text are my own.

1. Eli Cashdan, trans., *Menahoth, The Babylonian Talmud* (London: Soncino Press, 1948).

PART 5 *The New Genetics and Public Policy*

CHAPTER 20 *Religious Traditions in a Postreligious World*

Does *Halakhah* Have Insights for Nonbelievers?

JOHN LANTOS

Things are gonna slide, slide in all directions,
Won't be nothin', nothin', you can measure anymore.
The blizzard of the world has crossed the threshold
And it's overturned the order of the soul.
When they said repent, I wonder what they meant?
—Leonard Cohen, "The Future"

The purpose of this chapter is to ask two questions. First, what are we to make of the use of religion in our national discourse around bioethical issues, particularly issues related to stem cells and cloning? Second, how might religious thinkers think about their role as they participate in this discourse? Specifically, are they speaking as representatives on their own religious tradition and speaking primarily for adherents of that tradition? Or are they drawing insights from their own tradition that have universal relevance and so might shape public and civil discourse more broadly?

Recent developments in biology, reflected in popular culture, suggest that we are at an apocalyptic moment. Movies like Steven Spielberg's *Jurassic Park*, novels like Margaret Atwood's *Oryx and Crake*, and discourse such as that of President George W. Bush's Council on Bioethics all suggest that the stakes in current decisions about genetics research are so enormously high that one wrong move could plunge us into irreversible chaos. Such sentiments have an apocalyptic tone and a prophetic moral message: we must repent of our boundary-pushing hubris before it is too late. However, it is never quite clear what, exactly, is sinful about science or what precise form repentance ought to take in a secular, pluralist democracy. Are there avenues of scientific research or technological development that we should prohibit? If so, how do we identify them, and what exactly makes them taboo?

In the discussion, the role of religion in shaping government positions is a curious and ambiguous one. On the one hand, our nation enshrines the principle of separation of church and state. On the other hand, the United States is the most religious of the industrialized countries, with a higher percentage of citizens who believe in God, go to religious services regularly, and believe that religion plays a very important part in our lives than in most other Western countries.

This dichotomy has important implications for bioethics and health policy because many of the controversial issues in this field touch rather directly on fundamental religious beliefs. Both President Bill Clinton's National Bioethics Advisory Commission (NBAC) and President Bush's Council on Bioethics sought out religious leaders in their hearings on the ethical issues in genetics and stem cell research. In justifying this attention to religious concerns, Leon Kass wrote:

> Moral positions rooted in religious faith or in philosophy or in ordinary personal experience of life are equally relevant, provided that the arguments and insights offered enter our public discourse in ways that do not appeal to special privilege or authority. Respect for American pluralism does not mean neutering the deeply held religious or other views of our fellow citizens. On the contrary, with the deepest human questions on the table, we should be eager to avail ourselves of the

wisdom contained in the great religious, literary, and philosophical traditions.[1]

Read carefully, this suggests that religious works are analogous to literary or philosophic works as repositories of the wisdom of the ages. That does not seem quite right. At recent hearings before the national bioethics commissions, practitioners of particular religions were called upon to talk about what their denomination believed. No similar call was made to Aristotelians, Benthamites, postmodernists, or abstract expressionists. It would seem that the role of religious leaders and, by analogy, of religious thinking was somewhat different from that of philosophical or literary thinking. Thus, it seems important to ask how the presence of official representatives of officially recognized religions helps or hinders the process of consensus building in our secular, pluralist democracy.

One way to think of their role is as representatives of particular interest groups. In that sense, the priests would speak for the Catholics or the rabbis for the Jews in a way similar to that in which the National Association of Plumbing-Heating-Cooling Contractors speaks for the plumbers or the American Medical Association for the doctors. In this view, religious communities are simply one of many interest groups whose sensitivities must be considered as any other interest group's sensitivities ought to be considered.

But there is another way to think about the thoughts of theologians (just as, perhaps, there is another way to think about plumbers or doctors). Although they do speak out of a particular tradition and may represent a particular constituency, they also speak from and to a broader set of concerns. They have, as their focus, fundamental ideas about the spiritual well-being of humankind. They bring to that focus centuries of reflection about the meaning of life, about human nature, about our ultimate purpose on earth, about the meaning of human suffering, about yearnings for spiritual fulfillment. In that sense, they are not simply the representatives of a particular voting bloc. They are also bearers of a subtle, sophisticated, ancient, and time-tested wisdom that is at once both more arcane and more universal than the wisdom of science. This specifically religious wisdom ought to engage any citizen who is concerned with larger spiritual questions. (In this sense, plumbers or doctors also have a dual role. We all have an interest in good plumbing or good medical care in ways that are related to, but separable from, what plumbers or doctors might want or demand or think is good for them. Doctors and plumbers have interests of their own, but we all have interests in the social structures that allow them to do their work excellently.) Gilbert Meilaender summarized this view when he testified before the NBAC that they might be interested in the Protestant theology on bioethical matters "not only because it articulates the view of a sizable number of our fellow citizens but also because it seeks to uncover a vision of life that we all share."[2]

Because religion is less tangible than sanitation or health care, it may be harder to see how religious leaders can speak beyond their own dogma or constituency. And yet, in the President's Commission on Bioethics hearings, that is just what they did. They spoke not about what was good for Catholics or Protestants or Jews but about how the reflections and conclusions of theologians ought to guide public policy in these controversial areas.

One result of this uniquely American approach to the proper interface between religion and politics is that American health policy with regard to stem cells was shaped in a dramatically different way from the health policy of other countries. And, given that our country is predominantly Christian, the specific religious views that shaped policy were the Christian ones.

The Jewish point of view on stem cells or on the moral status of the embryo is quite different from the Catholic or the Protestant point of view. I would like to highlight three of these differences and examine the ways in which they might be of interest to the larger debate about appropriate public policy regarding stem cells and genetics.

First, there are differences on the question of when life begins. For many of the Christian commentators before the bioethics commission, there was confusion about this fundamental issue, but there was also agreement about the meaning of the question. The question was one of scientific fact as much as of moral significance. The debates, such as they were, focused on the moments in biological development when one might say that a new and unique individual existed. In the final report on cloning, the commission wrote:

> In short, how we respond to the weakest among us, to those who are nowhere near the zenith of human flourishing, says much about our willingness to envision the boundaries of humanity expansively and inclusively. It challenges—in the face of what we can know and what we cannot know about the human embryo—the depth of our commitment to equality. If from one perspective the fact that the embryo seems to amount to little may invite a weakening of our respect, from another perspective its seeming insignificance should awaken in us a sense of shared humanity. This was once our own condition. From origins that seem so little came our kin, our friends, our fellow citizens, and all human beings, whether known to us or not. In fact, precisely because the embryo seems to amount to so little, our responsibility to respect and protect its life correspondingly increases.[3]

Contrast this view of the moral status of embryos with that presented to the NBAC by two leading rabbis, Moshe Dovid Tendler and Elliot Dorff. Tendler writes: "The Judeo-biblical tradition does not grant moral status to an embryo before 40 days gestation. Such an embryo has the same moral status as male and female gametes, and its destruction prior to implantation is of the same moral import as the 'wasting of human seed.'"[4] And Dorff writes: "Genetic materials outside the uterus have no legal status in Jewish law, for they are not even a part of a human being until implanted in a woman's womb, and even then, during the first 40 days of gestation, their status is 'as if they were simply water.'"[5]

Jewish commentators do not deny the importance of biological science. They take the biology of these matters seriously. But their religious traditions lead them to different conclusions about the moral implications of particular facts from those reached by many Christian theologians and by the President's Council on Bioethics. An early embryo with no moral status is very different from an embryo that is imagined to be one of our own and "the weakest among us."

The different views derive from many of the same biblical sources, but they lead to very different conclusions and implications about the ways that we should go about making the moral commitments that we make or do not make in order to facilitate the flourishing of human communities.

For many Jewish commentators, the highest hopes for human flourishing rest upon the potential medical advantages of research in this area for patients who are currently alive and suffering from particular diseases. For many Christian commentators, whose views are represented by the conclusions of President Bush's council, the focus is on empathy for the embryo as "one of us," as a particularly weak and vulnerable member of the human community. Both views start with empathy. It leads them in different directions.

The rabbis drew upon a religious tradition in which precedents had been established centuries ago about the relevance of various stages of pregnancy for moral and legal decision making. For the Jewish theologians, the embryo does not have the same moral status as it does for the Christian theologians. Does this mean that Jews do not respect the sanctity of life as seriously as Christians do? I think not. Instead, it reflects a different way of balancing spiritual concerns about the sanctity of life with a vision of the actualities of any human life. The Jewish view was not developed with embryo research or stem cells in mind. Nevertheless, it has interesting implications for those new developments in biology.

The second area in which Jewish theologians testifying before the commission differed from some others had to do with the nature of our moral obligations to develop new medical treatments. Much of the testimony before the commission focused on fears that our attempts to relieve human suffering or to cure disease might lead to even greater suffering. Meilaender writes: "Stem cell research is offered to us as a kind of saving solution, and it is not surprising therefore that we should grasp at it. . . . However, we may sometimes need to deny ourselves the handiest means to an undeniably good end."[6]

This sort of concern has been central to modern bioethics. Advances in biology and medicine are always seen—and usually correctly seen—as a double-edged sword. We can no longer assume, as we once did, that the purity of the impulse to cure disease or relieve suffering will guarantee the desired moral outcome of an actual improvement in health or reduction in suffering. Sometimes, our efforts to relieve suffering lead to greater suffering for both individuals and societies. These sorts of unintended consequences need to be carefully considered.

Again, however, for the Jewish theologians, the implications of these fears were somewhat different from those for the Christian theologians. The differences derive, in part, from the different views of human suffering and the moral obligation of individuals to heal or repair the world. Because traditional Judaism acknowledges that the world as God created it is incomplete and in need of repair, Jews recognize the obligation of each person to try to improve what we have been given. In Judaism, this is not seen as hubristic as it is in some strains of Christianity. Instead, it is seen as morally essential.

Taken together, these two different issues—the moral status of the embryo and the obligation to repair a broken world—lead to fundamentally different conclusions about the way we ought to balance the risks and the benefits of basic science research involving embryonic stem cells. As Rabbi Tendler writes, "A fence that prevents the cure of fatal diseases must not be erected, for then the loss is greater than the benefit."[7]

A final difference has to do with the nature of individual identity. Many of the theoretical concerns about gene therapy, and particularly about reproductive cloning, focus on the nature of individual human identity. Jews have always been intensely concerned about both identity and about relatedness. One of the earliest Jewish concerns about artificial insemination by a donor (AID) focused not on the morality of the procedure itself, which was seen as unproblematic, but on the implications it might have for evaluating the permissibility of future marriages. The concern was that it might lead to a forbidden marriage between relatives, since the genetic identity of the child conceived by AID would not necessarily be known. This sort of concern suggests a recognition and acknowledgment that individual identity is, in part, genetic. By implication, one might worry that two people with an identical genetic makeup might be considered, in some sense, the same person. However, interestingly, the focus of this rabbinical concern is not on the implications for the individual himself or herself so much as it is on the implications for the community and the rules that govern relationships and marriages. The threat or the harm is not to the individual whose genetic identity is blurred but to others who must live in defined relationships to that person.

Each of these examples—sketched out briefly and schematically here—is meant only to suggest some ways in which the biologic, political, and spiritual concerns within Jewish moral theology might lead to different conclusions from the ones that dominate current American health policy. In particular, they suggest that the focus of spiritual concern be not on questions about the nature of the embryo or the genome itself but on the nature of the community that must respond to different life forms and different possibilities for human relationship.

In one sense, there are striking similarities between the worldviews. Both Jews and Christians focus on how we might best care for the most vulnerable among us. The Christian theologians, however, invest the embryo with the moral weight of a human being and use our response to the embryo as a test of our empathy and compassion. The Jewish theologians focus on the sick and suffering among the already born and on how we might fulfill our moral obligation to care for them with both compassion and scientific zeal. Both focus on questions of how we might best care for one another in a world where new technologies might change both the possibilities for improving the world and the possibilities for harming it and ourselves.

How, then, might these profound but profoundly different religious discourses be incorporated into public policy? Religious discourse clearly plays a curious and complex role in the secular, pluralist democracies. As I suggested at the beginning, it is both similar to and different from the discourse of other communities, interest groups, or political constituencies. There is an element of religious discourse that is relevant only to adherents of the religion. In that sense, what rabbis think is relevant only to the Jews, just as what the Vatican says is relevant only to Catholics. In another sense, however, and perhaps only in a pluralist, nontheocratic state, each religious tradition can and perhaps must try to speak beyond such parochialism, to speak to all of us.

Such a task, however, calls for speech of a different variety. Speech within a particular tradition by that tradition's religious leaders tends both to be highly encoded in the idiom and the discourse of that tradition and to be absolute and authoritarian. For observant Jews, for example, the authority of the leading halakhic thinkers represents a particular sort of utterance. It is not part of ordinary discourse. Leading halakhic thinkers are recognized to have an authority that few can question. They have unique and uniquely recognized expertise within the tradition. Others cannot and do not question it because they (we) understand how much less we understand of the relevant tradition than do the experts. Such privilege and authority exists among the leading theologians and those authorized to make decisions of practice and/or doctrine in any particular tradition.

However, in order for the insights of particular traditions to speak to those outside the tradition—that is, in order for the hope of President Bush's council that religious wisdom will be of benefit to us all to be fulfilled—religious leaders will have to understand and master a new type of discourse. This new type of discourse is one that translates or interprets the insights of the tradition for those who stand outside it. Such a task assumes a society that is willing to listen and a process of mutual respect, but a society in which no authority or tradition is automatically or primarily prioritized over any other. In such a society, we all have an obligation both to speak to one another and to listen to one another, recognizing that speaking and listening are both active and controversial tasks, since both require intrareligious humility and interreligious respect. Religious thinkers must begin to unpack their own traditions in ways that will make them accessible to others. Only thus might any religious wisdom have implications that go beyond its relevance for adherents of that tradition.[8]

1. Leon Kass, "Opening Remarks to President's Council on Bioethics," President's Council on Bioethics, http://www.bioethics.gov/about/chairman.html (8 September 2007).

2. Gilbert C. Meilaender, "Human Stem Cell Research: Religious Perspectives. Testimony before the National Bioethics Advisory Commission, June 2000," in *Ethical Issues in Human Stem Cell Research*, vol. 3, *Religious Perspectives* (Rockville MD: National Bioethics Advisory Commission, June 2000), E-3, http://bioethics.georgetown.edu/nbac/stemcell3.pdf (accessed 17 November 2013).

3. "Human Cloning and Human Dignity: An Ethical Inquiry," section of the report titled "The Moral Case against Cloning-for-Biomedical-Research," President's Council on Bioethics, http://www.bioethics.gov/reports/cloning report/research.html (8 September 2007).

4. Moshe Tendler, "Stem Cell Research and Therapy: A Judeo-Biblical Perspective," paper presented at the National Bioethics Advisory Commission, http://www.bioethics.gov/reports/past_commissions/nbac_stemcell3.pdf (8 September 2007).

5. Elliot N. Dorff, "Stem Cell Research," in *Ethical Issues*, C4.

6. Meilaender, "Human Stem Cell Research," E-4.

7. Moshe Dovid Tendler, "Stem Cell Research and Therapy: A Judeo-Biblical Perspective," testimony before the National Bioethics Advisory Commission, June 2000, in *Ethical Issues*, H-4.

8. On the philosophical grounds for this type of discourse and the pluralism that it requires, see Elliot N. Dorff, *To Do the Right and the Good: A Jewish Approach to Modern Social Ethics* (Philadelphia: Jewish Publication Society, 2003), chaps. 3 and 4.

CHAPTER 21 *How the Unconscious Shapes
Modern Genomic Science*

ROBERT POLLACK

Let me begin with a brief quote from an unpublished talk I gave at the opening of the Learning Center at Cold Spring Harbor Laboratory in 1988: "From the beginning I have known the Lab to be a place of education as well as research. In fact, that's how the unique moment in the 1970s came to pass, when molecular biologists all over the world voluntarily agreed to suspend certain experiments in gene splicing. When I first came to the Lab, it was at Joe Sambrook's invitation, to give a class in the 1969 summer course on culturing mammalian cells and to chair a session of the very first Tumor Virus Meeting later that summer." It may be a familiar story by now, but it bears repeating today: I learned from one of Stanford biochemist Paul Berg's students in my summer course that Paul was about to construct a recombinant DNA including the transforming gene of the tumor virus SV40. This experimental protocol seemed in principle to be capable of generating a molecule with a novel and unpredictable infectivity. That was worrisome. It was a worry that could be dealt with by experimentation, so I called Paul in California and asked him if he were worried too.

He was not, nor was he able to answer my worry, so he did the honorable thing and helped the National Institutes of Health to set up the experiments to test whether my worries were grounded. I am pleased to say the experiments absolved SV40 recombinant molecules of the novel danger I had imagined. As you might guess, my own research at Columbia is now totally dependent upon these very recombinant molecules.

I tell the story here for two reasons. First, because the work on recombinant DNA held up for years by these concerns included the work of colleagues here at the Lab, in particular Joe Sambrook's. I think I was right to make a fuss, but I think also that it must have been difficult for him and for others here at the Lab to see me unhindered in my work here while I raised issues that temporarily stopped them from pursuing their ideas.

That makes a nice story for some people but not for others. In his new biography of Jim Watson, Victor McElheny gives his narration of this incident the heading "Robert Pollack Has a Fit."[1]

To bring the story up to date, my call led to the Asilomar conferences, which led to the formation of the Recombinant DNA Advisory Committee, or RAC, a model of bioethics at work; and my friend and colleague, Dr. Ted Friedman, the newly retired chairman of the RAC, assures us that it is alive, well, and doing important work on behalf of both scientists and the public. But as the author Conan Doyle might have noted, there is something funny going on: there is a dog that has not barked. That is, I called Paul Berg in 1971, more than four decades ago, and from that time until now, there have been no reports that I have seen of any scientist, in any field, precipitating an internal moratorium on any line of active basic research, ever. Why not?

At this long remove from those days, two reasons come to mind. First, I was close enough to the work to understand it and its implications more fully than most observers might have been, but not so close nor so involved with it that my own lab would be immediately impacted by my call. Not right away perhaps, but before I gave my grants back to the NIH in the mid-1990s, I had plenty of time to think about the additional work I had given everyone, including myself. And second, even then

I was aware that fears—my own fears and the fears of others around me—were expected to be kept from the daily discourse of the lab, and even then I knew that that was wrong.

What, then, has kept other scientists from being open about their own fears? Are they simply fearless? Here is a quote from Albert Einstein that may help us to see another explanation. Many religiously observant scientists like me have used the last line of this paragraph to support our own choice, but as you will see, most of us have probably misunderstood his own intentions.

> Now even though the realms of religion and science in themselves are clearly marked off from each other, nevertheless there exist between the two strong reciprocal relationships and tendencies. Though religion may be that which determines the goal, it has, nevertheless, learned from science, in the broadest sense, what means will contribute to the goals it has set up. But science can only be created by those who are thoroughly imbued with the aspiration towards truth and understanding. This source of feeling, however, springs from the sphere of religion. *To this there also belongs the faith in the possibility that the regulations valid for the world of existence are rational, that is, comprehensible to reason. I cannot conceive of a genuine scientist without that profound faith* [my italics]. The situation may be expressed by an image: science without religion is lame, religion without science is blind.[2]

Einstein wrote of his belief that no scientist lacked his "profound religion" of science for a conference titled "Science, Technology and Religion" held at the Jewish Theological Seminary in September 1940, the month of my own birth. It was then quickly published across the street, in the November 1940 issue of the *Union Review*, by the students of the Union Theological Seminary, under Paul Tillich's guidance. As an aside, this welcome for a scientist at a Protestant seminary resonates with the happy fact that in a gesture of ecumenical liberality the Union Theological Seminary has given the Columbia Center for the Study of Science and Religion a handsome suite of offices.

Both Einstein's essay and Tillich's response were published in the season of the Blitzkrieg's greatest successes. Surely Einstein knew by then that if he was right in his judgment that no serious scientist failed to share his faith in science, then that faith in science was extremely dangerous in and of itself. By then a scientifically designed, medically authorized, technologically superb program to extinguish the lives of anyone who could claim at least one Jewish grandparent had already assured the deaths of his family at the hands of, among many other criminals, a considerable number of genuine scientists, many of whom now happily burned his books.

Here is a much larger context for my question than my own narrative: How could Einstein have kept these facts from himself when he wrote this paper? And if he, the greatest mind of the century, was capable of such strong faith in the goodness and purity of science even in the face of such astonishing evidence, what self-awareness can the rest of us hope to have?

Meaning, Denial, and Repression

Everyone alive needs to make some sense of life, to give it some meaning. In the past century, scientists and doctors have made four interlocking discoveries that have made the task of finding meaning much more difficult. The oldest discovery has had the deepest impact: it is that DNA-based natural selection generates life in all its diversity and orderliness—including a scientist with a brain of great capacity to understand life's structures and functions—while, by itself, natural selection contains no element of design or purpose.

The second discovery concerns the mind. Scientists have shown that the conscious mind is the product of cells in the brain, an expression of the capacity of genes in these brain cells to respond to the outside world as well as to selectively recalled memories of earlier interactions with it. Third, they have found that the brain that does this is a tissue made of cells like any other tissue, albeit one that can imagine it has—or is—an ineffable, nonmaterial soul. And most painful of all, they have found that the entropic tendency of large and complicated structures to degrade into smaller ones assures that death—including the death

of the inner voice we each hear when there is no one else in the room—
is irreversible.

Together these discoveries paint a coherent and clear picture of the
living world and of our place in it that is notable for its complete lack of
meaning. Everyone who learns of these discoveries has the double task of
finding a way to accept them, despite their cumulative power to exclude
design and purpose from the living world, and of helping to assure that
the science of the future will be made by men and women who have
found meaning in their lives despite these facts of nature.

It has not been easy. Scientists cannot simply avoid thinking about
these discoveries, as so many of the rest of us do. Many aspects of today's
medicine are based on precisely these discoveries, which is why medicine has come to reject any larger meaning or purpose to life beyond
the workings of genes and the capricious choices of natural selection.
Yet one must—or at least I think one must—see life as more meaningful
than that if one is to lead a life worth living. The alternative is denial.
This is the unconscious rejection of one or all of these facts of life. Denial
allows one to avoid confronting one's fears, but it can lead to fateful
errors of judgment.

In Einstein's case, perhaps there would have been no way to save
his family even if he were to lose his faith in the goodness of science;
we cannot ever know. In any case, it seems clear to me that in our own
world—the world of the medical sciences—a similar denial of unbearable
facts can lead to the fantasy of conquering death itself, an impossible
dream that is the root of many of the ethical issues we are now facing in
light of the new developments in genetics.

Denial as a Brain Function

The Freudian unconscious of ego, id, and superego does not map completely to current diagrams of the functional anatomy of the brain, but
unconscious matters of hunger, sexual desire, aggression, and fear
occupy portions of the inner brain, while outer, cortical regions of the
brain—especially the cortical regions behind the forehead—deal with
conscious ego-like matters of subjective thought, abstraction, language,

and planning. The unconscious superego's world of values, rules, standards, goals, rewards, and punishments is least centered.

The first data systematically demonstrating that the unconscious repression of difficult memories was an aspect of normal brain function came from studies of survivors of head wounds. Working to understand and help brain-damaged soldiers and civilians, the Russian psychiatrist Alexandr Luria was able to partially align the analytic model of the mind with the anatomy of the brain. His most dramatic conclusion was that the normal brain was indeed functionally as well as anatomically divided into inner and outer parts. The centers of the inner brain were concerned with unconscious processes, affects, and memories; the centers of the outer brain carried out abstract conscious thought, perception of the outer world, directed action, and judgment; and at their boundary, a set of centers called the limbic systems carried out the balancing acts of bringing together the past and the present.

Luria's work, like the work of many others since then, has shown the repressive capacity of the mind: its ability to prevent certain memories and fantasies from reaching consciousness and its ability to let these memories emerge in the form of fantasies and daydreams have an anatomical correlate in the limbic systems where the two brains and their worlds meet. Damage to one particular place on the boundary between the inner and outer brains, where the most frontal of the limbic centers meets the most internal portion of the frontal zones of the cortex, has a spectacular effect on the place of dreams in a person's life: victims become unable to distinguish their dreams from reality. These lesions leave people in a permanent dream world, unable to tell whether what they see and hear is happening in the outside world or in their imagination. They suffer from "a constellation of vivacious dreams, hallucinations, confabulation, and a breakdown of the distinction between thought and reality."[3] In unaffected people, this portion of the brain must be constantly choosing among fantasy, unconscious memory, and current reality.

Each part of the brain thus contributes to the one inner voice of consciousness. Even the conscious act of learning from an event—the minimal unit of scientific observation—is the sum of at least four different kinds of neural activity taking place simultaneously in the two brains and in their shared limbic boundary. At the conscious level, the right cortical hemisphere of the outer brain internalizes the sensory experience of the event in terms of self-definition—what does this mean to me?—while the left cortical hemisphere retains the event cognitively, in language, as a set of facts and observations.

Simultaneously, the limbic system attaches an emotional affect associated with the event; the hippocampal memory retains a trace of both the event and the affect; while the hypothalamic regions of the inner brain generate their unconscious responses to the event. Of these various expressions of the brain's functional anatomy, someone doing mental work—a scientist analyzing her data, for instance—is consciously aware of only the first two.

Denial in Science

The emotional affects and memories of the past will not be part of a scientist's conscious experience, but because they are registered as changes in brain circuitry, they necessarily will be part of each act of observing and understanding the natural world. Scientists may insist that no aspect of nature is hidden from them, but inevitably their own nature—the conscious manifestations of their unconscious fears and needs—shapes the questions they ask of nature and thus what they can discover about the body and the mind.

What happens if we apply this empirical, clinical model of the relation between conscious thought and unconscious memory to the question of how the agenda of a science may keep unwanted unconscious memories from emerging to trouble the consciousness of its scientists? The conscious part of science begins with an act of faith, the ancient Greek belief that the natural world works by mechanisms that we can understand, even though they may initially be hidden from view.

Today, as in Democritus's time and in Einstein's, science works within the Greek belief that despite the smallness of atoms, the largeness of the cosmos, the rapidity of atomic transmutation or chemical catalysis, and the imperceptible slowness of evolutionary change, the underlying reality of any aspect of nature will be consistent, understandable, and therefore knowable. Everyone who plays the game of science must come to it infused with the belief that the way the natural world works can be understood to any degree of detail by sufficiently clever experimental manipulation.

Since nature is clearly silent and uninterested in a scientist's curious faith, the first step toward understanding is for the scientist to come up with a hypothesis to explain how some natural phenomenon works. The hypothesis is then tested through experiments that compare its predictions to the actual behavior of nature. Good experiments must often use elaborate machinery, so science can be expensive. But a good experiment need not be complicated, because it is never simply a set of measurements; it is a test of the usefulness of a figment of the imagination and a moment of risk and drama.

If experiments confirm a hypothesis—for why a ball bounces, a cell dies, the moon turns, or a muscle contracts—then the scientist must expand the range of tests to determine whether the hypothesis explains a little or a lot. As in backgammon, the stakes in science always go up; the game is never more risky than when a hypothesis proven right in a small corner of the natural world is tested in a bigger one. Each successive confirmation carries with it the obligation to push a hypothesis into ever-larger realms of nature by more extensive and subtler experiments. When—usually sooner than later—careful experimentation confounds a hypothesis, it has reached the limits of its usefulness, and it must be redrawn or withdrawn.

One may think that a hypothesis that explained even a little would be treasured and preserved. But once a hypothesis has been bounded by contradiction, the faith of science demands that it be altered or entirely replaced so that the task of understanding nature more fully can go on. Solely on the conscious level, science is thus reduced to a mixture of ritual and game, complete with a game's obedience to its own rules, austere unworldliness, and willful naïveté.

The conscious part of science is what most scientists would insist is all there is to science: an agenda for understanding nature. However, based on what we know of some minds, we can expect that the minds of scientists and therefore perhaps even the mind of science—the communality of experience and motivation shared by most scientists—have both conscious and unconscious parts. Just as the conscious part of science is shaped by the set of simple and universal rules that govern the conscious activity of all scientists, engaging and pooling the efforts of many different people's conscious minds, the unconscious parts of the mind of science—in particular, the sciences that serve medicine—would be expected to emerge as fantasies and obsessions shared by scientists in these fields.

Although the notion that scientists may share their unconscious fears and conscious fantasies, dreams, and myths may seem disingenuous, meaningless, or just plain silly, recall that until not too long ago, many serious observers thought it was disingenuous, meaningless, or silly to imagine that an individual brain might contain—within its biological functions—any individual mind at all. The early behaviorist assumption that the mind is an illusory, ineffable by-product of the brain's mindless application of instinctive rules had to be set aside in light of what we now know about the brain's functional anatomy.

It is time to follow up on that conclusion, to set aside the notion that science can operate in the present moment without an unconscious component to its deliberations. Science is the product of the unconscious sources of imagination and introspection as much as it is the product of a set of rules. The emotions and memories shared by scientists in the same field are its inner voice, and there is no reason that these inner voices should not be dealing with the same unconscious, repressed memories as do any of a field's practitioners. The question is not whether but how the unconscious aspects of science, refracted in maturity through its methods, resurface in ways that deflect the course of science itself.

Though science may seem at first remote from unconscious memory or conscious fantasy and obsessive behavior, it remains a human enterprise, and the fantasies of infancy are likely to be the same, whatever a person's later career. When universal, negative experiences are dealt with in the same way by a group of people linked by language and culture, their shared fantasies can crystallize into a core of collective myth. Every time biomedical scientists look at a piece of the human body through the lens of science, the lens becomes a mirror. What they see in it is at once familiar and completely strange. The human mind and body, but especially the mind and body of the scientist, become uncanny; the German word *unheimlich* best conveys the way each becomes more strange as it becomes better understood. This uncanny element of the life sciences derives from the fact that we can neither fully accept nor consciously and rationally even understand our own death.

Inescapable Mortality

Nature makes us mortal; surely that affects the behavior of every thinking person. For the scientist, what better way to reduce the feared figure of our own mortality than to make it our experimental material? This is a modern version of the original Greek notion of science, born in a world that did not distinguish between science and religion. The underlying myth of science concerns one of their immortal but otherwise altogether human gods. Asklepios, the demigod of medicine, was the son of the immortal god Apollo and a human princess named Coronis. The centaur Chiron—a physician of consummate medical skill who also happened to be an early human-horse recombinant hybrid—taught Asklepios the arts of medicine, and Asklepios became so skilled at healing that he was able to resurrect the dead.

Hades, the immortal god who ruled the underworld, complained to Zeus, the father of the gods, that he feared the loss of future subjects if humans were no longer to die. Zeus's response—he killed Asklepios with a thunderbolt—explains our present mortality and leaves us with the fantasy that by rediscovering the skills of Chiron and Asklepios, we may yet one day escape death.

The rest of the myth as it has come down from the Greeks tells us that Apollo, the immortal god of song and light, took such offense at this act of Zeus that he slew the Cyclopes, the makers of Zeus's thunderbolts, in revenge. Revenge, however, came too late to help poor mortals then or now. This myth and the hope it expresses have survived for thousands of years longer than any of the gods it describes. Despite all the rewards of scientific understanding, Apollo—science itself—has not yet overcome mortality. But we all vest the same hope in Asklepios—consciously or unconsciously—just as the Greeks did each time we visit the doctor.

The original Greek myth is alive not only in the minds of patients; it also lives in the minds of many doctors and scientists. Sometimes a great scientist will let the dream of Asklepios surface, allowing it to peek out from behind that other ancient Greek mask, the rationality of science. For instance, in his autobiography, the French Nobel laureate François Jacob is ostensibly discussing what it feels like to carry out a series of experiments, and the American Nobel laureate Arthur Kornberg is writing an editorial explaining why the scientific endeavor is unique among human activities, when both emerge with unexpected confessions of faith:

> JACOB: "And with this idea that the essence of things, both permanent and hidden, was suddenly unveiled, I felt emancipated from the laws of time. More than ever, research seemed to be identified with human nature. To express its appetite, its desire to live. It was by far the best means found by man to face the chaos of the universe. To triumph over death!"
>
> KORNBERG: "The ultimate scientific languages used to report results are international, tolerate no dialects, and remain valid for all of time. . . . Science not only enables the scientist to contribute to the progress of grand enterprises, but also offers an endless frontier for the exploration of nature."

Only faith or obsession—if they are not the same—can expect a method for observing nature to give a vision of endlessness or of triumph over death. Hyperbole like Jacob's may be intended or read as metaphysical metaphor, but the underlying fantasy remains clearly expressed: omnipotence of thought will bring immortality. This notion does not stand up to rational analysis; that is why the conscious, operational agenda of science masks the fantasy in Kornberg's "endless frontier," the cloak of institutional immortality. But institutional immortality itself, born from the unconscious will that one's name not be scattered, is just a different version of the same fantasy, an ancient impulse not limited to the sciences.

Myths of immortality—personal or institutional—distort scientists' conscious behavior. They steer the game of science in directions that have less utility than the scientists themselves may believe but that point away from an explicit confirmation of the underlying fears that create these myths. The uncanny familiarity of death, always on the threshold of being rediscovered by the rules of science, obliges scientists unconsciously to subvert the rules of their game, turning away from some of their most important discoveries.

The denial of the fear of nature's terrible power of mortality, the projection of the suppressed wish not to die into a vision of nature as capable of bestowing immortality; these are the marks of a masked unconscious creating a biomedical science at war with its own stated purposes. When scientists say "give us your bodies, and we will cure you," they have found a way to deal with an otherwise unbearable ambivalence toward their own experimentally vulnerable existence. They protect themselves, not necessarily by curing anyone, but by gaining control of someone else's body, if not their own. Freud, incidentally, recognized this role of medical science as the one "higher superstition" he himself believed in.

Institutional Immortality

In its disciplined way of looking at the natural world, science requires its practitioners to act as if they were observers, not participants. The first and last scientific instrument, the one that must be used in every experiment, is the scientist's brain; scientists who choose the human body and mind as their playing field cannot fully meet this requirement without dislodging themselves from their own bodies and minds. The strain of trying to meet a standard of dispassionate curiosity without flying into pieces imposes an irrational gap between the student of the brain and the brain of the student, between the scientist and his or her body and mind. To deal with the emergence of this intolerable thought, medical scientists have created the myth that their instruments and procedures somehow free them from the boundaries of their minds and bodies. This is the myth of absolute rational control of the physician and scientist over their material, the notion that the metaphor of scientist as sculptor will not break down even when the sculptor and the sculpture are one and the same. This myth may work to keep thoughts repressed, but at the cost of requiring the belief in an invented, institutional immortality based on the dry fact of precedence.

Every discovery must have at least one discoverer, and many have more than one. As competing sculptors may race to clear the excess stone from blocks of marble to reveal their different visions of what lay hidden inside, competing scientists clear away layers of plausible models, racing to uncover a demonstrably accurate schematic explanation of a part of the natural world's inner workings. Consciously and conscientiously followed, these rules work; they permit at least some scientists to uncover the mechanisms and structures of the natural world, and they do permit a few to win the game.

The dream of winning takes on an obsessive quality in the medical sciences, where the subject of scientific study is the mind and body and the reality of mortality becomes unavoidable. The result is an obsessive hope: that a big enough win in the game of science will confer a form of immortality on the winner. Discoveries that set the agenda for the future work of a large group of other scientists do this after a fashion, permanently associating a scientist's name with an aspect of nature. Think of the Freudian slip and the Watson-Crick model of DNA. Players in a game that can confer even this sort of immortality—however rarely—cannot be playing only for conscious stakes. In the medical sciences, the belief in winning immortality of this sort can become problematic when it supports the denial of an unpleasant biological reality, especially when that reality emerges from precedent-setting discoveries themselves.

It is not that science and medicine wish to avoid finding cures. It is that they are too strongly motivated by an irrational, unconscious need to cure death to be fully motivated by the lesser task of preventing and curing disease simply to put off the inevitable end of their patients' lives and, by extension, their own. There is a way for the life sciences to end their denial of their own unconscious, freeing it from the obfuscations and inefficiencies it creates today out of its own fantasies. Choices are necessary, and it is at the moment when choices are made that the scientific method departs from the wholly conscious tool of scientific experimentation and enters the human world in which all choices are made in a personal and social historical context, replete with emotional affects and barely remembered feelings.

An enlightened medical science would acknowledge that there are limits to conscious thought and to life itself that cannot be transcended by any rational agenda. It would then be able to stop making promises that it cannot keep, whether to itself or to the rest of us who pay its way. Having next acknowledged the unconscious memories of its practitioners and the shared fantasies they have generated, it would then be ready to find ways to diminish the influence of these fantasies on its conscious agendas.

Interdependence instead of Immortality

Although the narratives of successful science—discoveries, we call them—are bounded by culture no less than any other narrative, the models they stem from, confirm, and alter are not. These models, the most recently adapted, current working hypotheses of science, float above all their previous narrative versions, persisting through time, never final. We live by such models because they mold the patterns of our thought. In Hamlet's soliloquies, Shakespeare gave us our way of seeing ourselves as having inner voices and developing through inner dialogue. In a similar way, the sciences continue to give us new and sometimes precarious perspectives from which to see ourselves. These, in time, become as completely taken for granted as the Shakespearean notion of a private monologue. In just this way, Freud's unconscious and Darwin's natural selection have not merely been added to our vocabulary. They have become aspects of the way we understand ourselves; it is left for scientists to learn that these insights of self-understanding apply to them as well.

For example, the denial of mortality is often accompanied by the denial of another aspect of the human genetic birthright, that we are intrinsically social beings. The mind is the product of social interactions; there would not be enough DNA in the world to encode a single mind. From birth on, minds develop in brains by the imitation of other minds, partly but not solely the minds of biological parents. Most of the few behaviors wired into our genes at birth maintain and thicken the bonds through which this imitation can proceed. The current biomedical model of a person as an autonomous object lacks a proper respect for these social interactions. It severs the patient from family and social context, and it devalues preventive—social—medicine to an afterthought or a charity.

This denial of the reality of the social bond, like the denial of mortality, is an avoidable mistake of science. These and other strains that have opened between scientific medicine and society are not simply matters of resource allocation. They are signs that the knowledge of death and the need for others in one's life cannot be suppressed any longer, that the dreams of science are no longer satisfying even the dreamers.

In the United States, the cost of medical care reached more than two trillion dollars in 2011, the highest percentage of GDP (17.7 percent) of any nation.[4] It is unlikely that the two intertwined mistakes of today's medical science can be corrected without a renewal of interest in preventive medicine. But what is to be prevented? Prevention has two meanings, depending on what is meant by a healthy person. If health is given a functional definition—you are healthy if you are free to work and think and play to the best of your born abilities—then preventive medicine—in the form of a vaccine, for instance—simply lowers the risk of developing a disease later in life.

If, on the other hand, one imagines there is an ideal of human form and function to which we all must aspire, then preventive medicine takes on a different, perhaps alluring, but in the end sinister purpose: the elimination of avoidable deviation from this ideal. Physicians have already begun to take on the role of gatekeepers, inadvertent agents of selection, eugenicists manqué, deciding on the relative value of different human lives. Should that become even more common, definitions of disease will once again become less a matter of biology than of politics.

I believe that it is our ethical obligation to help scientists accept the validity of their own inner voices and see their research as an expression of their innermost feelings. This will be difficult, but it is not impossible. In a sense, it is just an extension inward of the fundamental methods of science. The great physicist Richard Feynman of Cal Tech saw the possibility as an obligation: "It is our responsibility as scientists, knowing the great progress which comes from a satisfactory philosophy of ignorance, the great progress which is the fruit of freedom of thought, to proclaim the value of this freedom; to teach how doubt is not to be feared but welcomed and discussed; and to demand this freedom, as our duty to all coming generations."[5] Today, few scientists accept this obligation, and the public knows it. If there is a particular and specific way in which we can contribute to the general good as Jews and as scientists and ethicists, it is by embracing—as particularly Jewish—this notion of a "philosophy of ignorance." Every morning's prayers begin with "Reshit hokhmay yirat HaShem" (The beginning of wisdom is awe of Heaven). That awe suits the philosophy of ignorance perfectly, and it is, for that reason, a strong antidote to the arrogance that denial feeds. Absent awe, or modesty, the popular vision of scientists as white-coated practitioners of a pagan religion will remain all too accurate.

NOTES

This chapter is adapted from portions of my book, The Missing Moment *(Boston: Houghton Mifflin, 1999).*

1. Victor K. McElheny, *Watson and DNA: Making a Scientific Revolution* (New York: Perseus, 2003).
2. Albert Einstein, "Science and Religion," *Union Review* (Union Theological Seminary) 2 (1940): 5-7, 28.
3. These examples of lesions in the brain and their effects on the ability to dream are taken from the notes of an unpublished lecture given by the distinguished British psychiatrist David Solms to the New York Psychoanalytic Association in 1995.
4. http://www.oecd.org/els/health-systems/oecdhealthdata2013-frequently requesteddata.htm.
5. Richard P. Feynman, *What Do You Care What Other People Think? Further Adventures of a Curious Character*, ed. Ralph Leighton (New York: W. W. Norton, 1988), 240-48.

CHAPTER 22 *To Fix the World*

Jewish Convictions Affecting Social Issues

ELLIOT N. DORFF

Jewish responses to social issues, like those of any other group, are in part a function of the group's interests. As a small minority of the American population, Jews have historically supported laws, policies, and programs that guarantee the rights of individuals and minorities against those of the majority. This would explain, for example, why Jews overwhelmingly support a strong wall of separation, to use Jefferson's phrase, between religion and state, why acts of anti-Semitism and talk of the United States as a Christian nation both engender an immediate, nervous response in Jews, and why surveys routinely report that Jews as a group are among the most liberal in supporting individual rights and protections.

Another factor that undoubtedly affects how Jews respond to social issues is their socioeconomic status. Although the median Jewish household income is higher than that of American households generally, all Jews are definitely not rich; indeed, 22 percent report household incomes of less than $25,000 per year, and 5 percent are below the poverty line.[1] Still, the fact that Jews as a group are relatively well off sometimes leads them, as David Sidorsky pointed out almost forty years ago,[2] to take stands that other groups in that socioeconomic group take, with poorer groups of Jews taking other positions.

At the same time, in many ways it is still the case that, as Milton Himmelfarb famously quipped, "Jews earn like Episcopalians and vote like Puerto Ricans."[3] That is, although historically members of every other religio-ethnic group that became more prosperous invariably shifted from supporting Democrats to Republicans, Jews, in contrast, stubbornly remain as liberal as they had been when living in poverty. So, for example, even though President George W. Bush was, by most accounts, among the staunchest supporters among American presidents of the state of Israel, only 24 percent of American Jews voted for him in 2004. (Ironically, Puerto Ricans, although still among the poorest subgroups in America, have increasingly supported Republicans against their economic interests due to their strong Catholic or Evangelical faith.)[4] This voting pattern among Jews clearly stems from factors other than group interest.

In this chapter, I will suggest that Jews' positions on social issues stem from their faith commitments about both how to make policy and what that policy should be. That is, they come from Jews' *methods* of interpretation of their own tradition and from the *content* of Jewish conceptions and values as the various groups of American Jews understand them and live by them. Thus, Orthodox Jews are the most likely to support the Republican agenda because of their version of Jewish faith, and non-Orthodox Jews, who constitute some 90 percent of the American Jewish community, support Democratic candidates and policies due to their approach to Jewish tradition.[5] After describing how these positions flow from the denominations' diverse understandings of the authority and methods of Judaism, I shall list a number of core Jewish commitments that also influence how Jews respond to issues of social policy, including the issues of stem cell research and genetics discussed in this volume.

Methods

James Packer, in his 1958 book *Fundamentalism and the Word of God*, characterizes the varying views of Christians on how to interpret the Bible as follows: "For evangelicals, the final court of appeal is to be found in Scripture as interpreted by itself; Romanists [i.e., Roman Catholics], some Anglo-Catholics [i.e., Anglicans/Episcopalians], and Orthodox [Christians] locate it in Scripture as interpreted (and in some measure amplified) by official ecclesiastical sources; and Liberal Protestants look to Scripture as evaluated in terms of extra-biblical principles by individual Christian[s]."[6] Jews reading this will immediately say that Scripture cannot be interpreted by itself, that we human beings inevitably bring our own perspectives to any text, including Scripture, and that one's perspective is shaped by one's own biography, personality, and historical context. Hence the rich Jewish heritage of midrash, in which, as the Rabbis say, "there are seventy faces to the Torah" (seventy being just a big number), and Jews through the ages have delighted in discovering and arguing about all of them.[7] In fact, Jews worry about those who think, first, that some text (whether it be the Torah, the whole Hebrew Bible, some legal code, the New Testament, or the Koran) is the direct word of God and, second, that they have the only correct interpretation of it, for the combination of these beliefs can be downright dangerous. After all, it is precisely such assurance of divine knowledge that has justified some of history's worst atrocities, all too many of them committed against us Jews.

Moreover, as I argue in my book *To Do the Right and the Good,* there are historical, philosophical, and theological reasons for pluralism among Jews and also in Jews' relationships with non-Jews.[8] That is, to assert that I am right does *not* necessarily mean that you are wrong.

In modern times, Reform Jews, with their emphasis on individual autonomy, advocate a methodology whereby each individual interprets God's Covenant with Israel for himself or herself.[9] This, though, leads to major questions about the authority of the Torah itself, for if I can make it mean whatever I want it to say, why do I need it in the first place? It also raises deep questions of Jewish identity, specifically, what makes my interpretation of the Torah Jewish? Just because I am a Jew? If Jews disagree about how to interpret and apply a verse from the Torah, as they almost always do, how can there be any identifiably Jewish interpretation? Nothing less than the identity and coherence of the tradition are at stake here.

Orthodox Jews, on the other hand, will warm to the second mode that Packer mentions—the meaning asserted by the officials in charge of such things. For them, rabbis over the centuries have determined what we should believe and do through their official and authoritative interpretations of both the Written and Oral Torahs, and they continue to do so today. This, though, raises deep questions about the source of their authority, what to do when they disagree, whether their rulings can and should be affected by factors outside the texts of the law (like morality, Jewish narratives, economics, politics, advances in science and technology, and social changes), and whether, in the end, their judgments can and should be changed or ignored.

Conservative Judaism—and so that my biases are clear, let me state openly that I am a Conservative rabbi—makes its decisions on the basis of both of Packer's last two factors as applied to Judaism, namely, the Rabbinic tradition and the response to that tradition by contemporary rabbis and lay Jews. That is, Conservative Jews understand Judaism historically, and in the past as well as in our own time, Jewish law is the product of the *interaction* between what the rabbis say and what the people do. Because Conservative Judaism does not embrace the last of Packer's methodologies alone, it does not reduce Judaism to individual choice, and because it does not embrace the second one alone, it does not reduce Judaism to tradition alone either. In the area of law, this has replaced the neat formulations of the Reform and Orthodox movements (either follow your Jewish conscience or follow the Shulhan Arukh or whatever other code you find authoritative) with a much more messy methodology that requires *judgment* as one balances a number of competing claims—tradition, modern social and economic conditions, morality, customs, theology, science, technology, and so on. On the other hand, what one loses in neatness and clarity through this Conservative approach one gains in both authenticity to the tradition, which also balanced these factors in every era, and in reality, for today is not exactly like ages past, and Judaism cannot survive if it is interpreted as if it were.

These differences among Jewish denominations have important ramifications for how Jews of various sorts approach matters of public policy.

Reform Jews, with their emphasis on individual autonomy, tend to support liberal causes. Just as individuals should determine the religious parts of their lives, they maintain, so too individuals should decide as many other parts of their lives as possible. At the same time, government should provide for the poor, education, and other social needs in the name of justice and equality. As with most liberals, Reform thinkers have never clearly delineated how they would balance the conflicting claims of individual liberty and equality.

Orthodox Jews, with their emphasis on tradition, have increasingly supported politically and socially conservative positions in such matters as faith-based funding for schools and social service programs, government controls on abortion, and laws banning homosexual marriage. Even though, as heirs to the classical Rabbis of the Mishnah, Midrash, and Talmud, no Jews can be fundamentalist in the sense that some Protestants are, Orthodox Jews are as close as Jews come to the kind of certainty that fundamentalists assert in what they believe to be God's will. The Written and Oral Torah that has come down to us is the word and will of God, and so our task is to apply it to modern circumstances. New developments, Orthodox interpreters assert, are relevant only to the extent that they pose new challenges and opportunities for living life as the classical texts demand; the norms of Judaism must not be changed as a result of new findings in science or new technology, let alone new moral or social sensitivities.

Some parts of the Orthodox community (specifically, ultra-Orthodox [*haredi*] Jews and some sects of Hasidim) also have a deep distrust of the modern world and live in self-secluded communities in places like Kiryas Joel in Monroe, New York. They take a very narrow view of their involvement in matters of public policy, namely, that the state and federal governments should give them as much as they need and interfere with their communities as little as possible. This has perhaps especially been evident in the legal battles that Kiryas Joel has had over its educational system—battles that have been the cause of a U.S. Supreme Court decision.[10]

Conservative Jews, ever in the middle, have generally opted for individual freedom on matters affecting people's private lives, arguing that the state should not intervene in matters of individual conscience but for a thick sense of community in providing social services, including government funding of the poor, education, and scientific endeavors such as embryonic stem cell research and genetic cures for diseases. This follows from the Conservative attempt to blend tradition with modernity, for Jewish tradition requires communal support for the poor, education, and health care. On the other hand, the recognition that revelation can be interpreted in multiple ways, the involvement of Conservative Jews in modern society to a much greater extent than most Orthodox Jews are, and the minority status of Jews in America together make Conservative Jews support individual freedoms on personal matters more than Orthodox Jews do.

Thus, in another chapter of *To Do the Right and the Good*, I argue, along with Thomas Jefferson, that governments should make laws only on those moral issues where Americans overwhelmingly agree, such as prohibitions against murder, theft, rape, and fraud, and positive duties such as paying taxes and providing education, food, clothing, and shelter for your children.[11] Where sizable segments of Americans disagree about moral issues, however, the government should stay out of the fray and allow Americans to make their own decisions, guided, if they so choose, by their families and religions, unless there are secular grounds for adopting a particular law or policy. So, for example, I believe that for both heterosexual and homosexual couples, states should create only civil unions or domestic partnerships in order to accomplish their legal goals and leave it to the various religions to decide whether they will consecrate given unions as marriage. Although I do not discuss that example in that chapter, I do address abortion there. Even though I personally agree with classical Jewish texts and most of their modern interpreters that abortion should not be used as a form of ex post facto birth control but only when the life or health of the woman is at stake or the fetus has a lethal or devastating illness, and even though I do my utmost to convince other Jews that their heritage demands that they act accordingly and that this is the wise and moral stance to take,

I do not think that the state should ban abortions to those who view the moral situation differently. On the other hand, to cite the other example I explore in that chapter, because I think that assisted suicide will have deleterious effects on health care in this country quite apart from my Jewish and moral convictions that it is wrong—that the legal right to assist a suicide will all too quickly become the duty to gain assistance to commit suicide so as to minimize costs—I would argue on secular grounds that it should not be legalized.

How does this approach affect the topics of this book—stem cell research, genetic identity, genetic testing, and genetic engineering? Catholics and some Evangelical Christians who oppose embryonic stem cell research should certainly not be required to engage in it or to use its therapies once they are developed. In light of the facts that American law does not construe an embryo as a full human being and that the majority of Americans support embryonic stem cell research, the American government should not only allow embryonic stem cell research but support it financially because of its significant promise to create cures for devastating and even lethal diseases.

Primarily through the Health Insurance Portability and Accountability Act (HIPAA) of 1996, American law has already taken steps to protect individuals' genetic identity from insurance companies and employers in order to prevent the kind of society depicted in the movie *Gattica*, in which all your choices in life are determined by your DNA, which is diagnosed at birth. At the same time, children and their representatives have the right to determine the identity of their biological parents for purposes of learning about potential medical problems and identifying who has parental responsibilities toward them.

Genetic testing should likewise be predominantly done only with the consent of the people involved, which includes adults for themselves and parents for their minor children. There are some cases, though, when the government may have a legitimate right to test someone's DNA. These include law enforcement officials trying to determine culpability for a crime; indeed, I would argue that legal authorities have the duty to consider such tests legally binding to determine not only guilt but also innocence so that prisoners wrongly convicted could be freed on the basis of DNA evidence, despite the legal system's interest in not revisiting cases that have been decided.

Genetic engineering of plants, which has produced much more food than we otherwise could, should be allowed, but only with safeguards to protect the environment. After all, if we develop one strain of an ideal tomato—one that does not spoil quickly and is as tasteful and juicy as one could hope for—and then some blight devastates that strain, we will be living in a world without tomatoes. That is, biodiversity must be preserved, and harm to other parts of the environment must be anticipated and avoided.

The same is true for genetic engineering of animals, but with animals we also have to worry about causing as little pain to them as possible in any genetic engineering we do with them. Moreover, because of the sentient nature of animals, bioengineering may be justifiable only if the goal is to provide food or therapies but not for other less important uses, such as cosmetics.

Content

In addition to the Jewish view of multiple meanings in texts of revelation and the consequent methodologies used by the various modern movements as described above, the core teachings of Judaism explain a significant part of why Jews support what they do. A good way to get at these teachings is to consider classical Judaism's view of the ideal world, the kind of world we should try to fix our world to become. In my book *The Way into Tikkun Olam (Repairing the World)*, I explore eight parts of that ideal vision:

1. Children.
2. The Land of Israel.
3. The ingathering of the exiles.
4. Prosperity.
5. Health.
6. Justice.

7. Knowledge of the word of God.
8. Peace.[12]

In the following, I will indicate how some of these core Jewish goals affect Jewish attitudes toward public policy, in particular regarding the topics of this book.

Children

The very first blessing and commandment in the Torah is to "be fruitful and multiply," and one of God's blessings for each of the Patriarchs is the promise of children.[13] At the same time, each of the four Matriarchs has trouble having children, indicating how difficult it can be for many couples to conceive and bear children.[14] Because Jews go to college and graduate school at much higher rates than Americans generally,[15] most non-Orthodox Jews today delay marriage until their late twenties or thirties and do not begin to try to have children until their thirties. Because biologically the optimal ages for procreation for both men and women are the late teens and twenties, this means that Jews suffer from infertility problems at higher rates than the general population. They therefore use assisted reproductive techniques to a great extent. They thus support public policies that permit birth control and abortion during the years that they do not want to have children and techniques to assist reproduction when they do. They also strongly support scientific research in this area, as in most others.

Jews also have been heavily involved in testing for genetic diseases and in the search for their cures. This in part stems from the strong Jewish concern for health, which will be discussed below, but it also is a function of specific genetic diseases that affect particularly the Ashkenazi Jewish community, that is, those with ancestors from northern or eastern Europe. As discussed in some of the essays in this volume, the small numbers of Jews, their restriction to only certain places to live, and the norm of endogamy have together meant that some genetic diseases that would normally be filtered out in a larger population persisted in the Jewish community. Beginning with testing for Tay-Sachs in the 1970s, young married Jews intending to begin having children—and sometimes before marriage—are routinely tested for a panel of genetic diseases that are more common among Jews than they are in the general population because of the factors noted above.

Even that is not enough, as the experience of my son and daughter-in-law proved. They were tested for the usual panel of Ashkenazi diseases, and the tests indicated that neither one of them was a carrier of any of the diseases. Their first child, however, has fragile X syndrome. That is not any more common among Jews than it is in the general population, and that is why it is not included in the panel of Ashkenazi Jewish diseases. It is, however, among the most common of genetic diseases, and so Israel tests for it. Once they knew that my daughter-in-law is a carrier, they investigated using preimplantation genetic diagnosis (PGD) to test embryos for future children so as to avoid the disease, and they were told that it is very difficult to test for fragile X in embryos. They found a clinic in Chicago, though, that claimed some success in identifying embryos with fragile X and those free of the disease, and they thankfully have a second child who does not have it.

I have told this story because my son and daughter-in-law are not by any means unusual among Jews. On the contrary, many Jewish couples are involved in genetic testing both before they try to have children and as part of PGD if one of them is a carrier of a genetic disease.

Furthermore, both because of this unique genetic situation and because of the Jewish commitment to medical research and cures, Jews, including my son and daughter-in-law and now their parents, are heavily involved in the search for cures of these genetic diseases. This includes the use of genetic engineering and stem cell research in efforts to find such cures.

Prosperity

The Torah asserts that "there will never cease to be needy ones in your land, which is why I [God] command you: open your hand to the poor and needy kinsman in your land."[16] As I describe elsewhere,[17] Jewish laws found in both the Torah itself and later Rabbinic literature delineate a number of ways in which individuals and communities must help the poor, including the non-Jewish poor. Even though, with very few exceptions, Jews have historically been quite poor themselves, Jews have had a remarkable record of helping the destitute among them, whether Jewish or not—to the extent that Maimonides says, "We have never seen nor heard of an Israelite community that does not have a charity fund."[18] Jewish law, though, also prescribes limits to how much charity one should give so that donors do not become recipients, and it delineates duties of the poor themselves to seek to find work so that they can get off the communal dole. Along these lines, Maimonides's famous ladder of charity ranks helping a poor person to help himself or herself through job training and loans as the highest form of charity.

This deep and long-standing tradition of providing for the immediate needs of the poor in as private and as dignified a way as possible while also helping the poor find their way out of poverty has meant that the vast majority of Jews have supported a thicker safety net than most other subgroups of the American population do while at the same time heavily supporting job training and incentives for those who can support themselves. These efforts have included support for government programs along these lines, private efforts, and Jewish communal activities. As a past president of Jewish Family Service of Los Angeles, an agency with an annual budget of $28 million, most of it from government funds but some from Jewish communal funds and private donations, I proudly point out that we offer over sixty programs on a nondenominational basis to help people with needs of all sorts and that we have been offering such services since 1854, seven years before the city of Los Angeles was incorporated.

Because food insufficiency is a major part of poverty throughout the world, and because genetic engineering of foodstuffs has already produced much more food than we could possibly have produced by traditional methods, Jews tend to support the use of genetic engineering for this purpose. At the same time, as indicated above, the Jewish tradition is also very much concerned with preserving the environment and with health and safety, and so our efforts to provide food for the hungry through genetic engineering must be balanced with significant measures to protect the environment and to guarantee the safety and health of those who eat genetically engineered foods.

Health

It is not an exaggeration to say that Jews have had a virtual love affair with medicine for the last two thousand years. Many rabbis have also been physicians until recent times, when advances in clinical practice and research have required many more years of education to become a doctor. Jewish texts see physicians as the agents and partners of God in the ongoing act of healing, and so doctors are greatly respected. Furthermore, as Judaism sees it, our bodies belong to God; we have them on loan during the course of our lives in trust from God. We therefore have a fiduciary duty to God to take care of them through the preventive methods at our disposal—including proper diet, exercise, hygiene, and sleep—and through curative interventions when needed. As I discuss more thoroughly elsewhere,[19] the duty to seek to cure disease devolves especially on physicians because of their special training, but it also applies to the community as a whole in that the community must provide health care for those who cannot otherwise afford it—which, these days, includes almost everyone. Most derive this obligation from Leviticus 19:16, "Do not stand idly by the blood of your neighbor," which the Talmud interprets to mean that we must rescue people from drowning and highway robbers—and, by extension, from illness as well.[20] Nahmanides, a thirteenth-century rabbi, additionally bases this duty to care for the sick on Leviticus 19:18, "Love your neighbor as yourself," for if you were ill and in need of care you could not afford, you would want others to provide it for you, and so you need to do the same when others are in need.[21] The Talmud also uses that verse to justify curing others even when it involves inflicting a wound.[22] The Rabbis additionally impose a duty to visit the sick, in part to emulate God, who visited Abraham when he was recovering from his circumcision.[23]

These precedents have made Jews disproportionately involved in health care as both providers and researchers, and it has led Jews to support health-care delivery and research in all its forms. Further, because the

Jewish tradition did not adopt the form of naturalism that Catholics and some Evangelicals did, Jews find no problem in using artificial means as well as natural ones for birth control and all other medical objectives. Whatever works best should be used. Pain is not seen as instructive, salvational, or good for some other reason; we should avoid or diminish it as much as possible. This has led rabbis to endorse hospice care. Rabbis differ widely as to whether machines and medications, and, especially, artificial nutrition and hydration, must be used on those patients who will not otherwise live and, if applied, whether it may be removed if the patient is not improving. On the other hand, because the mandate to heal is very strong in Judaism, and because the Jewish tradition understands fetal development in stages, representatives of all three major denominations of American Judaism have endorsed aggressive stem cell research, including the use of embryonic stem cells. As indicated above, Jews have also strongly supported genetic testing to avoid diseases in their offspring and genetic engineering to cure such diseases when that becomes possible.

Justice

"Justice, justice shall you pursue" (Deuteronomy 16:20) rings through Judaism and, indeed, all of Western civilization as a cardinal principle. For Judaism, that entails both procedural and substantive measures.[24] Procedures must be used to ensure that only the guilty are punished, and since we do not always know who is indeed guilty, the Rabbis made the death penalty virtually inoperative.[25] As indicated above, Judaism would certainly support reversing convictions on the basis of new DNA evidence or any other convincing corroboration of a prisoner's innocence, judging the disruption to the justice system that such overturning of convictions causes as a secondary concern that must yield to achieving justice for the particular person who was wrongly convicted. It would also support using the latest in technology and sufficient resources to provide apt and appropriately equipped defense attorneys to make sure that the innocent are exonerated in the first place, for the presumption of innocence is even stronger in Jewish law than it is in American law.[26]

At the same time, Judaism insists that adults take responsibility for their own illegal actions, including compensation where that is possible and punishment where appropriate.

In addition to these measures of procedural justice, Judaism mandates that we take specific measures to ensure substantive justice. This includes all the methods described above and others as well to make sure that the sick, the poor, and the otherwise downtrodden are cared for. In our context, the demand for justice would also demand that the benefits of stem cell research and genetic testing and engineering be afforded to all members of society who need them. This is simply a corollary of the general Jewish conviction that all members of society are entitled to have their basic needs met, including their health-care needs.

The Influence of Ideology

In all of these ways, then, the content of the Jewish vision of an ideal society affects how Jews think of a range of social issues, including those involving the topics of this book. The methodology that Jews use to interpret their own tradition, including both traditional awareness of the multivocal nature of authoritative texts and modern denominational interpretations and applications of those texts, together with the content of Jewish convictions, some of which are described above, have led Jews to respond to social issues in characteristically Jewish ways. Sometimes Jews ignore their minority status and their own best interests. It is precisely in such cases—when Jews "vote like Puerto Ricans," where the influence of their convictions is most in evidence. It also, though, affects how they see matters when their pragmatic interests accord with their religious commitments. Sometimes faith plays the dominant role and sometimes a lesser one; but in virtually all decisions that Jews make about social issues, their distinctive lens on the world as Jews makes a real difference in how they live their lives and try to fix the world.

This chapter, without its discussions of the topics of this book, first appeared in the Journal of Ecumenical Studies *44, no. 1 (Winter 2009): 57–69. Reprinted with permission.*

1. [No author indicated], *The National Jewish Population Survey 2000–2001* (New York: United Jewish Communities, 2003), 6, 23–25, http://www.ujc.org/njps (accessed 6 June 2014).

2. David Sidorsky, "The Autonomy of Moral Objectivity," in *Modern Jewish Ethics: Theory and Practice*, ed. Marvin Fox (Columbus: Ohio State University Press, 1975), 153–73, esp. 169.

3. Quoted in Norman Podhoretz, "The Christian Right and Its Demonizers," *National Review*, 3 April 2000, http://connection.ebscohost.com/c/articles /3044635/christian-right-demonizers (accessed 6 June 2014).

4. Andrea Elliott, "The Political Conversion of New York's Evangelicals," *New York Times*, 14 November 2004.

5. Approximately 21 percent of Jews affiliated with a synagogue in the United States identify as Orthodox, but less than half (46 percent) of America's Jews belong to a synagogue, so that means that about 10 percent of the total Jewish population sees itself as Orthodox. See *National Jewish Population Survey*, 7. The more recent Pew report on American Jews asserts that Orthodox Jews still are about 10 percent of American Jewry; see http://www.pewforum.org/2013/10/01/jewish -american-beliefs-attitudes-culture-survey/ (accessed 6 June 2014). Orthodox Jews tend to support the Republican Party, and Jews of the other denominations or no denomination support the Democratic Party; see http://www.pewforum.org /2013/12/03/infographic-survey-of-jewish-americans/ (accessed 6 June 2014).

6. James I. Packer, *Fundamentalism and the Word of God: Some Evangelical Principles* (Grand Rapids MI: Wm. B. Eerdmans, 1958), 46–47. For an excellent discussion of this and other characteristics of Evangelical Christianity in comparison to Catholicism and Mainline Protestantism, see Richard J. Mouw, "The Problem of Authority in Evangelical Christianity," in *Church Unity and the Papal Office: An Ecumenical Dialogue on John Paul II's Encyclical "Ut Unum Sint,"* ed. Carl E. Braaten and Robert W. Jenson (Grand Rapids MI: Eerdmans, 2001), 124–41.

7. *Numbers Rabbah* 13:15–16.

8. Elliot N. Dorff, *To Do the Right and the Good: A Jewish Approach to Modern Social Ethics* (Philadelphia: Jewish Publication Society, 2002), chaps. 2 and 3, 36–95.

9. A relatively traditional statement of this theory, but one that still leaves it to "the Jewish self" to decide what to believe and do, is that of Rabbi Eugene Borowitz in his book *Renewing the Covenant* (Philadelphia: Jewish Publication

Society, 1991), esp. 284–99. He and I had a respectful but spirited and, I think, illuminating dialogue about his theory in print. See Elliot N. Dorff, "Autonomy vs. Community," *Conservative Judaism* 48, no. 2 (Winter 1996): 64–68; Eugene Borowitz, "The Reform Judaism of *Renewing the Covenant*: An Open Letter to Elliot Dorff," *Conservative Judaism* 50, no. 1 (Fall 1997): 61–65; and Elliot Dorff, "Matters of Degree and Kind: An Open Response to Eugene Borowitz's Open Letter to Me," *Conservative Judaism* 50, no. 1 (Fall 1997): 66–71. This entire exchange is discussed and reprinted in my book *The Unfolding Tradition* (New York: Aviv Press [Rabbinical Assembly], 2005), 463–80 (in the 2011 revised edition, 507–22).

10. Board of Education of Kiryas Joel School District v. Grumet, 512 U.S. 687 (1994).

11. Dorff, *To Do the Right*, chap. 4, 96–113. I quote Thomas Jefferson to this effect on p. 105 there. Note also that Jefferson and Madison specifically included the Ninth and Tenth Amendments in their Bill of Rights to ensure the rights of individuals "retained by the people" against the government but not delineated in the first eight amendments and the rights of states to govern anything not explicitly assigned by the Constitution to the federal government.

12. Elliot N. Dorff, *The Way into Tikkun Olam (Repairing the World)* (Woodstock VT: Jewish Lights, 2005), chap. 11, 226–49.

13. Genesis 1:28; Abraham: Genesis 15:1–7, 17:1–7; Isaac: Genesis 26:4; Jacob: Genesis 28:14, 35:9–10.

14. Sarah: Genesis 15:2–4, 18:1–15; Rebekah: Genesis 25:21; Rachel: Genesis 29:31, 30:1–8, 22–24, 35:16–20. Leah is first able to have four children but then cannot have others, at least for a while (Genesis 29:31–35, 30:9), until God opens her womb again so that she can deliver three more children (Genesis 30:17–21). Rachel is ultimately able to produce two children but dies in childbirth with the second one (Genesis 30:22–24, 35:16–20).

15. *National Jewish Population Survey*, 6: "More than half of all Jewish adults (55%) have received a college degree, and a quarter (25%) have earned a graduate degree. The comparable figures for the total U.S. population are 29% and 6%." Based on the 2010 census and subsequent statistics, the Pew Research Center reported that in 2012, 33 percent of Americans between the ages of twenty-five and twenty-nine had earned a college degree (37 percent of women and 30 percent of men); see http://www.cbsnews.com/news/record-numbers-earning -college-degree/ (accessed 6 June 2014). It did not record the percentage of Americans with graduate degrees, however, or the percentages of Jews with college or graduate degrees, so the 2001 statistics are the latest available for comparison of the levels of education of American Jews to Americans as a whole.

16. Deuteronomy 15:11.

17. Dorff, *To Do the Right*, chap. 6, 126–60; and Dorff, *The Way into Tikkun Olam*, chap. 5, 107–30.

18. M.T. *Laws of Gifts to the Poor* 9:3.

19. Elliot N. Dorff, *Matters of Life and Death: A Jewish Approach to Modern Medical Ethics* (Philadelphia: Jewish Publication Society, 1998), chap. 2, 14–34 (esp. 27 and n. 42 on 333), and chap. 12, 279–324. See also Dorff, *The Way into Tikkun Olam*, chap. 7, 144–62.

20. B. *Sanhedrin* 73a.

21. Nahmanides, *Kitvei HaRamban*, ed. Bernard Chavel (Jerusalem: Mosad HaRav Kook, 1963 [Hebrew]), 2:43. This passage comes from Nahmanides's *Torat Ha-Adam (The Instruction of Man), Sha'ar Sakkanah (Section on Danger)* on B. *Bava Kamma*, chap. 8, and is cited by Joseph Karo in his commentary to the *Tur*, the *Beit Yosef, Yoreh De'ah* 336. Nahmanides uses similar reasoning in his comments on B. *Sanhedrin* 84b.

22. B. *Sanhedrin* 84b.

23. B. *Sotah* 14a.

24. For a fuller description of both the procedural and substantive requirements of justice as Judaism defines it, see Dorff, *To Do the Right*, chaps. 5 and 6, 114–60.

25. M. *Makkot* 1:10.

26. In both systems, culprits can plead guilty to civil charges, but in American law, culprits can also plead guilty to criminal charges, while in Jewish law, "a person may not make himself an evil person" (B. *Sanhedrin* 9b, 25a; see also B. *Yevamot* 25b; B. *Ketubbot* 18b). See Aaron Kirschenbaum, *Self-Incrimination in Jewish Law* (New York: Burning Bush Press, 1970).

CONTRIBUTORS

REBECCA ALPERT (rabbi, Reconstructionist Rabbinical College; PhD in religion, Temple University) is professor of religion at Temple University. She is the author of *Whose Torah? A Concise Guide to Progressive Judaism.*

JEFFREY BURACK (MD, Harvard Medical School; MPP, Harvard John F. Kennedy School of Government; BPhil in philosophy, Oxford) is medical director of the East Bay AIDS Center in Oakland, California; associate clinical professor of bioethics and medical humanities at the University of California, Berkeley; and associate clinical professor of medicine at the University of California, San Francisco. He chairs the Ethics Committee at Alta Bates Summit Medical Center in Berkeley and serves on the editorial board of the Society of Jewish Ethics. He has contributed to the Jewish Publication Society's series Jewish Choices, Jewish Voices.

ELLIOT N. DORFF (rabbi, Jewish Theological Seminary of America; PhD in philosophy, Columbia University) is rector and Sol and Anne Dorff Distinguished Professor of Philosophy at American Jewish University and, every other year since 1974, visiting professor at UCLA School of Law. He has served on three federal government commissions—on health care, sexual ethics, and research on human subjects—and he currently serves on the state of California's commission charged with setting the ethical guidelines for stem cell research within the state. He has served on the national advisory board of the American Association for the Advancement of Science's Dialogue on Science, Ethics, and Religion, and he serves on the Broader Social Impacts Committee for the Smithsonian's Natural History Museum. In 1997 and 1999 he was invited to give testimony on behalf of the Jewish tradition on cloning and on stem cell research before the President's National Bioethics Advisory Commission. He is a past president of the Society of Jewish Ethics and the Academy of Jewish Philosophy and a past chair of the Jewish Law Association, which has recently elected him as its honorary president. He was awarded the Lifetime Achievement Award of the *Journal of Law and Religion,* and he has four honorary doctoral degrees. He has been elected a fellow of the Hastings Center, the first bioethics institute in the United States. He has written the rabbinical rulings on infertility and end-of-life issues for the Conservative Movement's Committee on Jewish Law and Standards and has served as its chair since 2007. He is a past president of Jewish Family Service of Los Angeles, and he serves on the board of the Jewish Federation Council of Los Angeles. This is the fourteenth book that he has either edited or coedited on Jewish thought, law, and ethics, and he has additionally written over two hundred articles and twelve books on those subjects, including *Matters of Life and Death: A Jewish Approach to Modern Medical Ethics.*

ROBERT GIBBS (PhD, University of Toronto) is professor of philosophy at the University of Toronto. His research interests include Jewish thought, German idealism, French postmodern literary theory, social theory, philosophy of law, existentialism, pragmatism, and the phenomenological tradition. His major publications are two books, *Correlations in Rosenzweig and Levinas* (1992) and *Why Ethics? Signs of Responsibilities* (2000).

SHIMON GLICK (MD, State University of New York Downstate Medical Center, and professor emeritus of medicine at the Faculty of Health Sciences, Ben Gurion University of the Negev, Beer Sheva, Israel) has served as the dean, the head of the Moshe Prywes Center for Medical Education, and the head of the Lord Rabbi Immanuel Jakobovits Center for Jewish Medical Ethics, all at the Faculty of Health Sciences, Ben Gurion University of the Negev.

RONALD M. GREEN (PhD, Harvard University) is the Cohen Professor for the Study of Ethics and Human Values at Dartmouth College. He has taught at Harvard, Dartmouth, and Stanford. He is the author of over 150 articles and 9 books in bioethics, organizational ethics, and philosophy of religion. A recent book is *Babies by Design: Ethics and Genetic Choice* (Yale University Press, 2007).

JOHN LANTOS (MD, University of Pittsburgh) is currently professor of pediatrics at the University of Missouri, Kansas City, and director of the Children's Mercy Hospital Bioethics Center. He is the author of hundreds of peer-reviewed articles and six books, including *Do We Still Need Doctors?* (Routledge, 1997) and *The Lazarus Case: Life and Death Decisions in Neonatal Intensive Care* (Johns Hopkins, 2001). He is an associate editor of *Pediatrics,* on the Executive Committee of the American Academy of Pediatrics Section on Bioethics, a past president of the American Society of Bioethics and Humanities, and an ethics advisor to numerous NIH-sponsored clinical trials.

YOSEF LEIBOWITZ (rabbi, Yeshiva University; PhD, University of California, Berkeley), formerly the rabbi of an Orthodox synagogue in Berkeley, California, is the founder and head of the Yad Ya'akov Foundation for Jewish Education in Israel.

AARON L. MACKLER (rabbi, Jewish Theological Seminary of America; PhD in philosophy, Georgetown University) is associate professor of theology at Duquesne University in Pittsburgh, where he teaches in the Center for Healthcare Ethics. Formerly president of the Society of Jewish Ethics, he currently serves as a member of hospital ethics committees in the Pittsburgh area. Dr. Mackler's publications include *Introduction to Jewish and Catholic Bioethics: A Comparative Analysis* (Georgetown University Press, 2003) and an edited volume, *Life and Death Responsibilities in Jewish Biomedical Ethics* (Finkelstein Institute, Jewish Theological Seminary of America, 2000). He served as staff ethicist for the New York State Task Force on Life and the Law and as a member of the Rabbinical Assembly's Committee on Jewish Law and Standards (1991–2011), for which he served as chair of the Subcommittee on Bioethics.

JUDITH S. NEULANDER (PhD, Indiana University) is the codirector of the Judaic Studies Program at Case Western Reserve University. Neulander is a folklorist whose doctoral dissertation discredited academic claims of a crypto-Jewish folk survival in New Mexico (2001), subsequently supported by independent DNA research conducted at Stanford and New York Universities by

Wesley K. Sutton for his PhD in biological anthropology (2004). With funding from the National Institutes of Health and the Center for Genetic Research Ethics and Law, Neulander and Sutton established the first genetic-ethnographic database for historical crypto-Jews in Portugal and the first investigation of Jewish-by-disease labeling in New Mexico. For her crypto-Jewish research, Neulander received the prestigious Don Yoder Award from the American Folklore Society; she authored the crypto-Jewish entry for the *Encyclopedia of American Folklife* and coauthored with Sutton in *Annals of Human Biology*.

LOUIS E. NEWMAN (PhD in Judaic studies, Brown University) is the John M. and Elizabeth W. Musser Professor of Religious Studies and the Humphrey Doermann Professor of Liberal Learning at Carleton College. He is the founding president of the Society of Jewish Ethics. His other writings on Jewish bioethics include "Woodchoppers and Respirators: The Problem of Interpretation in Contemporary Jewish Ethics" and "Text and Tradition in Contemporary Jewish Bioethics," both of which can be found in his collected essays, *Past Imperatives: Studies in the History and Theory of Jewish Ethics* (SUNY Press, 1998).

ROBERT POLLACK (PhD in biology, Brandeis University) is professor of biological sciences, codirector of the Center for the Study of Science and Religion, and director of University Seminars at Columbia University. Additionally, he is a member of the Earth Institute faculty, a lecturer at the Center for Psychoanalytic Training and Research, and an adjunct professor at Union Theological Seminary. He was dean of Columbia College from 1982 to 1989. He is the author of more than a hundred research papers on the oncogenic phenotype of mammalian cells in culture. In addition, he has written many opinion pieces and reviews on aspects of molecular biology, medical ethics, and science education. He has edited two books on these matters for Cold Spring Harbor Laboratory Press, and he has published three books since 1994: *Signs of Life*, *The Missing Moment*, and *The Faith of Biology, the Biology of Faith*. His fourth book, a narration of the work of his wife, the artist Amy Pollack, *The Course of Nature*, was just published.

TOBY L. SCHONFELD (PhD in philosophy, University of Tennessee, Knoxville) is the director of the Master of Arts in Bioethics Program at the Center for Ethics, associate professor of medicine, and codirector of the Program for Scholarly Integrity at Emory University. After receiving her doctorate in philosophy with a concentration in medical ethics, she served as the director of the Center for Humanities, Ethics, and Society and associate professor of Health Care Ethics and vice-chair of the Health Promotion, Social and Behavioral Health Sciences Department in the College of Public Health at the University of Nebraska Medical Center. She is a past president of the Society of Jewish Ethics. Dr. Schonfeld has published articles on genetics and feminism in the *Cambridge Quarterly of Healthcare Ethics, Perspectives in Biology and Medicine,* and the *Journal of the Society of Christian Ethics* and has dozens of presentations and publications in women's health more generally. With D. Micah Hester, she is the coeditor of *Guidance for Healthcare Ethics Committees* (Cambridge, 2012), an important educational text for members of health-care ethics committees.

PAUL ROOT WOLPE (PhD, University of Pennsylvania) is the Asa Griggs Candler Professor of Bioethics, the Raymond F. Schinazi Distinguished Research Chair in Jewish Bioethics, a professor in the Departments of Medicine, Pediatrics, Psychiatry, and Sociology, and the director of the Center for Ethics at Emory University. Dr. Wolpe also serves as the first senior bioethicist for the National Aeronautics and Space Administration (NASA), where he is responsible for formulating policy on bioethical issues and safeguarding research subjects. He is coeditor of the *American Journal of Bioethics* (AJOB), the premier scholarly journal in bioethics, and editor of *AJOB Neuroscience*, and he sits on the editorial boards of over a dozen professional journals in medicine and ethics. Dr. Wolpe is a past president of the American Society for Bioethics and Humanities; a fellow of the College of Physicians of Philadelphia, the country's oldest medical society; and a fellow of the Hastings Center, the oldest bioethics institute in America. He was the first national bioethics advisor to Planned Parenthood Federation of America. He is the author of over 125 articles, editorials, and book chapters in sociology, medicine, and bioethics and has contributed to a variety of encyclopedias on bioethical issues. He is the coauthor of the textbook *Sexuality and Gender in Society* and editor and a key author of the end-of-life guide *Behoref Hayamim: In the Winter of Life.*

NOAM ZOHAR (PhD, Hebrew University) is an associate professor in the Department of Philosophy and the Department of Jewish Thought, Bar Ilan University, and is the director of the Graduate Program in Bioethics there. He is a member of the Sheba Medical Center Ethics Committee and of Israel's National Bioethics Council. Author of books and articles in the fields of bioethics, Rabbinics, and philosophy of *halakhah*, his works include *Alternatives in Jewish Bioethics* (1997), *Quality of Life in Jewish Bioethics* (ed., 2006), and *The Jewish Political Tradition* (ed., with Michael Walzer and others, 2000, 2003, forthcoming).

LAURIE ZOLOTH (PhD, Graduate Theological Union, University of California, Berkeley) is president of the American Academy of Religion, former president of the American Society for Bioethics and Humanities, and former vice president of the Society for Jewish Ethics. She was the principal investigator (PI) of the grant from the American Association for the advancement of science, which helped to produce this book; co-PI of a National Institute of Health (NIH) Ethical, Legal, and Social Implications Project on the Human Genome and Identity; and chair of the Howard Hughes Medical Research Institute Ethics Advisory Board. As a Charles McCormick Deering Professor of Teaching Excellence at Northwestern University, she was the founding director of the Center for Bioethics, Science, and Society and the Brady Program in Ethics and Civic Life. She is professor of Bioethics and Humanities at the Feinberg School of Medicine and professor of religion and a member of the Jewish Studies faculty at Weinberg College of Arts and Science. She is on the editorial board of the *Journal of Narrative Ethics*. Her book, *Health Care and the Ethics of Encounter*, on justice, health policy, and the ethics of community, was published in 1999. She is also coeditor of four books: *Notes From a Narrow Ridge: Religion and Bioethics*, with Dena Davis; *Margin of Error: The Ethics of Mistakes in Medicine*, with Susan Rubin; *The Human Embryonic Stem Cell Debate: Ethics, Religion, and Policy*, with Karen LeBacqz and Suzanne Holland; and *Oncofertility: Ethical, Social, Legal and Medical Issues* with Teresa K. Woodruff, Lisa Campo-Engelstein, and Sarah Rodriguez. She is a member the NIH National Recombinant DNA Advisory Committee (RAC) and the Ethics Committee of the American Society of Reproductive Medicine as well as 3 NIH data safety monitoring boards, and Ethics committee for NASA, and National Science Foundation Centers Of Excellence in both Synthetic Biology and Nanotechnology. She received both ASBH's award for Service to the Field in 2007, and the NASA National Public Service Award for six years of outstanding service as the first ethicist on NASA's National Advisory Board. In 2005 she was awarded the Graduate Theological Union's Alumni of the Year award for service to the field of religion and social ethics. She was a founding board member and currently serves on the boards of the International Society for Stem Cell Research and the Society for Neuroethics. Professor Zoloth has served on the national advisory boards of the American Association of the Advancement of Science's Dialogue on Science, Ethics and Religion; the American Association for the Advancement of Science's Working Group on Human Germ-Line Interventions and on Stem Cell Research Advisory Board and has testified for state and federal commissions on issues in emerging scientific research.

INDEX TO CLASSICAL SOURCES

GENERAL INDEX

AAAS (American Association for the Advancement of Science), xi–xii, xvi–xvii, 14, 241, 311

Abaye (rabbinic sage), 350, 352, 363, 372

abortion: in American law, 10; embryos outside the womb and, 33–36, 51–52n30; forty/fifty-four days of gestation, initial, 12, 27, 30–31, 33–37, 50–52n25–26, 50n21, 51n30, 57, 68; genetic testing and, 148, 199, 202, 209–10; in Israeli law, 245; Jewish views on, 27–29, 31, 408–9, 411; "man inside of a man," Noahide prohibition on murder of, 21n23; Roman Catholic position on, xiv, 11, 51n26; therapeutic, 28, 43, 47–48n11; using aborted fetuses for stem cell research, 26–30, 44, 49–50n15–16; views on legal and moral status of embryos, 9–12, 27, 31. *See also* beginning of life; embryos

Abraham (biblical patriarch), 138, 139–40, 142–44, 159n7, 164, 179, 182–83, 251, 335, 347, 414

absence of Jewish law on particular subject, methodologies for dealing with, 31–33

acquisition, possession, and control: commodification of the self, 38, 96, 281; problem of, 88–90; science, sense of rational control in, 397–400

adult or somatic stem cells, 4–5, 6, 7, 15–18, 40–41, 44, 45

Africa: colonialist vision of, 162, 167, 172, 173; lost tribe tales of, 168–71; sickle cell anemia and malaria in, 150. *See also* Falasha; Lemba, identity, and genetic mapping

aggadah, 94, 95, 291, 348

agricultural metaphor for medicine, 296–97

Ahad Ha'am, 140

AID (artificial insemination by donor), 383
AIDS/HIV, 104, 237–38, 265
Akiba (rabbinic sage), 83–84, 90–92, 99n5, 100n11, 282n4, 295–96
alchemy, 94–95
Aleinu, 339–40
Alfasi, 99n3
allele, defined and described, 107
Allen, Woody, 141
allergies, 262
Allport, Gordon W., 112, 129–30
Alpert, Rebecca, 109, 137, 421
Alzheimer's disease, 65, 208
amal (struggle and effort), valuing, 253–54, 326–28
American Association for the Advancement of Science (AAAS), xi–xii, xvi–xvii, 14, 241, 311
anavah. See humility before God
Anderson, W. French, 280, 283–84n17
anencephaly, 290
animals: creation of, 73–74; genetic engineering of, 235, 410, 413–14; as research models, 6, 15, 17, 26
anorexia, 261
anthropological uses of genetic mapping, 107–9
anti-Semitism: African Jews and, 171–73; folk taxonomies and, 129; genetic disease prevalence and stigmatization of Jews, 154–57, 204–5, 212, 263; Jewish genetic distinctiveness influenced by, 195–96; misuse of term, 157n1
Antonovsky, Aaron, 246
Arabs not distinguishable by DNA from Jews, 114

Aristotle and Aristotelianism, 50n26, 86, 368, 379
artificial insemination by donor (AID), 383
artificial reproductive techniques (ARTS), 13
Asch, Adrienne, 88
Asher ben Yehiel, 99n3
Ashi (rabbinic sage), 359, 368, 373
Ashkenazi Jews: breast cancer and *BRCA1/2* mutations in, 131–33, 154, 196–97; colorectal cancer and, 154; defined, xiv, 194–95, 228n5; disease burden of, xiv–xv, 194–96, 197–98, 217; founder effect and, 149–50, 195, 211; genetic testing and, 196–97, 205, 206, 207, 210, 211, 212; Isserles, Moses, glosses of, 99n3; Kohen modal haplotype and, 263; matrilineal origins of, 159n16; Niemann-Pick types A and B, 121; split from Sephardi, 108, 131; as study population, 149; TSD (Tay-Sachs disease) and, 148, 158n12
Asilomar conferences, 240, 388
Asklepios, 396–97
atherosclerosis, 245–46
Atlanta Jewish Gene Screen, 206
Atwood, Margaret, 378
Azulai, Chayim, 272n20

Bacon, Sir Francis, 88
Baylis, Françoise, 248
beginning of life, 68–78; biblical creation in the divine image, 69–77; flood story, "image of God" in, 77–78; forty/fifty-four days

of gestation, initial, 12, 27, 30–31, 33–37, 50–51n25–26, 50n21, 51n30, 57, 68; implications for stem cell research, 77, 78; Jewish views of, 12, 27, 30–31, 68, 208, 381–82; rabbinic texts on, 68–69; role of religion in public life and, 380–82; Roman Catholic position on, xiv, 51n26, 208; "simply fluid/liquid/water," early embryos regarded as, 12, 27, 31, 33–36, 38, 39, 57, 68

behavior, genetic interventions to alter, 161, 246, 253, 302, 310, 327–28

Ben Azzai, 275

Ben-Amos, Dan, 111–12

Benjamin of Tudela, 170

Berg, Paul, 387–88

Better Than Well (Elliott), 328

Billy Budd (Melville), 94

Bioethics Commission, 55

biology, culture, and religion, inter-twining of, in Jewish identity, 137, 139–44, 149, 184–89

bipolar affective disorder, 324

Black Death, 104

"Black Jews." *See* Falasha; Lemba, identity, and genetic mapping

black-Jewish race relations in America, 176–77

blastocyst, 5, 12, 13–14, 306n4

Bleich, J. David, 30, 49n16

blessings: Noahide, 77–78; on occasion of hearing good news, 304–5, 308–9n28; for physically disabled persons, 42, 297–98

Bloom's syndrome, 197

Blum, Lawrence, 221

the B'nei Bateira, 274

bodies, God's ownership of, 25, 27, 317–18, 320, 414

bone marrow transplants, 7

Bordenave, Kristine, 117

Bornstein, Abraham, 62, 63

Botox injections, 267, 326

The Boys from Brazil, 245

Brave New World (Huxley), 244, 281

BRCA1/2 mutations, 131–33, 196–97, 199, 204, 207, 208, 262

breast augmentation, 266

breast cancer, 131, 154–55, 194, 196, 199, 203–4, 205, 261–62, 265

Brodkin, Karen, 176

Brother Daniel case (Israeli Supreme Court, 1962), 138, 144

b'tzelem Elokim. See divine image

bubbameises, 186–87

Buber, Martin, 142

Burack, Jeffrey H., 310, 311, 421

Burke, Wylie, 155

burn victims and stem cell research, 7

Bush, George W., 55, 243, 378, 382, 385, 404

Calf, Golden, 100n10, 143, 174

calf, half-grown, creation and eating of, in Talmud, 86, 87, 90, 91, 93, 100n10

Callahan, Daniel, 88

Canavan disease, 197

cancer. *See specific types*

care, ethic of. *See* genetic decision making

Caro, Joseph. *See* Karo, Joseph

Casey decision *(Planned Parenthood v. Casey;* U.S. Supreme Court, 1992), 10

Caspi, Avshalom, 246
Catholicism. *See* Roman Catholicism
CCR5 mutation, 104
CD4 receptor, 238, 265
Center for Genetic Research Ethics
  and Law, 129
charity, duty of, 412–14
Chaucer, Geoffrey, 94
Chicago Center for Jewish Genetic
  Disorders, 197–98, 205
Chicago Jewish News, 186
children and Jewish convictions on
  social issues, 411–12
Chmielnicki massacres, 149
cholera, 173
Christian Scientists, 307n16, 308n27
Christians: public bioethics, con-
  tributions to public debate on,
  380–85; scriptural interpretation,
  methods of, 404–5. *See also* Prot-
  estantism; Roman Catholicism
circumcision, 178–79, 249, 269
civic witness, 96–97
CJD (Creutzfeldt-Jakob disease),
  heritable or familial, 121–27
CJLS (Conservative Movement's
  Committee on Jewish Law and
  Standards), 23, 28, 47–48n11
Clinton, William Jefferson, 317, 378
cloning: of embryonic stem cells,
  18–19; therapeutic versus reproductive, 67n15
Clouser, Dan, 259, 262
cocreation, 316–17
Cohen, Cynthia, 96
Cohen, Leonard, 377
Cold Spring Harbor Laboratory,
  Learning Center at, 387–89

Cole-Turner, Ronald, 202
Collins, Francis, 212, 243
colonialist vision of Africa, 162, 167,
  172, 173
colorectal and colon cancer, 154–55,
  196
commodification of the self, 38, 96,
  281
Conan Doyle, Arthur, 388
confidentiality and disclosure issues,
  203, 204, 209, 288–89, 409
congestive heart failure and stem cell
  research, 7
consequences, responsibility for,
  332–34
Conservative Judaism, xv, 23, 406,
  407–8
Conservative Movement's Committee on Jewish Law and Standards
  (CJLS), 23, 28, 47–48n11
control. *See* acquisition, possession,
  and control
conversion: Abraham, covenant with,
  142; consanguineous marriage
  of siblings following, 179–80;
  crypto-Judaism and, 125, 126, 165;
  female founders of Ashkenazi
  gene pool and, 159n16; identification of converts as such,
  prohibitions on, 179; identity,
  tensions in concepts of, 177; Jewish culture versus Jewish religion,
  connection to, 144; Jewish lineage
  by, 108, 138, 142, 143, 207; of Khazars, 181, 183; legal definition of
  Jewishness and, 109, 138, 178, 179;
  recent increases in, 141; reconver-
  sion, 126, 138

coronary artery disease, 245–46

cosmetic surgery, 252–53, 266, 267, 314, 326

Council on Bioethics, 55, 243, 247, 248, 378, 380, 382, 385

Cousins Club, 175–76

creation narrative: animals, 73–74; biblical creation in the divine image, 69–77; blessing on occasion of hearing good news and, 304–5; cocreation and genetic enhancement, 316–17; dominion of humans, 75–77; humans, 74–75; inanimate objects, 72; plants, 72–73; stem cell research affecting, 12–16, 25; women's knowledge of/power over, 83–84

creativity, as human duty, 249–50

Creutzfeldt-Jakob disease (CJD), heritable or familial, 121–27

Crick, Francis, 399

CRISPR/Cas, 239

crypto-Judaism: conversion and, 125, 126, 165; lost tribes/lost Jews and, 165; Portuguese crypto-Jews in colonial America, 115. *See also* New Mexican crypto-Jews

Cultural Zionism, 140

culture, biology, and religion, intertwining of, in Jewish identity, 137, 139–44, 149, 184–89

Culver, Charles M., 54n43, 259, 262, 271n8

curiosity versus studiousness, 96–97

cystic fibrosis, 106, 197, 208, 239

da'at Torah, 291, 306–7n9

Dana-Farber Cancer Institute, 212

dangerous or forbidden knowledge. *See* knowledge

Darwin, Charles, 88, 104, 400

Davis, Dena, 158n8

decision making. *See* genetic decision making

defining disease, 257–70; application to enhancement versus therapy issue, 264–67; as condition of the person, 260; distinct sustaining causes, 263–64; distinguished from rational beliefs and desires, 260–61, 373n9; elements of a malady, 260–64; importance of, 257–60, 270; Jewish thinking about, 267–70; suffering or increased risk of suffering, 261–63, 268

degenerative diseases and stem cell research, 7–8

Democritus, 393

demut, 283n5

denial: as brain function, 391–93; of human interdependence, 400–402; meaning, scientific discovery and loss of, 391; of mortality, 398, 399, 400, 402; in science, 393–96

depression, 246

"depth theology," 32

Dershowitz, Alan, 171

diabetes, 7, 27, 90, 290

diagnostic genetic testing, 203

Dialogue on Science, Ethics, and Religion (DoSER) program, AAAS, xi–xii

Dickinson, Emily, 94

dihydrolipoamide dehydrogenase deficiency, 197

disability and disabled persons: blessing for, 42, 297–98; defined, 262; enhancement versus therapy, 42–43; genetic decision making and caring for, 218, 223, 225–26; genetic testing of embryos and, 209–10; as image of God, 25; Jewish valuation of, 42

disappearance, non-disappearance, and memory in stories of African Jews, 171–72, 175–76

disclosure and privacy issues, 203, 204, 209, 288–89, 409

discrimination, genetic, 196–97, 212, 288–89, 301, 321

disease and Jewish identity: Ashkenazi Jews, disease burden of, xiv–xv, 194–96, 197–98, 217; endogamy, xiv–xv, 149, 155, 159n15, 195, 263, 411; ethical dilemmas associated with, 150–54; eugenics linking, 145, 198; Falasha, 173; in folk taxonomies, 119–27, 128–33; founder effect, 149–50, 195; history of anti-Semitism and, 195–96; mere chance versus genetic distinctiveness, 195–96, 211; statistical impossibility of using disease as Jewish genetic marker, 132, 155; stigmatization and, 154–57, 204–5, 212, 263. *See also specific diseases, e.g.* Tay-Sachs disease

disease, defining. *See* defining disease

diversity and difference, human, 103, 110, 184–86, 322, 324

divine image (*b'tzelem Elokim*): aborted fetuses used in research and, 29; animal research and, 26; biblical creation in the divine image, 69–77, 275; concept of, 275–78; flood story, "image of God" in, 77–78; genetic enhancement and, 274–82, 297, 321–22; humans made in, as fundamental theological principle, 25, 55–57; "image" and "likeness," 71; implications for stem cell research, 77, 78; intellect viewed as constituting, 57, 276; referred to humans as God's representatives on earth, 282–83n5

divine Name, embryos as emblematic of, 55–65; concept of, 57–58; countervailing values, 64–65; moral and legal status of embryos, 55–57; objective criteria for, 58–60; subjective criteria for, 60–64; woman as author or scribe, 66n8

divine Name, omitted from Torah scroll, 353–59, 364–68

DNA: defined and described, 233–34; image of God, not constituting, 278–79; in justice system, 409–10, 415; replication, transcription, and translation of, 344–46; Watson-Crick model of, 399

Documentary Hypothesis, 69

Dor Yeshorim, 148, 204, 205–7

Dorff, Elliot N., xiii, xv, 3, 12, 15, 23, 99n3, 103, 193, 218, 233, 306n8, 317, 322, 381, 403, 410–11, 421–22

DoSER (Dialogue on Science, Ethics, and Religion) program, AAAS, xi–xii

dreams, brain lesions preventing people from distinguishing reality from, 392

Duchenne muscular dystrophy, 265

Dury, John, 171

duty, obligation, and responsibility: of charity, 412–14; for consequences, 332–34; creativity as, 249–50; freedom and, balancing, 185–89; to future generations, 334–36; Jewish emphasis on, 331–32; new medical treatments, duty to develop, 382–83; power and, 286–87, 299; suffering, obligation to reduce, 252, 299

E. coli, genetic manipulation of, 235, 241

Eckstein, Josef, 204

ectopic pregnancy, 52n35

Edelson, Paul, 148

Editas Medicine, 239

efficacy and safety issues, 236–37, 240–41, 243, 306n5, 332–34, 336–37

EGSO (ethnogeographic stories of origin), 188

Einstein, Albert, 79n4, 188, 389–90, 391, 393

Elad the Danite, 169–70

election, doctrine of, 139–40, 142–44

Eliezer (rabbinic sage), 90–92

Elliott, Carl, 328

embryonic stem cells, 5–6, 7, 9, 15, 18–19, 26–27, 306n4, 409

embryos: created specifically for medical research, 14–15, 36–39, 44–45; forty/fifty-four days of gestation, initial, 12, 27, 30–31, 33–34, 35, 36, 37, 50–52n25–26, 50n21, 51n30, 57, 68; frozen, 9–12, 14, 30, 34, 36, 37, 40, 44, 61, 63, 67n12; genetic testing and PGD, 39, 198–99, 207–11, 245, 412; legal and moral status of, xiv, 9–12, 27, 31, 55–56, 208, 380–82; outside the womb, 33–36, 51–52n30; removing cells from, 39–40; sex selection, 209; as "simply fluid/liquid/water," 12, 27, 31, 33–36, 38, 39, 57, 68, 382; size, location, and motility of, 12, 21n23. *See also* abortion; divine Name, embryos as emblematic of

endogamy and presence of genetic disease, xiv–xv, 149, 155, 159n15, 195, 263, 411

enhancement versus therapy, 243–54, 310–40; *amal* (struggle and effort), valuing, 253–54, 326–28; consequences, responsibility for, 332–34; cosmetic surgery, 252–53, 266, 267, 314, 326; distinguishing, 42–43, 244–47, 264–67; divine image and, 274–82, 297, 321–22; duty, Jewish emphasis on, 331–32; equal access issues, 241–42, 254, 281, 322–23; ethical questions associated with, 240; expressivist critique of, 321; future lineages/populations, effects on, 241–42, 334–36; giftedness, human sense of, 329–31;

enhancement versus therapy (*cont.*) goals of Jewish reflection on, 312, 337–39; human diversity, value of, 322, 324; humility before God and, 311, 315–16, 321, 324–26, 330–32; intelligence, interventions designed to enhance, 246–47, 266, 302–3, 312–14; in Jewish law and tradition, 249–54; as medical goal, 247–48; naturalism versus cocreation, 316–17; preventive enhancement, 265, 269–70; risks associated with, 266–67; secular critiques of, 314–16; self-fulfillment and, 328–29; stem cell research and, 42–43, 54n43; stewardship, concept of, 317–18, 320; substantive Jewish moral vision and, 319–21; theological response to, 290; theoretical nature of discussion of, 243–44, 274–75; unanticipated effects, 336–37. *See also* defining disease

equal access to medical technology: genetic intervention/enhancement and, 241–42, 254, 281, 322–23; genetic mapping and, 150–54; stem cell research and, 41–42, 95–96

ethic of care. *See* genetic decision making

ethics and scientific research. *See* Jews and genes

Ethiopian Jews. *See* Falasha

ethnogeographic stories of origin (EGSO), 188

eugenics: disease and Jewish identity, linking, 145, 198; genetic

intervention and, 235, 244, 284n19, 290, 306n8, 310; genetic testing and, 198, 202, 208, 210; preventive medicine and, 401; stem cell research and, 42, 97

exceptionalism, genetic, 245

expressivist critique of genetic enhancement, 321

Falasha: black-Jewish race relations in America and, 176, 177; colonialist vision of Africa, 162, 167, 172, 173; disease and Jewish identity, 173; Lemba compared, 164, 180; marks of Jewishness for, 165–66, 173–74; narrative construction of story, 164; Operation Moses, 162; *The Return* (Sonia Levitin), 160, 162, 167, 172–74, 189; seeking to migrate to Israel, 162, 167. *See also* Lemba, identity, and genetic mapping

familial dysautonomia, 90, 151–52, 197

Familial Dysautonomia Foundation, 151–52

familial hyperinsulinism, 197

Fanconi anemia, 197, 265

Faust, 94

feminist scholarship, 219, 220, 269

fertility, low human rate of, 107. *See also* infertility treatment

Feynman, Richard, 402

Fisher, Berenice, 220–22, 224, 225

Flavor-Saver Tomato, 235

flood story, "image of God" in, 77–78

folk taxonomies, 110–33; academic logic versus, 111–12; anti-Semitism

and, 129; construction of, 118–20;
on disease and Jewish identity,
119–27, 128–33; FOAF ("Friend-
of-a-Friend") tales, 118, 120, 122,
123–24, 126, 127; prejudice and
overgeneralization of Cultural
Others, 112, 122, 127–30; "rabbi in
Colorado" element, 119, 120, 124,
125, 126; religion and genetics,
false links between, 112–13, 128–
30, 132. *See also* "lost tribes" of
Israel; New Mexican crypto-Jews
forbidden or dangerous knowledge.
  See knowledge
forty/fifty-four days of gestation, initial, 12, 27, 30–31, 33–37,
  50–51n25–26, 50n21, 51n30, 57, 68
founder effect, 149–50, 195, 211
fragile X syndrome, 198, 412
Frankel, Mark S., xi
Frankenstein (Shelley), 94, 333
Freedman, Benjamin, 318, 331, 332,
  333
freedom and responsibility, balancing, 185–89
Freud, Sigmund, 391, 398, 399, 400
Friedmann, Theodore, 235, 237, 388
"Friend-of-a-Friend" (FOAF) tales,
  118, 120, 122, 123–24, 126, 127
frozen embryos, 9–12, 14, 30, 34, 36,
  37, 40, 44, 61, 63, 67n12
Fundamentalism and the Word of God
  (Packer), 404–5
future lineages/populations, effects
  of genetic interventions on, 241–
  42, 334–36

Gattica (movie), 409

Gaucher disease, 152–54, 195, 197
Gauss, Karl, and Gaussian curves,
  188
Gearhart, John, 4, 9
Geertz, Clifford, 285
Gelsinger, Jesse, 36, 236–37
Gemara. *See* Talmud
gemilut hesed, 252
gene therapy. *See* genetic
  intervention
gene transfer. *See* genetic
  intervention
genes, defined and described, 233–34
genetic decision making, 215–27;
  application of ethic of care,
  222–24; concept of ethic of care,
  219–22; context and individual
  circumstances, importance of,
  218–19, 223, 225–27; objections to
  ethic of care approach, 224–26;
  pikku'ah nefesh and, 218, 226;
  principle- or rule-based approach
  in Jewish bioethics, 217–19; TSD
  and, 215–17, 218, 223, 225–26; utilitarianism, 224–25
genetic discrimination, 196–97, 212,
  288–89, 301, 321
genetic engineering: of animals
  and plants, 235, 410, 413–14; of
  humans. *See also* enhancement
  versus therapy; eugenics; genetic
  intervention
genetic enhancement. *See* enhancement versus therapy
genetic exceptionalism, 245
genetic intervention, 233–42; in animals
  and plants, 235, 410, 413–14; concept of, 234–35; equal access issues,

genetic intervention (*cont.*) 241–42, 254, 281, 322–23; ethical issues related to, 240–42; future lineages/populations, effects on, 241–42, 334–36; genes, DNA, and RNA, 233–34; germ line modification, 240, 243, 244, 290, 334, 336, 346; in human patients, 235–40, 282n1; powers released by, 287–90, 299, 301, 303; safety and efficacy issues, 236–37, 240–41, 243, 306n5, 332–34, 336–37; theoretical nature of discussion of, 243–44. *See also* defining disease; enhancement versus therapy; repairing Torah scroll and gene therapy; theological response to genetic intervention

genetic mapping, 103–9; anthropological uses of, 107–9; continuing scientific questions about, 107; HGC (Human Genome Project), 103–5, 106, 147, 160, 324; human difference, challenge of, 103, 110, 184–86; identity affected by, 177–78; Jewish identity and, 107–9, 184–89; SNPS (single nucleotide polymorphisms), 108, 160–61; understanding and curing disease through, 104–7. *See also* disease and Jewish identity; folk taxonomies; Lemba, identity, and genetic mapping

genetic testing, 193–213; abortion and, 148, 199, 202, 209–10; ambiguous associations of, 193–94, 196–97, 198, 212; Ashkenazi Jews and, 196–97, 205, 206, 207,

210, 211, 212; benefits of, 198–99; *BRCA1/2* mutations and, 196–97, 199, 207; burdens of, 202–5, 215–16; clinical testing on adults, 203–7; defined and described, 201; diagnostic, 203; of embryos, 198–99, 207–11; Jewish cooperation with, 194, 196–98, 205–6, 210, 212, 268–69, 411–12; Jewish identity and, 207; Jewish tradition and ethical issues raised by, 201–2; in justice system, 409–10, 415; privacy and disclosure issues, 203, 204, 209; programs and panels for Jewish genetic testing, 197–98; recent advances in, 199, 203; research on, 211–13; social convictions of Jews and, 409–10, 411–12; stigmatization and, 204–5; for susceptibility, 203–7; TSD (Tay-Sachs disease) and, 148–49, 194, 195, 197, 208. *See also* genetic decision making

genetics and Jews. *See* Jews and genes

genetics, defined and described, 103, 105

The Genetics of the Jews (Mourant), 195

genome and genomics, defined and described, 103–7, 234

genome map, 234–35

genome sequence, 234–35

germ line modification, 240, 243, 244, 290, 334, 336, 346

Gert, Bernard, 259, 262

Gibbs, Robert, 342, 422

giftedness, human sense of, 329–31

Gilligan, Carol, 220
Gilman, Sander, 193
Glick, Shimon, 243, 422
glycogen storage disease, 197
goats, genetic manipulation of, 235
Golden Calf incident, 100n10, 143, 174
golem, 87, 89, 189n7
Goodman, Richard M., 125
graft-versus-host problem, 8, 15
Graves, Joseph, 130
Greek classical culture, 69, 298, 393, 396–97
Green, Ronald M., 53n43, 257, 422–23
Greenberg, Irving, 307–8n17
Gross, Michael, 153–54
Gustafson, James, 285
Gutiérrez, Ramón, 113

HAART (highly active antiretroviral therapy), 238
Hadad-yis'i, 282–83n5
Hadassah, 212
Hadrian (Roman emperor), 91
halakhah and genetics. See Jews and genes
halakhic-theological discourse, 56, 65, 66n3
Halevi, David, 61
Halevi, Yehuda, 140, 181–84
half-grown calf, creation and eating of, in Talmud, 86, 87, 90, 91, 93, 100n10
Hananel (rabbinic sage), 354, 355, 359, 365, 366, 368, 373
Hanina (rabbinic sage), 87
haplotype, defined and described, 107

Harry Potter, 100n20
Hastings Center, 41, 53n38, 314–15
Hauerwas, Stanley, 88
health and healing. See medicine, practice of
health and life, preservation of. See pikku'ah nefesh
Health Insurance Portability and Accountability Act (HIPAA; 1996), 409
heart disease and heart attacks, 7, 208, 245–46, 261, 290
Hebrew language and genetic code, 173–74, 347
hemophilia, 178–79, 269
hepatitis, 8
herem, 138
Herodotus, 124
Heschel, Abraham Joshua, 276
hesed, 36
hesed shel emet, 252
HGC (Human Genome Project), 103–5, 106, 147, 160, 324, 345
hidden, inner code of Jewishness, 173–75, 177
highly active antiretroviral therapy (HAART), 238
Hillel (rabbinic sage), 157, 276–77
Himmelfarb, Milton, 404
HIPAA (Health Insurance Portability and Accountability Act, 1996), 409
HIV/AIDS, 104, 237–38, 265
HLA (human leukocyte antigen) molecule, 239
Hogan, Brigid, 98n2
Holocaust/Shoah, 105, 144, 156, 163, 166, 175, 193, 208, 390

homosexuality, 208, 259, 263, 264, 407, 408

Hordes, Stanley M., 115, 117

human difference and diversity, 103, 110, 184–86, 322, 324

Human Genome Project (HGC), 103–5, 106, 147, 160, 324, 345

human giftedness, 329–31

human interdependence, 400–402

human leukocyte antigen (HLA) molecule, 239

human life, sanctity and preservation of. See *pikku'ah nefesh*

humility before God *(anavah)*: enhancement versus therapy and, 311, 315–16, 321, 324–26, 330–32; genetic intervention and, 249, 250–51, 267, 291–98; ignorance, philosophy of, 402; stem cell research and, 26, 34

Huntington's chorea, 104, 106, 203, 204, 239, 261

Huxley, Aldous, *Brave New World*, 244, 281

hypertension, 52n35, 261

iatrogenic genetic disease, 266–67

identity: commodification of the self and, 38, 96, 281; fixed versus mutable concepts of, 177–78; genetic knowledge challenging, 289; human difference, challenge of, 103, 110, 184–86; public policy on bioethics, role of religion in, 383–84; scientific research affecting, 4. *See also* Jewish identity

ignorance, philosophy of, 402

image of God. *See* divine image; divine Name, embryos as emblematic of

Imagining Russian Jewry (Zipperstein), 171

immortal cell lines, 6, 8

immortality, institutional, 397–400

India, Jewish groups in, 108

induced pluripotent stem cells, 4–5

infertility treatment: AID (artificial insemination by donor), 383; donation of excess embryos for stem cell research, Jewish law on, 30–36, 44; health risks associated with, 37, 52–53n35; IVF donations and development of stem cell research, 9–11, 13–14; Jewish convictions regarding children and, 411; privacy, sex selection, and *Kohen* status, 209

information and privacy issues, 203, 204, 209, 288–89

inner voice, importance of listening to, 391, 392, 395, 400, 401–2

institutional immortality, 397–400

intelligence and intellect: as constituting divine image, 57, 276; genetic interventions designed to enhance, 246–47, 266, 302–3, 312–14; modafinil, 312–13, 315, 337; *Nr2b* gene, 313–14, 327, 336, 337

interbreeding, Levitical prohibition of, 307n10

interdependence, human, 400–402

intermarriage, xv

International Jewish Discourse Project on Religion and Ethics at the Frontier of Genetic Medicine, 311

International Society for Intelligence
  Research, 247
International Society of Stem Cell
  Research (ISSCR), 15
intervention, genetic. *See* genetic
  intervention
in vitro fertilization (IVF): for PGD
  (preimplantation genetic diagnosis), 39, 198–99, 207–11, 245, 412;
  stem cell research and, 9–11, 13–14
Iranian Jews, 132
Isaac (biblical patriarch), 142, 182–83,
  251
Isaac (rabbinic sage), 354, 356, 357,
  365, 366, 373
Isaac ben Samuel (rabbinic sage),
  354, 365, 366, 373
Isaac ben Samuel ben Martha (rab-
  binic sage), 350, 363, 372
Ishmael (rabbinic sage), 295–96
Islam, 12, 95, 99n3, 146, 163, 181
Israel: abortion in, 245; Falasha seek-
  ing to migrate to, 162, 167; genetic
  testing in, 209, 210, 212, 245, 412;
  Law of Return in, 109, 138, 144;
  Lemba and, 189n6; Palestinian
  Arabs in, 153–54, 156; Supreme
  Court (Brother Daniel case), 138,
  144; treatment of Gaucher dis-
  ease in, 153–54
ISSCR (International Society of Stem
  Cell Research), 15
Isserles, Moses, 99n3
IVF. *See* in vitro fertilization

Jacob (biblical patriarch), 142, 182–83,
  251, 252, 322, 335
Jacob, François, 397

Jakobovits, Immanuel, 33
Jare da Bertinoro, Obadiah, 170
Jefferson, Thomas, 403, 408, 418n11
Jehovah's Witnesses, 307n16
Jeremiah of Difti (rabbinic sage), 359,
  368, 369, 373
Jewish Family Service of Los Angeles, 413
Jewish Federation of Metropolitan
  Chicago, 197
"Jewish geography," 175–76
Jewish identity, 137–57; biology, culture, and religion, intertwining of,
  137, 139–44, 149, 184–89; doctrine
  of election and covenant with
  Abraham, 139–40, 142–44; ethical
  dilemmas and, 145–46, 150–54;
  genetic mapping, issues raised
  by, 107–9, 184–89; genetic testing and, 207; in Halevi's account
  of Khazars, 181–83, 184; Jewish
  definitions of who is a Jew, 138–39;
  "Jewish genes," concept of, 146–48;
  knowledge, burdens and limits of,
  81–82; Lemba and issues of, 178–81,
  184–89; by matrilineal descent,
  108, 109, 122–23, 138–39, 159n16;
  patrilineal Jews, 138. *See also* conversion; disease and Jewish identity
Jewish revolt (132–35 CE), 91
Jewish United Fund (JUF), 197–98
Jewish Woman (periodical), 152
Jews and genes: AAAS and, xi–xii,
  xvi–xvii, 14, 241, 311; absence of
  Jewish law on particular subject,
  methodologies for dealing with,
  31–33; conflict between tradition/
  ethics and scientific research, 3–4;

86–88; power of, 287–89, 291–94, 301, 303–4; real illness, impact on dealing with, 90, 96; resources for theological response to, 287–89; Talmud on, 83; teaching/research versus practice, 90–93, 96–97; tensions inherent in, 93–95; women and, 84–85

Kohanut status, 139–40, 180, 189n2, 209

Kohen modal haplotype, 107–8, 109, 161, 163, 178, 180, 187

Kook, Rav, 140

Kornberg, Arthur, 397

The Kuzari: In Defense of a Despised Faith (Yehudah Halevi), 181–84

Lantos, John, 377, 423

lashon kodesh, 174

Law of Return (Israel), 109, 138, 144

Learning Center at Cold Spring Harbor Laboratory, 387–89

Lehmann, Lisa, 196

Leibowitz, Yosef, 68, 423

Lemba, identity, and genetic mapping: Africa, colonialist vision of, 162, 167; "discovery" of, 162–64; Falasha compared, 164, 180 (*See also* Falasha); genetic claims for, 109, 147, 166–67, 177; hidden, inner code of Jewishness, 173–75, 177; identity, what constitutes, 177–78; Jewish identity and, 178–81, 184–89; Khazars compared, 181–84; Kohen modal haplotype and, 109, 163, 178, 180, 187; "lost tribes" as literary and epistolary trope and, 163, 164–66, 168–71;

memory, disappearance, and non-disappearance, 171–72, 175–76; narrative construction of story, 164; Parfitt's studies of, 108–9; persecution and anti-Semitism, 171–73; race relations and, 176–77; religious practices of, 167, 173–74; return to Israel, 189n6

leukemia, 236, 237, 239, 282n2, 333, 336, 337

Levinas, Emmanuel, 85, 99n5

Levitin, Sonia, *The Return,* 160, 162, 167, 172–74, 189

Libyan Jews, 122, 124, 125, 127

Liebman Jacobs, Janet, 119, 121–25, 127

life and health, preservation of. See *pikku'ah nefesh*

Linnaeus, Carl, 110–11, 128, 130

liposuction, 267, 326

Lipschutz, Alejandro, 113

liver disease, 7–8, 239

Loew, Judah (the Maharal of Prague), 250

Loewenthal, Ron, 117

"lost tribes" of Israel: indigenous populations of Americas identified as, 113, 164, 171; as literary and epistolary trope, 163, 164–66, 168–71. *See also* Falasha; Lemba, identity, and genetic mapping; New Mexican crypto-Jews

Luria, Alexander, 392

Lurianic Kabbalah, 140

Luzzato, Chaim, 140

Mackler, Aaron L., 274, 423

"mad cow" epidemic, 122

Madison, James, 418n11

the Maharal of Prague (Judah Loew), 250

Maimonides, 99n3; on Abraham's calling by God, 140, 158n7; on charity, 413; on disease prevention and health promotion, 249, 253, 268, 298–99; on human intellect, 276; on legal and moral status of embryo, 31. *See also Index to Classical Sources*

maladies, defining. *See* defining disease

malaria, 150, 260

"man inside of a man," Noahide prohibition on murder of, 21n23

many-to-one relationships in genetics, 324

maple syrup urine disease, 197

mapping the genome. *See* genetic mapping

Maria the Jewess (alchemist), 95, 100n22

matrilineal descent in Jewish communities, 108, 109, 122–23, 138–39, 159n16

McElheny, Victor, 388

MDNA disease, 241

meaning, scientific discovery and loss of, 390–91

medicine, practice of: agricultural metaphor for, 296–97; Jewish tradition and, xiv, 194, 267–68, 317, 414–14; Maimonides on, 249, 253, 268, 298–99; new medical treatments, duty to develop, 382–83; preventive medicine, 401; as service to God, 299, 300

Mehlman, Maxwell J., 282n3

Meilaender, Gilbert, 81, 88, 380, 382

Meir (rabbinic sage), 354, 356, 358, 365, 366, 373

memory, disappearance, and non-disappearance in stories of African Jews, 171–72, 175–76

Menassah ben Israel, 171

Mendel, Gregor, 104

Mendelssohn, Felix, 324

mending Torah scroll. *See* repairing Torah scroll and gene therapy

mental anguish, Jewish principle of, 210, 218

mental illness, 258, 259, 260–61, 264. *See also specific types*

messianism, Jewish, as process, 331

mezuzah, 349

Midrash, xvi; genetic intervention and, 249, 251, 269, 291, 293, 295, 297, 302, 305; lost tribes in, 168; social issues, Jewish convictions regarding, 405, 407; stem cell research and, 94, 95. *See also Index to Classical Sources*

mikveh, 251–52, 277, 306n8

Milton, John, 94

Mishnah: genetic intervention and, 278, 280, 291–92, 294, 301, 348; genetic mapping and, 169; social issues, Jewish convictions regarding, 407; stem cell research and, 31, 32, 52n33, 83, 90, 98n3. *See also Index to Classical Sources*

mitzvah d'rabbim, 29, 49n15

mitzvah ha-ba'a b'aveirah, 29

mitzvot, 36, 83, 277, 331

Mizrahi Jews, 205, 207, 210

onanism (wasting seed), 37, 52n33–34
one-to-many relationships in genetics, 324
Operation Moses, 162
organ and tissue transplants, 7, 8–9, 29, 40, 42, 51n27, 299
Orthodox Jews, xv; business world, rabbinical decisions in, 99n3; genetic testing and, 148, 194, 195, 204, 205, 209; Jewish identity and, 138; percentage of American Jews identifying as, 417n5; social issues and, xv, 404, 406, 407, 408; stem cell research and, 23, 25, 30, 46n7, 47n11; Temple, prayers for restoration of, 189n4; on Torah, 346
Oryx and Crake (Atwood), 378
Oshaia (rabbinic sage), 87
ovarian cancer, 37, 52n35, 154, 196, 199

Packer, James, 404–5, 406
Pale of Settlement, 149, 175
Paradise Lost (Milton), 94
The Paradox of Choice (Schwartz), 328–29
Parens, Erik, 43
Parfitt, Tudor, 108–9, 177, 189n12
Parkinson's disease, 7, 90
parthenogenesis, 14–15, 36–39, 44–45
Pascal, Blaise, 14
Patai, Raphael, 95, 100n22, 115
patrilineal Jews, 138
pemphigus vulgaris (PV), 117–18
performance-enhancing drugs, 266
personalized genetic medicine (PGM), 104–5

PGD (preimplantation genetic diagnosis), 39, 198–99, 207–11, 245, 412
PGM (personalized genetic medicine), 104–5
philosopher's stone/sorcerer's stone, 94, 100n20
philosophy of ignorance, 402
pikku'ah nefesh: genetic intervention and enhancement, issues of, 252, 267–68, 317; in genetic testing and decision making, 205, 218, 226; social convictions of Jews and, 414–15; stem cell research and, xiv, 25, 48, 56, 317
plants: creation of, 72–73; genetic engineering of, 235, 410, 413–14
pleuropulmonary fibrosis, 52n35
Pliny, 124
pluripontent genes, 324
polio vaccinations, 265
Pollack, Robert, 141, 147, 387–89, 424
polymorphic alleles, 107
poor, duty to help, 412–14
Portuguese crypto-Jews in colonial America, 115
possession. *See* acquisition, possession, and control
Pratt, Mary Louise, 116, 118, 128
preimplantation genetic diagnosis (PGD), 39, 198–99, 207–11, 245, 412
preservation of human life and health. See *pikku'ah nefesh*
Press, Nancy, 155
Prester John, 171
preventive enhancement, 265, 269–70

preventive medicine, 401

principle- or rule-based approach in Jewish bioethics, 217–19

privacy and disclosure issues, 203, 204, 209, 288–89, 409

prophylactic mastectomy, 204, 262

proteomics, 106, 344, 345–46, 348, 362

Protestant work ethic, 326

Protestantism: naturalism and, 415; New Mexican crypto-Jews and, 114, 115; role of religion in public bioethics debate and, 380; scriptural interpretation by, 405; social issues and, 405, 407; stem cell research and, 11, 12

Prouser, Joseph, 48n13

Provigil, 312

p'sak din, 291

pseudepigraphy, 122

public policy and genetics, 377–416; on beginning of life and moral status of embryo, 380–82; on duty to develop new medical treatments, 382–83; on identity of individuals, 383–84; religion, role of, 377–85. *See also* social issues, Jewish convictions regarding; unconscious, shaping of genetics by

PubPeer (science blog), 19

Putnam, Hillary, 96

PV (pemphigus vulgaris), 117–18

Rab (rabbinic sage), 350–52, 354, 359, 360, 362–66, 368–70, 372

Raba (rabbinic sage), 359, 369, 370, 373

Rabbah bar bar Hanah (rabbinic sage), 354, 355, 365, 366, 373

Rabbah the Younger (rabbinic sage), 359, 360, 368, 369, 370, 373

rabbinic tradition, 57, 68–69, 79n2, 83, 86, 98–99n3. *See also* Midrash; Mishnah; Talmud; Tosefta and Tosafists; *Index to Classical Sources*

rabbinical system and democracy of scholarship, 179, 190n14

RAC (Recombinant-DNA Advisory Committee), NIH, 239, 388

race: African Jews and black-Jewish relationship in America, 176–77; New Mexican crypto-Jews and, 111, 113–18, 119, 121–26; as term, 157–58n1

Rambam. *See* Maimonides

Ramban. *See* Nahmanides

Rashi, 50n16, 62, 66n9. *See also Index to Classical Sources*

Rawls, John, 273n31

Recombinant-DNA Advisory Committee (RAC), NIH, 239, 388

Reconstructionists, xv, 138, 141, 143

Reform Jews, xv, 23, 138, 140, 141, 143, 145, 405, 406, 407

Reisner, Avram Israel, 50n16, 322

religion: false links between genetics and, 112–13, 128–30, 132; intertwined with culture and biology in Jewish identity, 137, 139–44, 149, 184–89; role in bioethics debate, 377–85

religious Zionists, 140

Remennick, Larissa, 210

repairing Torah scroll and gene therapy, 342–74; missing words (text 2), 353–59, 364–68, 372–73; number of mistakes that can be corrected (text 1), 350–53, 362–64, 372; parallels between, 344–49; person/human body, scroll traditionally compared to, 343, 348–49; torn texts (text 3), 359–61, 368–71, 373–74; tractate *Menahot* texts on scroll mending, 343–44, 349–50, 361–62

repression. *See* denial; unconscious, shaping of genetics by

Reshit hokhmay yirat HaShem, 402

responsibility. *See* duty, obligation, and responsibility

The Return (Sonia Levitin), 160, 162, 167, 172–74, 189

Reubeni, David, 170–71

RIKEN, 17–18, 19

risks, allowing for, 236–37, 240–41, 243, 306n5, 332–34, 336–37

RNA, defined and described, 233–34, 345

Robert, Jason S., 248

Roche, Ellen, 36

Rodriguez, Damian, 333, 334, 337

Roe v. Wade (U.S. Supreme Court, 1973), xiv, 10, 11

Roman Catholicism: on embryo's moral status, xiv, 11, 208; on genetic testing of embryos, 207–8; naturalism and, 415; neo-Thomism, 368; role of religion in bioethics debate and, 379, 380, 385; scriptural interpretation by, 405; social issues and, 404, 405,

409, 415; stem cell research and, 11–12, 36, 51n26, 80, 96

Romans, 48n14, 86, 91, 101n11, 159n16, 164, 190n14, 277

Rosenfeld, Arziel, 306n8

Rosh Hashanah, 274

Rosner, Fred, 306n8

Roth, Joel, 52n31

Roth, Sol, 325

Rothman, David J., 248

Rothman, Sheila M., 248

rule- or principle-based approach in Jewish bioethics, 217–19

Sachs, Bernard, 217

safety and efficacy issues, 236–37, 240–41, 243, 306n5, 332–34, 336–37

salutogenesis, 246

Sambatyon River, 168, 170–71

Sambrook, Joe, 387, 388

Sandel, Michael, 329–32

schizophrenia, 129, 259

Schofer, Jonathon, 187

Schonfeld, Toby L., 215, 424–25

Schwartz, Barry, 328–29

SCIDS (severe combined immunodeficiency), X-linked, 236, 237, 239, 279, 282n2, 333

Science (periodical), 19

scientific research into genetics. *See* Jews and genes

SCNT (somatic cell nuclear transfer), 44

screening, genetic. *See* genetic testing

scroll of Torah, repairing. *See* repairing Torah scroll and gene therapy

self-fulfillment, 328–29

Semmelweis, Ignaz, 18

Sephardic Jews: defined, 195; genetic diseases of, 195; Isserles, Moses, glosses of, 99n3; New Mexican crypto-Jews and, 117, 121, 124, 125, 126, 127, 131–33; split from Ashkenazi, 108, 131

severe combined immunodeficiency, X-linked (SCIDS), 236, 237, 239, 279, 282n2, 333

sex selection of embryos, 209

Sforno, Obadiah ben Jacob, 283n8

Shakespeare, William, 400

Shapiro, Rami, 339–40

Shattuck, Roger, 14, 21n26, 93–94, 95

sheheheyanu, 308n28

Shimon ben Elazar (rabbinic sage), 354, 356, 365, 366, 373

Shimon of Shezur (rabbinic sage), 354, 356, 365, 366, 373

Shoah/Holocaust, 105, 144, 156, 163, 166, 175, 193, 208, 390

shofar, sounding, 274

Shweder, Richard, 331, 332

sickle cell anemia, 150, 239

Sidorsky, David, 403

Silver, Lee M., 272n11

Silvers, Anita, 321

Simeon ben Yohai (rabbinic sage), 294

Simon, Herbert, 329

"simply fluid/liquid/water," early embryos regarded as, 12, 27, 31, 33–36, 38, 39, 57, 68, 382

single nucleotide polymorphisms (SNPS), 108, 160–61, 263

"sluggish schizophrenia," 259

smallpox vaccinations, 186

SNPS (single nucleotide polymorphisms), 108, 160–61, 263

social interdependence, 400–402

social issues, Jewish convictions regarding, 403–16; children, 411–12; content of Jewish conceptions and values, 404, 410–11; factors affecting, 403–4; justice, 415–16; medicine and health, importance of, 414–15; methods of interpretation of tradition and, 404–10; poor, duty to help, 412–14

social justice, 174, 322–23. *See also* equal access to medical technology

Soloveichik, Joseph B., 250

somatic cell nuclear transfer (SCNT), 44

somatic or adult stem cells, 4–5, 6, 7, 15–18, 40–41, 44, 45

sorcerer's stone/philosopher's stone, 94, 100n20

sotah, 64, 65

special needs children. *See* disabled persons

spinal injuries and stem cell research, 8

spinal muscular atrophy, 197

Spinoza, Baruch, 140, 145

Steinberg, Avraham, 21n23, 25, 46–37n7, 47n11, 49n15, 51n30, 52n34

Steinberg, Bernard, 321, 324

stem cell research, 3–45; aborted fetuses, use of, 26–30, 44; animal models, 6, 15, 17, 26; benefits, potential and actual, 6, 7–9; cloning embryonic cells, unproven research on, 18–19; controversy surrounding, stem cell research (*cont.*)

4; creating embryos specifically for medical research, 36–39, 44–45; creation/reproduction narrative affected by, 12–16; definition of stem cells, 5–6; divine image and beginning of life, implications of, 77, 78; embryonic stem cells, 5–6, 7, 9, 15, 18–19, 26–27, 306n4, 409; embryos outside the womb, treatment of, 33–36; equal access to medical technology, 41–42, 95–96; fundamental theological principles, 24–26; immortal cell lines, 6, 8; induced pluripotent stem cells, 4–5; infertility treatments, donation of excess embryos from, 9–11, 13–14, 30–36, 44; isolation and extraction process, 5–6, 9; legal and moral status of embryos and, 9–12, 27, 31, 55–56; methodologies for dealing with absence of Jewish law on particular subject, 31–33; official Jewish rulings on, 23; paying for egg and sperm donations for, 14–15, 52n31; *pikku'ah nefesh* and, xiv, 25, 48, 56, 317; rabbinical support for, 12; removing cells from embryos, 39–40; size, location, and motility of embryo as factors, 12, 21n23; social convictions of Jews and, 409, 415; somatic or adult stem cells, 4–5, 6, 7, 15–18, 40–41, 44, 45; telomerase, 6–7; therapy versus enhancement, 42–43, 54n43; totipotent or pluripontent nature of stem cells, 5; transforming adult cells into totipotent cells, unproven research on, 16–18; transplantation therapies, 8–9, 42. *See also* abortion; beginning of life; divine Name, embryos as emblematic of; embryos; knowledge

stewardship, concept of, 317–18, 320

Stewart, Martha, 89

Stoicism, 368

storage disorders, 121

The Stranger (Camus), 94

stroke, 7, 52n35, 262

struggle and effort *(amal)*, valuing, 253–54, 326–28

studiousness versus curiosity, 96–97

suffering, in Jewish thought: defining disease and, 261–63, 268; obligation to reduce, 252, 299; as part of human life, 88–89; value attached to, 268, 308n20, 318

Supreme Court, Israeli, 138

Supreme Court, U.S., xiv, 10–11, 407

susceptibility testing, 203–7

Sutton, Wesley, 114–15, 117

SV40 tumor virus, 387–89

Talmud (Gemara), xvi, 83, 98n3, 179, 184, 187, 348, 407. *See also entries for* Babylonian Talmud *and* Jerusalem Talmud *in Index of Classical Sources*

Tanhuma (rabbinic sage), 275, 282n4

Tay-Sachs disease (TSD), 148–49, 216–17; abortions due to, 28; creation/reproduction narrative and selection of embryos without, 13; defining disease and, 263, 265; founder effect and, 149; genetic decision making and, 215–17, 218, 223, 225–26; genetic intervention in, 290; genetic testing for, 148–49, 194, 195, 197, 208; genetics of, 158–59n12, 217; HGP and, 106; mere chance versus genetic distinctiveness in occurrence of, 195; in non-Jewish populations, 159n15; prognosis for, 216; single mutation causing, 104, 194

TB (tuberculosis), 150, 154

tefillin, 349

Tell Fekheriyeh, 282–83n5

telomerase, 6–7

Tendler, Moshe, 23, 49n16, 99n3, 204, 212, 269, 317, 381, 383

testing, genetic. *See* genetic testing

thalassemia, 153

theological response to genetic intervention, 285–305; blessing on occasion of hearing good news and, 304–5, 308–9n28; disabled, blessing for, 297–98; ethics as reflection of theology and world-view, 285–86; genes, power to alter, 289–90, 294–98, 301; humility before God, importance of, 291–98; knowledge, power of, 287–89, 291–94, 301, 303–4; medicine, Jewish views on practice of, 296–97, 298–99, 300; power and responsibility, relationship between, 286–87, 299; powers released by genetic tech-nologies, 287–90, 299, 301, 303; resources for, 290–98; service to God as guideline for, 299–304

therapy versus enhancement. *See* enhancement versus therapy

Thomas, Mark, 108, 187

Thompson, James, 4

Thompson, Stith, 119–20

tikkun olam, 250, 319, 336

Tillich, Paul, 389

Tineus Rufus (Tyrannos), 100n11

tissue and organ transplants, 7, 8–9, 29, 40, 42, 51n27, 299

To Do the Right and the Good (Dorff), 405, 408, 418n11

to'ar, 71

Torah scroll, repairing. *See* repairing Torah scroll and gene therapy

Tosefta and Tosafists, 62–63, 98n3. *See also Index to Classical Sources*

tradition, Jewish. *See* Jews and genes

transplantation therapies, 7, 8–9, 29, 40, 42, 51n27, 299

Tronto, Joan, 220–22, 224, 225

TSD. *See* Tay-Sachs disease

Tsien, Joe, 313, 336

tuberculosis (TB), 150, 154

tumor virus SV40, 387–89

Tyrannos (Tineus Rufus), 100n11

tza'ar ba'alei hayyim, 26

tzedekah, 152

tzelem, 71, 275, 278, 279, 281, 282, 283n5, 321. *See also* divine image

Übermensch, 169

ultra-Orthodox Jews, 206, 268, 407

unconscious, shaping of genetics by, 387–402; denial, as brain function, 391–93; denial, in science, 393–96; faith in science, Einstein's remarks on, 389–90, 391; human inter-dependence, acceptance versus denial of, 400–402; inner voice, unconscious (*cont.*)

importance of listening to, 391,
392, 395, 400, 401–2; institutional immortality, desire to win,
and sense of rational control,
397–400; meaning, scientific
discovery and loss of, 390–91;
mortality, problem of, 390–91,
396–402; philosophy of ignorance, embracing, 402; SV40
tumor virus controversy, 387–89

Union Review (periodical), 389

U.S. Supreme Court, xiv, 10–11, 407

Usher syndrome, 197

utilitarianism, 224–25

Vacanti, Charles, 17

Vacanti, Martin P., 17

vaccinations, 186, 265

The Vanishing American Jew (Dershowitz), 171

Varmus, Harold, 236

Vatican Council I, 51n26

v'nishmartem m'od l'nafshoteikhem, 205

Walker-Warburg syndrome, 197

Wallenberg, Eliezer, 47n11

Washofsky, Mark, 23

Watson, Jim, 388, 399

*The Way into Tikkun Olam (Repairing
the World)* (Dorff), 410–11

Weinberger, Leor, 237, 238

Weisbard, Alan, 51n27, 308n27

white identity and New Mexican
crypto-Jews, 111, 113–18, 119, 121–26

WHO (World Health Organization),
247–48

"The Wife of Bath" (Chaucer), 94

Winkler, Mary, 89

Wolpe, Paul Root, 201, 312, 313, 425

women: as author or scribe of divine
Name in embryos, 66n8; feminist
scholarship, 219, 220, 269; knowledge, access to, 84–85

Woo Suk Hwang, 19

World Health Organization (WHO),
247–48

Wurzburger, Walter, 307n9

Wyschogrod, Michael, 140

X-linked SCIDS, 236, 237, 239, 279,
282n2, 333

Yamanaka, Shinya, 15

Yanni (rabbinic sage), 94

Yavneh, as center of Jewish teaching,
274

Yiddls of Prague, 171

Yohanan ben Zakkai, Rabban, 274

Yohanan (rabbinic sage), 86, 251,
306n8

Yom Kippur, 91

Zeiri (rabbinic sage), 359, 368, 369,
370, 373

Zipperstein, Steven, *Imagining Russian Jewry*, 171

Zliberstein, Yitzhak, 210

Zohar, Noam, 55, 425–26

Zoloth, Laurie, xiii, xv, 3, 14, 15, 18–
19, 80, 103, 109, 160, 193, 233, 239,
241, 426

zucchini, 84, 85, 88, 91, 92, 99–100n7

zygote, 5, 9, 38–39